AF429430

Organic Name Reactions:

Principles, Mechanisms and Applications

Organic Name Reactions:

Principles, Mechanisms and Applications

Sanjay B. Bari

M. Pharm., Ph. D. D.I.M.F.I.C.

Principal and Professor
H. R. Patel Institute of Pharmaceutical Education and Research,
Shirpur, Maharashtra, India.

Vinod G. Ugale

M. Pharm., Ph.D.

Assistant Professor
R. C. Patel Institute of Pharmaceutical Education and Research
Shirpur, Maharashtra, India.

PharmaMed Press

An imprint of Pharma Book Syndicate

A unit of BSP Books Pvt. Ltd.

4-4-309/316, Giriraj Lane,
Sultan Bazar, Hyderabad - 500 095.

Organic Name Reactions : Principles, Mechanisms and Applications

by Sanjay B. Bari, Vinod G. Ugale

Published by

PharmaMed Press

An imprint of Pharma Book Syndicate

A unit of BSP Books Pvt. Ltd.

4-4-309/316, Giriraj Lane, Sultan Bazar, Hyderabad - 500 095.

Phone: 040-23445688 Fax: 91+40-23445611

E-mail: info@pharmamedpress.com

www.pharmamedpress.com/pharmamedpress.net

ISBN: 978-93-89974-30-0

PREFACE

Knowledge of basic and applied organic chemistry is very indispensable on many fronts. At broader sense, in depth knowledge of fundamental principles, basic structures, and mechanisms of organic chemistry is vitally important; but it is not possible to cover each topic in great depth in one text book of organic chemistry. Nor would this be desirable, even we make it possible! Nevertheless, students will often wish to pursue individual topics further in detail by using multiple reference books. In our opinion, organic chemistry has been grown tremendously but not randomly; the exploration follows certain broad lines, but students come to us knowing no more chemistry than in their past; they must be led carefully along the this path in **'Wohler's jungle'**; if we wish that they should not get lost. The vitally focused and difficult part of organic chemistry is through understanding of organic name reactions. Study of organic name reactions is an important part of theoretical and applied chemistry. The breaking and making of covalent bonds usually occurs in several discrete steps before conversion into final products. To understand and apply these theoretical principles in experimental synthetic research is possible only when students have through knowledge of stepwise organic name reactions. Keeping in view of all facts concern with available literature and insufficient compiled data of name reactions, an effort has been made to guide undergraduate and graduate students about organic name reactions. It has been our experience that students often have a hazy collection of material that creates confusion and enhances fear about organic chemistry; hence this book has been written to cover only one entire topic of organic name reactions at one place with pictorial presentations that gives complete understanding of subject. We have taken one organic name reaction at a time; we explain it as fully and clearly as we can. It has been our aim to provide thorough knowledge about organic name reactions with its mechanisms and latest applications in synthetic research or in pharmaceutical industry. In this book, we have covered not more than a tiny fraction of this enormous field of chemistry; but what we can anticipate for is to perform a good job for undergraduate and graduate students. The knowledge gained from this book enable readers to predict the products from nearly analogous substances or reactants and what is more significant is to distinguish a pattern in apparently diverse reactions.

Aim and Scope

The book comprises of organic name reactions of imperative synthetic importance and presents thoroughly updated recent data on name reactions. Although this context will complement the knowledge of many organic chemistry reference books but it is highly suitable for learning and understanding the detailed basic concepts of most organic name reactions. This project is initiated only to provide complete knowledge of organic name reactions with mechanisms and updated synthetic and/or latest industrial applications. All the reactions have been dealt in great details. Basic mechanisms of reactions along with considerations of reactivity and orientations are also elaborated in detail. As manuscript will cover 200 organic name reactions, this will be most up-to-date major text book of its kind. It offers students and industrial chemists a valuable resource for conducting experiments and performing wide range of applications, from pharmaceuticals to pesticides. Each reaction listing is clearly organized into discrete uniform sections that allow readers to quickly gather the information they need to understand and conduct their own experimental procedures. **Organic Name Reactions: Principles, Mechanisms and Applications** tender several features that help the readers to obtain information quickly and conduct their experiments effectively:

a) Reaction summaries provide basic information about each name reaction,

b) Schematic reaction index offers a quick overview of each reaction,

c) The principle parts of every reaction include categories of reactions and organize all subtypes of organic reactions according to their type of transformations (For example: oxidation, reduction, molecular rearrangement, etc.)

Till date the chemistry of organic compounds and methods for their synthesis form the bedrock of modern medicinal, chemical, and pharmaceutical research. Organic Name Reactions: Principles, Mechanisms, and Applications will be the state-of-art resource and valuable addition to bookshelves of pharmaceutical scientists and faculties along with industrial organic chemists.

-Authors

1. Acyloin Condensation

Principle

The carboxylic acid esters undergoes bimolecular reductive coupling upon refluxing with aprotic solvents such as ether, benzene, toluene or xylene to afford α-hydroxy ketone is known as acyloin condensation. Symmetrical α-hydroxy ketones (aliphatic analogs of benzoins) are commonly known as acyloin; the name is derived by adding the suffix '-*oin*' to the name of corresponding acid.

The reaction is more favored when R is an alkyl group. With longer alkyl chains, higher boiling solvents can be used. Di-esters are used to prepare cyclic acyloins. When the acyloin condensation is carried out in the presence of chlorotrimethylsilane, the enediolate intermediate is trapped as bis-silyl derivative which is hydrolysed in acidic condition to the acyloin. Reaction occurs between two moles of ester (intermolecular condensation) or one mole of di-ester (intramolecular condensation). Rearrangement is promoted by either acid or base; the thermal acyloin rearrangement can be accelerated by high pressure.

General Reaction

Mechanism

Step 1: Reaction proceeds through free radical mechanism.

A reaction occurs in presence of metallic sodium; a direct transfer of electron towards carbonyl carbon atom takes place to give an intermediate (**I**) which rapidly dimerize to produce unstable intermediate product (**II**). Rapid loss of both alkoxy groups from intermediate (**II**) gives 1,2-diketone.

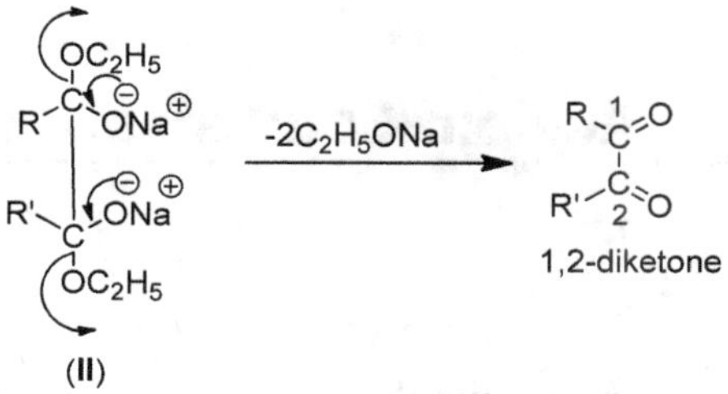

$-2C_2H_5ONa$ → 1,2-diketone

(II)

Step 2: 1,2-diketone is highly reactive, undergoes reduction with metallic sodium to give sodium salt of enediol. Finally addition of carboxylic acid affords 1,2-diol which after tautomerization resulted into the stable product acyloin (α-ketol or α-hydroxy ketone).

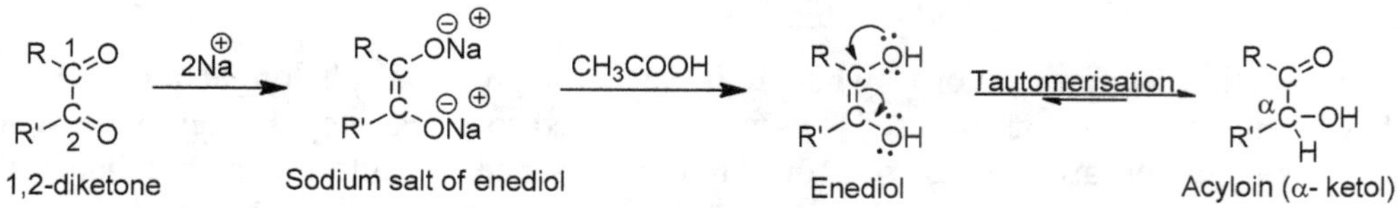

1,2-diketone → $2Na$ → Sodium salt of enediol → CH_3COOH → Enediol → Tautomerisation → Acyloin (α-ketol)

Applications

Intramolecular acyloin condensations of diesters have been widely used for synthesis of medium and large ring compounds with better yield.

   a) *Preparation of cyclic acyloins.*

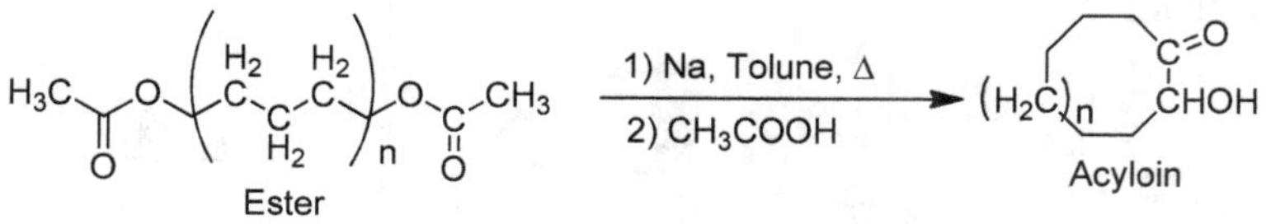

Ester → 1) Na, Tolune, Δ; 2) CH_3COOH → Acyloin

   b) *Preparation of catenane.*

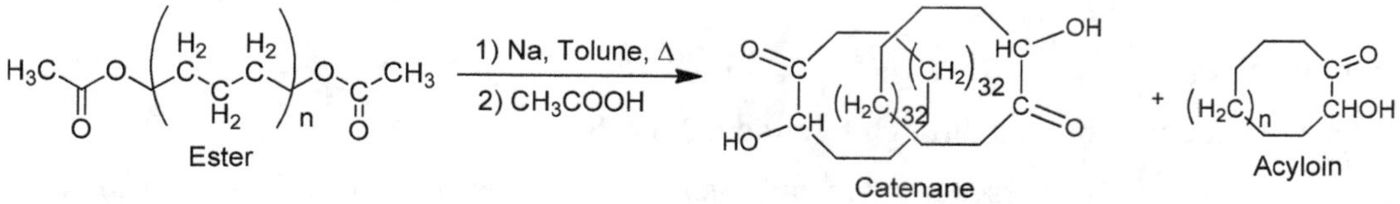

Ester → 1) Na, Tolune, Δ; 2) CH_3COOH → Catenane + Acyloin

2. Alder-Ene Reaction (Conia Reaction)

Principle

Pericyclic Reactions

A reaction in which simultaneous bond breaking and bond formation takes place in a single step through a cyclic transition state is known as Pericyclic reaction. These reactions are said to be concerted and do not produce any reaction intermediates. These reactions are also referred as no mechanism reactions. Polar reagents, solvents or catalysts do not affect the pericyclic reactions. Pericyclic reactions are affected by heat or light and are highly stereoselective.

Alder-ene reaction is an example of pericyclic reaction. The reaction involves four-electron system containing an alkene $>C=C<$ (π-bond) and an allylic C-H (σ-bond); the double bond shifts from alkene to form new C-H and C-C σ-bonds. Alder-ene reaction resembles to Diels-Alder reaction. The allylic system reacts similarly to a diene, while in alder-ene reaction the other reactant is an enophile, compared to dienophile in Diels-Alder reaction. The Alder-ene reaction requires higher temperature because of higher activation energy and stereo-electronic requirement for breaking of allylic C-H bond (σ-bond). The reaction involves addition of an enophile to an alkene *via* allylic transposition hence also termed as hydro-allyl addition reaction. These enophiles may be an aldehydes, ketones or imines to produce β-hydroxy or β-amino olefins. These compounds may be unstable under reaction conditions, so that at elevated temperature ($>400°C$) reverse reaction takes place called as retro-ene reaction. **General reaction:**

Ene   Enophile   Adduct

Methylenecyclohexane   Furan-2,5-dione (Enophile)   3-(cyclohex-1-en-1-ylmethyl)dihydrofuran-2,5-dione

Mechanism

Reaction involves four-electron system containing an alkene $>C=C<$ (π-bond) and an allylic C-H (σ-bond); the double bond shifts from alkene to form new C-H and C-C σ-bonds.

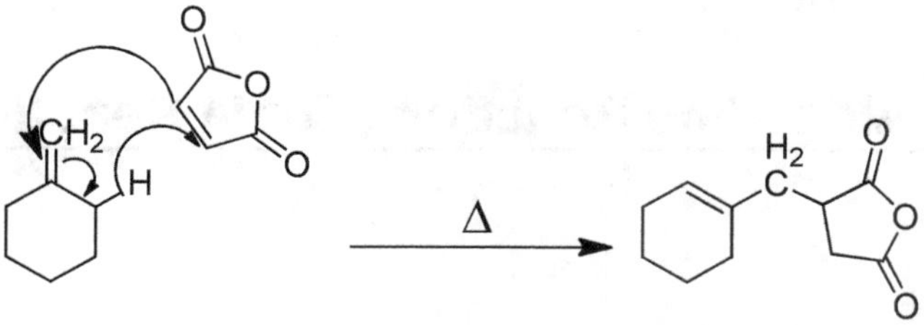

Applications

a) *Preparation of N-(1,3-diphenylbut-3-en-1-yl)-4-methylbenzenesulfonamide.*

(*E*)-*N*-benzylidene-4-
methylbenzenesulfonamide

Prop-1-en-2-ylbenzene

CHCl$_3$: THF (4:1)
RT, 0.25h

N-(1,3-diphenylbut-3-en-1-yl)-4-
methylbenzenesulfonamide

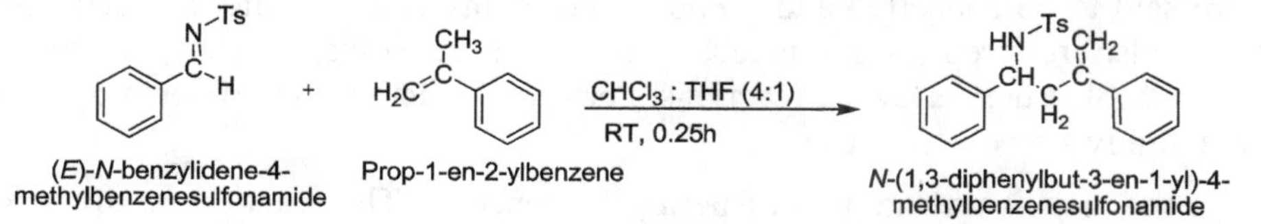

b) *Preparation of ethyl 3-(furan-2-yl)-2-hydroxypropanoate.*

(*E*)-(furan-2-
ylmethylene)hydrazine

KOH, Pt/Clay, 35%
(Kishner reduction)

2-methylene-2,3-dihydrofuran

0 °C, CH$_2$Cl$_2$, 89%
Alder-ene reaction

Ethyl 3-(furan-2-yl)-2-
hydroxypropanoate

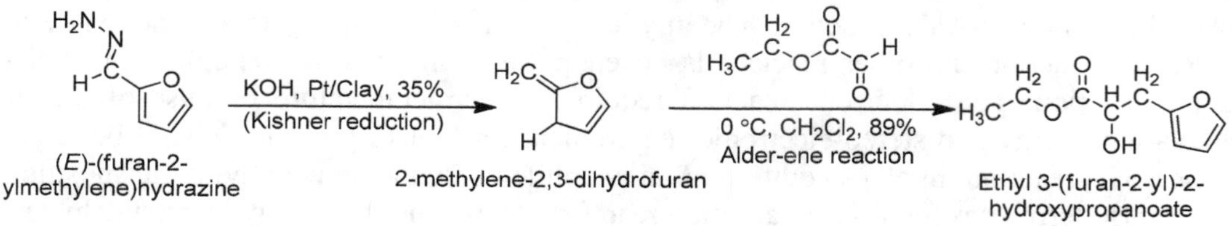

3. Alder-Rickert Reaction

Principle

Alder-Rickert reaction was first invented by Alder and Rickert in 1936. The reaction is extension of a Diels-Alder reaction. Diels-Alder cycloadducts extrude the cleavable groups to give more stable aromatic compounds under thermal conditions or in presence of acid or base is known as Alder-Rickert reaction. In addition, Diels-Alder cycloadducts may be converted into aromatic compounds *via* rearrangement or oxidation.

General Reaction

In this reaction, R' and R'' groups of reactants are eliminated. A formation of stable aromatics will be the driving force for Alder-Rickert reaction.

Application

a) The reaction is applicable in synthesis of different derivatives of aromatics.

4. Allan-Robinson Condensation

Principle

Allan and Robinson discovered condensation reaction of *o*-hydroxyaryl ketones and an anhydride of aromatic acid in 1924. The synthesis of flavones or isoflavones compounds by condensation between *o*-hydroxyaryl ketones and an anhydride of aromatic acid in presence of a sodium salt of corresponding acid is called as Allan-Robinson condensation or Allan-Robinson's flavone synthesis.

General Reaction

o-hydroxyaryl ketones   Flavones or isoflavones derivatives

Mechanism

o-hydroxyaryl ketones and sodium salt of corresponding acid undergoes enolization to give reactive intermediate which on addition of an anhydride eliminate carboxylate ion. The intermediate on reaction of carboxylate ion followed by enolization gives cyclic compound which on conjugate addition results flavones or isoflavones.

Applications

The reaction is widely applicable for the synthesis of structurally different flavonoids and isoflavones.

a) Synthesis of 7-hydroxy-6-methoxy-3-(2,4,5-trimethoxyphenyl)-4H-chromen-4-one.

2,4,5-trimethoxybenzyl cyanide   2-methoxybenzene-1,3,5-triol

1) $ZnCl_2$, Et_2O
2) Pyridine, ethoxalyl chloride

7-hydroxy-6-methoxy-3-(2,4,5-trimethoxyphenyl)-4H-chromen-4-one

b) Synthesis of 5,7-dimethoxy-2,8-dimethyl-3-phenyl-4H-chromen-4-one.

2-phenyl-1-(2,4,6-trihydroxy-3-methylphenyl)ethanone

CH_3COCl, Pyridine

$(CH_3)_2SO_4$/Acetone
K_2CO_3

5,7-dimethoxy-2,8-dimethyl-3-phenyl-4H-chromen-4-one

5. Aldol Condensation

Principle

Aldol condensation reaction reported by Kane (1838) is one of the most important C-C bond formation reactions for aldehydes and ketones. In this reaction, two molecules of α-hydrogens containing aldehydes (or ketones) reacts in presence of base to form β-hydroxyaldehyde (aldol) or β-hydroxyketone. The reaction is well-known as aldol condensation reaction. Aldols on heating undergo dehydration to form α,β-unsaturated carbonyl compounds. Under kinetic control, the mixed Aldol addition can be used to prepare adducts that are otherwise difficult to obtain selectively. With an unsymmetrically substituted ketone, such a non-nucleophilic, sterically-demanding, strong base will abstract a proton from the least hindered side. Proton transfer is avoided with lithium enolates at low temperatures in ethereal solvents, so that addition of a second carbonyl partner (ketone or aldehyde) will produce the desired aldol product.

General Reaction

Ethanal + Ethanal $\xrightarrow{\text{NaOH}}$ 3-hydroxybutanal (an aldol)

Propanal + Propanal $\xrightarrow{\text{NaOH}}$ 3-hydroxy-2-methylpentanal (an aldol)

Mechanism

Step 1: Abstraction of α-hydrogen of aldehyde by base results in formation of enolate ion. α-hydrogen as being acidic easily abstracted by base to produce carbanion.

α hydrogen $\longrightarrow$ Carbanion

Step 2: Nucleophilic attack of carbanion on a second molecule of aldehyde. The carbonyl carbon is electrophilic in nature and is attacked by a nucleophilic carbanion.

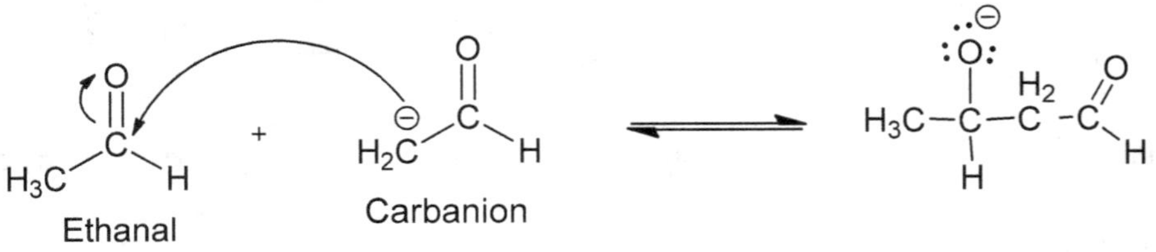

Step 3: Protonation (Formation of aldol). Heating results in the elimination of water from aldol to yield an α,β-unsaturated aldehyde. Dehydration may be affected by mineral acids.

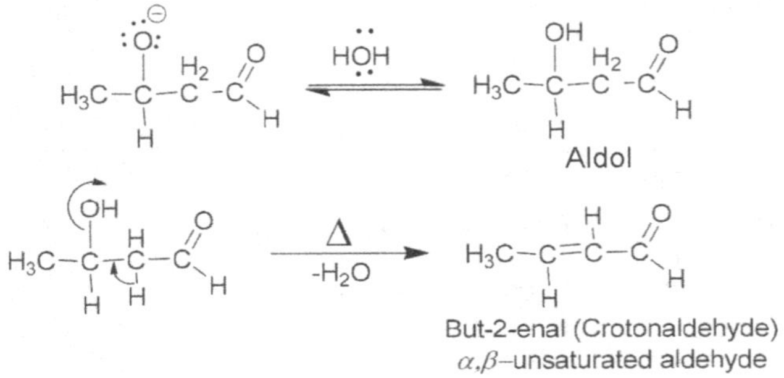

The ketones containing α-hydrogen's also undergo condensation to produce β-hydroxy ketones (ketols). 2-propanone (acetone) in presence of a base undergoes condensation to form 4-hydroxy-4-methylpentan-2-one.

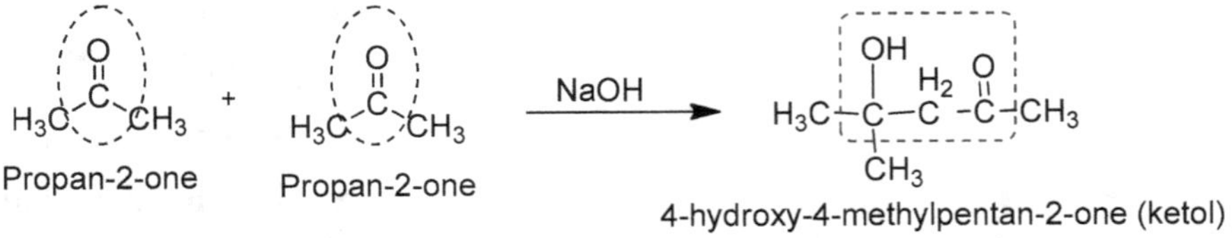

Ketols in acidic conditions or heating undergo dehydration to yield α,β-unsaturated ketones.

Crossed Aldol Condensation

To avoid formation of a mixture of products, the aldol condensation with different aldehydes is performed in such a way that one aldehyde contains α-hydrogen and other does not have any α-hydrogen. During reaction carbanion is formed from the α-hydrogen containing aldehyde.

Benzaldehyde (No α hydrogen)   Ethanal (α hydrogen)   3-hydroxy-3-phenylpropanal

Applications

a) *Both simple and crossed aldol condensations are used for synthesizing saturated and unsaturated aldehydes and alcohols of synthetic importance, for example: Synthesis of but-2-enal and butan-1-ol.*

3-hydroxybutanal (an aldol)   But-2-enal (Crotonaldehyde)   Butan-1-ol

b) *Synthesis of glucose from glycoaldehyde.*

3 Glycoaldehyde   →   Glucose

c) *Synthesis of Vitamin A*: The intermediate β-ionone required for the synthesis of vitamin A was synthesized by the condensation of citral (aldehyde) and acetone gives ψ-ionone; subsequent treatment of ψ-ionone with boron trifluoride gives β-ionone.

Citral   +   Acetone   → Ψ-ionone

Ψ-ionone $\xrightarrow{BF_3}$ β-ionone

6. Allylic Rearrangement

Principle

Claisen reported Allylic rearrangement in 1912. A migration of carbon-carbon double bond in three-carbon system (allylic; $>C=C-CH_2$), often occurring on the nucleophilic substitution of allylic systems in which a nucleophile adds to the double bond ($>C=C<$) along with the cleavage of allylic leaving group is termed as allylic rearrangement. As allylic cation is relatively stable, the allylic rearrangement is competed with regular S_N1 and S_N2 reactions; therefore, the allylic rearrangement is also known as S_N1' or S_N2' reaction. The allylic rearrangement is influenced by light, enzymes, solvents, lewis acids and transition metal catalysts (tungsten, rhodium, cobalt and palladium). The compounds which have a functional group X other than an unsaturated linkage on a carbon atom α to a double bond are known as allylic compounds. These compounds, when subjected to acid or base catalysis, undergo functional group (X) migrations to yield new compounds.

General Reaction

Allylic systems + $Y^{\ominus}$ ⟶ New allylic systems +

Allylic compounds (I) $\xrightarrow{\overset{\oplus}{H} \text{ or } \overset{\ominus}{OH}}$ New allylic compounds (II)

But-2-en-1-ol $\xrightarrow{\overset{\oplus}{H} \text{ or } \overset{\ominus}{OH}}$ But-3-en-2-ol

Mechanism

(a) S_N1 Mechanism

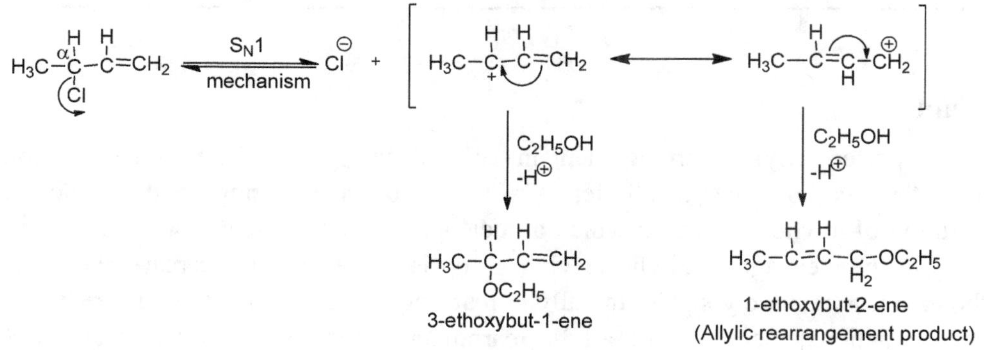

[Chemical mechanism diagram showing S_N1 mechanism: starting material $H_3C-CHCl-CH=CH_2$ (with α-carbon labeled) reacting to form $Cl^{\ominus}$ plus a resonance-stabilized allylic carbocation. Reaction with C_2H_5OH, $-H^{\oplus}$ gives two products.]

Product 1: 3-ethoxybut-1-ene ($H_3C-CH(OC_2H_5)-CH=CH_2$)

Product 2: 1-ethoxybut-2-ene
(Allylic rearrangement product) ($H_3C-CH=CH-CH_2-OC_2H_5$)

(b) S_N2 Mechanism

$C_2H_5O^{\ominus}$ (Nucleophile) + $H_2C=CH-CHCl_2$ (3,3-dichloroprop-1-ene) $\xrightarrow{\text{$S_N2$ mechanism}}$ $C_2H_5-O-CH_2-CH=CH-Cl$

1-chloro-3-ethoxyprop-1-ene
(Allylic rearrangement product 60%)

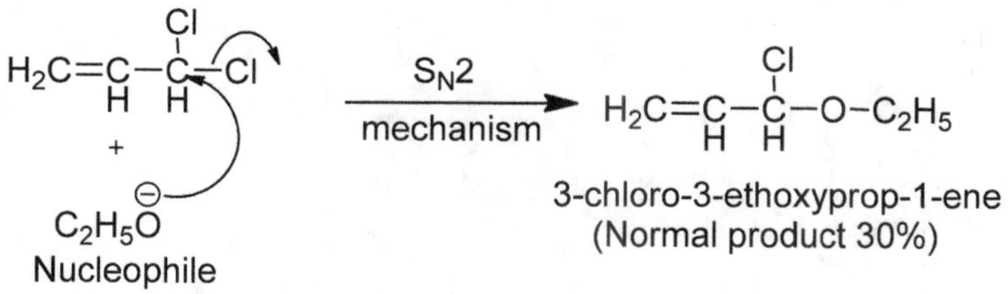

$H_2C=CH-CHCl_2$ + $C_2H_5O^{\ominus}$ (Nucleophile) $\xrightarrow{\text{$S_N2$ mechanism}}$ $H_2C=CH-CHCl-O-C_2H_5$

3-chloro-3-ethoxyprop-1-ene
(Normal product 30%)

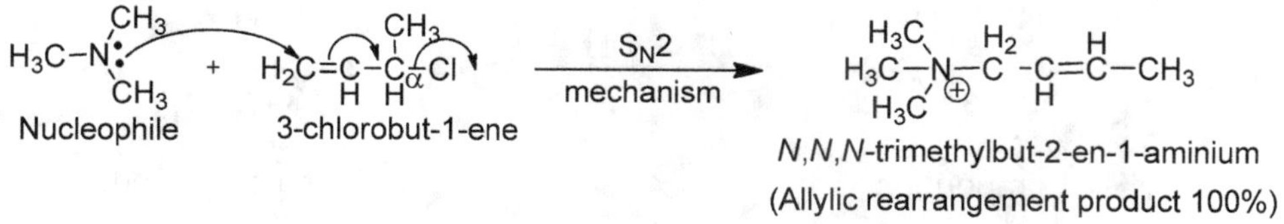

$H_3C-N(CH_3)_2$ (Nucleophile) + $H_2C=CH-CHCl-CH_3$ (3-chlorobut-1-ene) $\xrightarrow{\text{$S_N2$ mechanism}}$ $(H_3C)_3N^{\oplus}-CH_2-CH=CH-CH_3$

N,N,N-trimethylbut-2-en-1-aminium
(Allylic rearrangement product 100%)

4,4-dichloro-3-phenylcyclobut-2-enone + $C_2H_5O^{\ominus}$ (Nucleophile) $\xrightarrow{\text{$S_N2$ mechanism}}$ 2-chloro-4-ethoxy-3-phenylcyclobut-2-enone
(Allylic rearrangement product 100%)

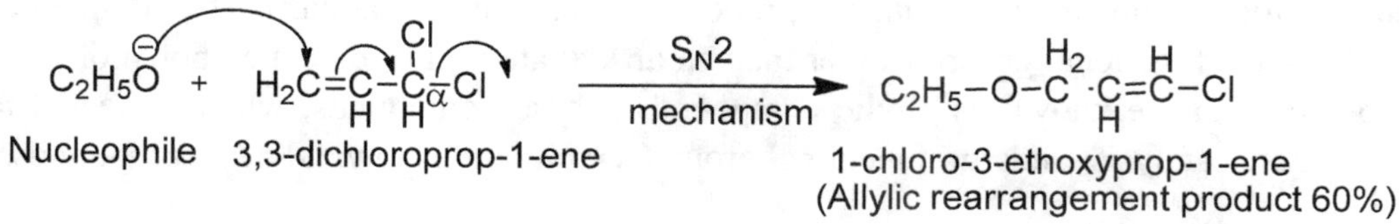

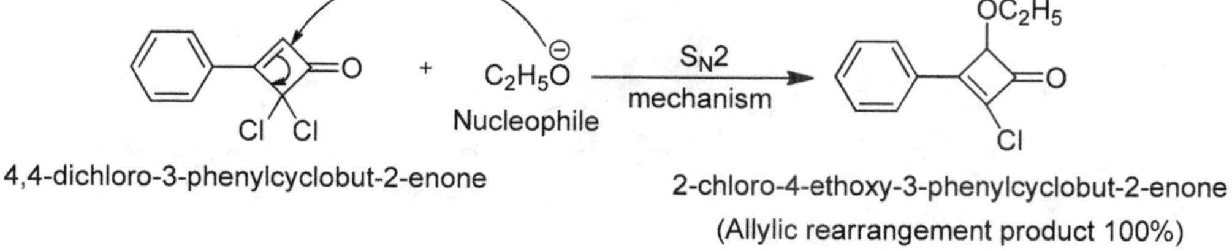

But-3-en-2-ol → 3-chlorobut-1-ene + 1-chlorobut-2-ene (via $SOCl_2$, -HCl)

But-3-en-2-ol → 1-chlorobut-2-ene (via $SOCl_2$, -HCl; $-SO_2$)
(Allylic rearrangement product 100%)

Applications

The allylic rearrangement is useful in synthesis of organic molecules.

a) Preparation of 1-ethoxybut-2-ene and 3-ethoxybut-1-ene from 3-chlorobut-1-ene.

3-chlorobut-1-ene → 3-ethoxybut-1-ene + 1-ethoxybut-2-ene (via C_2H_5OH)

b) Preparation of 3-methylbut-2-en-1-ol from 2-methylbut-3-en-2-ol.

2-methylbut-3-en-2-ol → 3-methylbut-2-en-1-ol (via $\overset{\oplus}{H}$ or $\overset{\ominus}{OH}$)

Principle

Amadori rearrangement was first reported in 1925. The transformation of *N*-glycosides of aldoses into *N*-glycosides of corresponding ketoses under acidic condition (conversion of aldimines into ketoamines) is known as Amadori rearrangement. The rearrangement is involved in the formation of osazone from glucose. This reaction also appears in the formation of complex glycation end products has very important role in biological processes. Amdori Glucosamine rearrangement is different from glycosylation which involves the formation of glycosides.

General Reaction

Mechanism

Step 1: Reaction involves protonation of ring oxygen in glycosylamine which on subsequent ring opening results the intermediate compound.

Step 2: Keto-enol tautomerism followed by ring closure to form 1-amino-1-deoxyketose

Enol form

Keto form

1-amino-1-deoxyketose

Applications

In the *Maillard reaction* pathway, Amadori rearrangement appears to be involved in the manifestations of the pathological effects of Diabetes, Alzheimer's disease, and aging processes. The reaction has been extensively applied for the preparation of glycoproteins by reaction of reducing sugars with side chains of lysine and arginine residues in proteins. In addition, the Amadori rearrangement has been also used for the preparation of amino polysaccharides.

a) Synthesis of 1-amino-1-deoxy-D-fructose from β-D-glucopyranosylamine.

β-D-glucopyranosylamine

1-amino-1-deoxy-D-fructose

8.   Angeli-Remini Reaction

Principle

The preparation of hydroxamic acids from aldehyde and benzosulfohydroxamic acid is known as Angeli-Remini reaction. The reaction was given by Angeli in 1896. As of today, many other methods have been developed to prepare different hydroxamic acids.

General Reaction

Benzaldehyde + N-hydroxybenzenesulfonamide $\xrightarrow{NaOCH_3}$ Hydroxamic acids

Mechanism

Step 1: Attack of nucleophilic $^-OCH_3$ group results in the deprotonation of N-hydroxy benzene sulfonamide to form hydroxyl (phenylsulfonyl) amide. Hydroxy (phenylsulfonyl) amide attack at carbonyl carbon of benzaldehyde to form intermediate.

Step 2: Intermediate ion upon hydride shift gives hydroxamic acid.

Applications

a) *The reaction is utilized for the synthesis of different substituted hydroxamic acids.*

   For example: Formation of N-hydroxy-3-phenylpropanamide from alkoxyamine resin.

Alkoxyamine resin $\xrightarrow[\text{2) TFA/H}_2\text{O in CH}_2\text{Cl}_2]{\text{1) Ph(CH}_2)_2\text{COCl, iPr}_2\text{NEt, CH}_2\text{Cl}_2}$ N-hydroxy-3-phenylpropanamide

9. Anschutz Anthracene Synthesis

Principle

The reaction was intially reported Anschutz during 1886. A synthesis of anthracene from vinyl bromide and benzene in the presence of aluminum chloride as a catalyst is popularly referred as Anschutz Anthracene synthesis. In addition, methyl phenyl carbinol or 1,1-diphenyl ethane can be converted into anthracene derivative in presence of lewis acid (*For example*: $AlCl_3$).

General Reaction

1. Benzene is converted into 9,10-dimethyl-9,10-dihydroanthracene in presence of $AlCl_3$.

Benzene   bromoethene   9,10-dimethyl-9,10-dihydroanthracene

2. Ethane-1,1-diyldibenzene is converted into 9,10-dimethyl-9,10-dihydroanthracene in presence of $AlCl_3$.

Ethane-1,1-diyldibenzene   9,10-dimethyl-9,10-dihydroanthracene

Mechanism

The reaction mechanism is quite similar to that of Friedel-Crafts Alkylation.

Step 1:

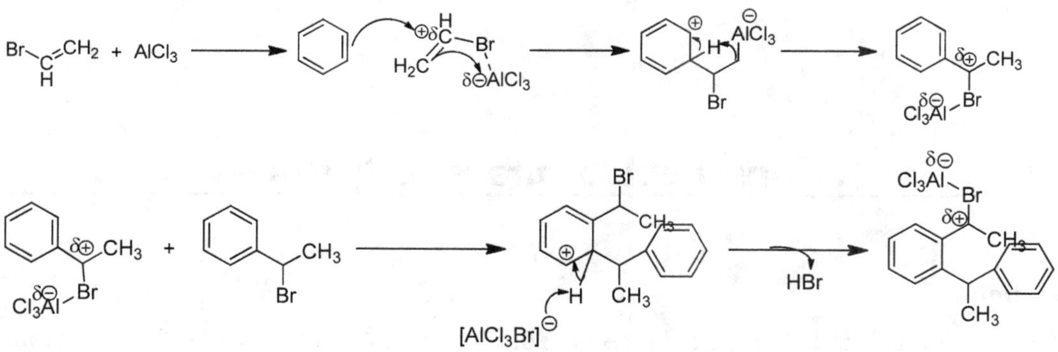

Step 2:

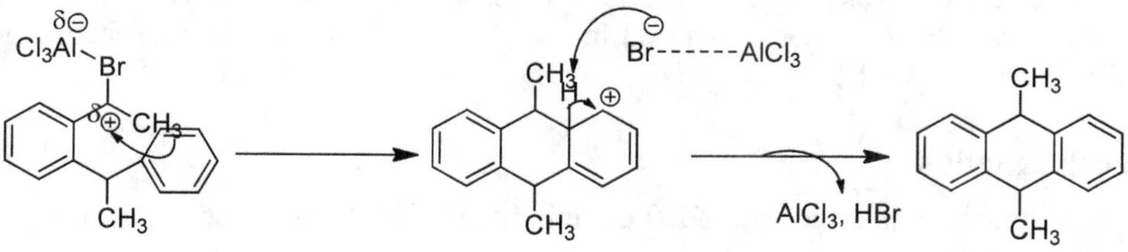

Applications

a) *Reaction is helpful in the formation of 9, 10-dimethyl-9,10-dihydroanthracene from benzene and ethyne in presence of aluminum chloride.*

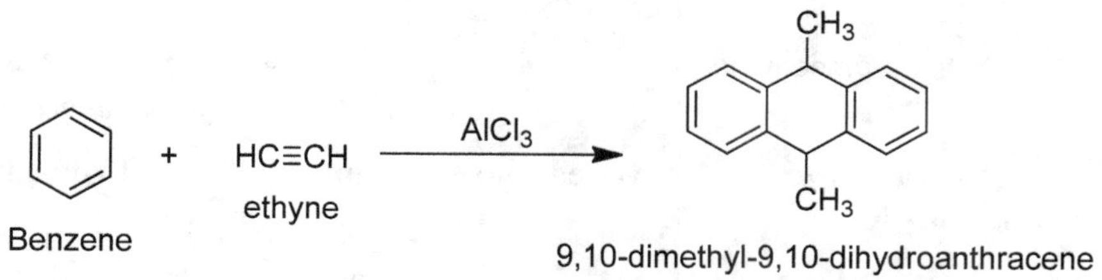

Benzene + $HC{\equiv}CH$ (ethyne) $\xrightarrow{AlCl_3}$ 9,10-dimethyl-9,10-dihydroanthracene

b) *9, 10-dimethyl-9,10-dihydroanthracene is obtained from benzene and chloroethene in presence of lewis acid aluminum chloride.*

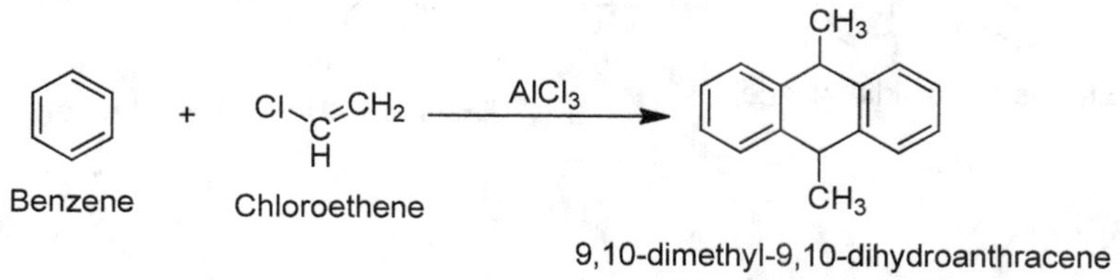

Benzene + Chloroethene $\xrightarrow{AlCl_3}$ 9,10-dimethyl-9,10-dihydroanthracene

10. Appel Reaction

Principle

The reaction was first reported in 1966 by Lee after initial work of Horner for halogenation of alcohol using triarylphosphine dihalides in 1959. Primary and secondary alcohols are converted into corresponding chlorides in presence of triphenylphosphine and carbon tetrachloride is popularly referred as Appel reaction. A mixture of triphenylphosphine and carbon tetrachloride is known as Appel reagent.

General Reaction

CX_4, PPh_3

Primary or Secondary alcohol

R = Aryl, Alkyl;
R$_1$ = Aryl, Alkyl, H

Corresponding halides

For example: Transformation of propan-2-ol to 2-chloropropane in presence of Appel reagent

CCl_4, PPh_3

Propan-2-ol

2-chloropropane

Mechanism

Step 1: Formation of Appel's salt

Appel's salt

Step 2: Appel reagent reacts with alcohol to form corresponding chlorides

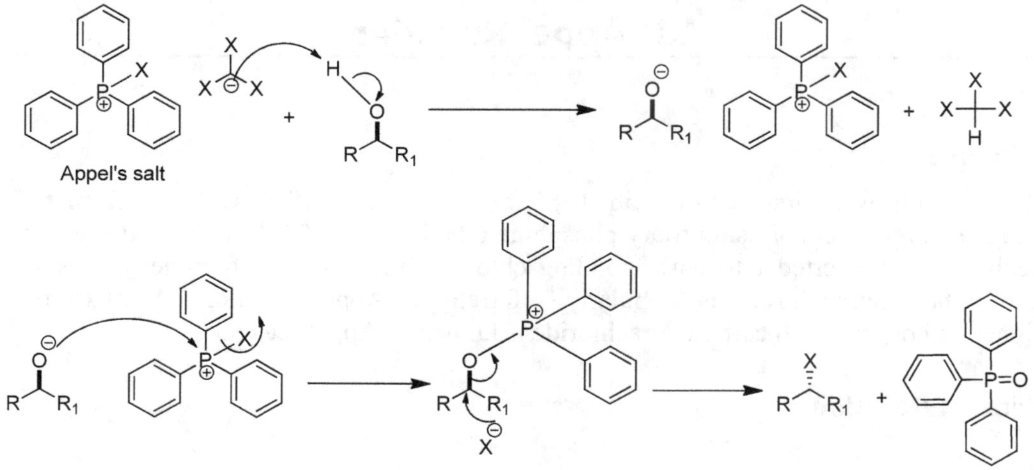

Appel's salt

Applications

The reaction has general applications in preparation of alkyl chlorides.

   a) Conversion of (S)-octan-2-ol to (S)-2-chlorooctane.

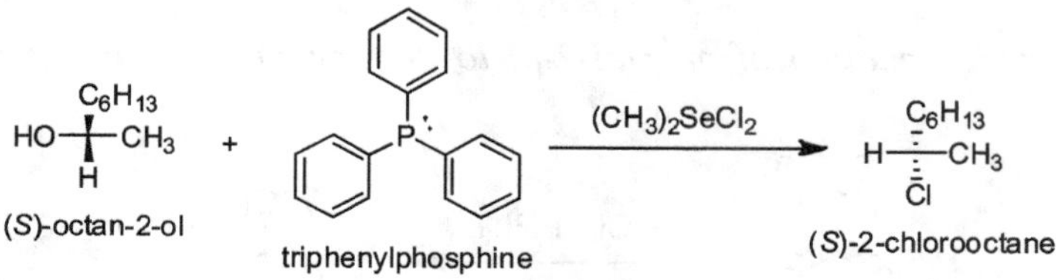

(S)-octan-2-ol   triphenylphosphine $(CH_3)_2SeCl_2$ (S)-2-chlorooctane

11. Arndt-Eistert Synthesis

Principle

The formation of homologated carboxylic acids or their derivatives by reaction of activated carboxylic acids with diazomethane and subsequent Wolff rearrangement of intermediate diazoketones in presence of nucleophiles such as water, alcohols, or amines is known as Arndt-Eistert synthesis. This is one of the vitally used methods for one carbon homologation of carboxylic acids using diazomethane. The reaction was first reported by Arndt and Eistert in 1935. The extension of carboxylic acid by one CH_2 unit by the reaction of acyl chloride with diazomethane is also an example Arndt-Eistert synthesis.

General Reaction

Carboxylic acid   Higher homologue of carboxylic acid

1. The first step involves the conversion of carboxylic acid into acid chloride in presence of thionyl chloride.

$$R-COOH \ + \ SOCl_2 \longrightarrow R-COCl \ + \ SO_2 \ + \ HCl$$

Acid   Thionyl chloride   Acid chloride

2. When acid chloride is reacted with the excess of diazomethane results the formation of a diazoketone. If the quantity of diazomethane is not excess, the halomethylketone is formed by the action of hydrogen chloride on diazomethane. The excess diazomethane can be removed by addition of small amounts of acetic acid or after vigorous stirring. Most of α-diazoketones are stable; isolated and purified by chromatographic techniques.

$$R-COCl \ + \ CH_2N_2 \longrightarrow R-COCHN_2 \ + \ CH_3\text{-}Cl$$

Acid chloride   Diazomethane   Diazoketone

$$R-COCl \ + \ CH_2N_2 \longrightarrow R-COCHN_2 \ + \ HCl$$

Acid chloride   Diazomethane   Diazoketone

$$R-COCHN_2 \ + \ HCl \longrightarrow R-COCH_2Cl \ + \ N\equiv N$$

3. Finally, diazoketones undergoes decomposition and rearrangement in presence of catalysts (Silver oxide, colloidal silver, Pt or Cu, silver benzoate, and triethylamine) to give the final products. Thus, an acid is formed in the presence of water and alcohol; reaction of ammonia with diazoketone gives aimde, primary amines on reaction yields substituted amides.

$$R-\overset{\overset{O}{\|}}{C}-CHN_2 \quad \xrightarrow{Ag_2O}$$

Diazoketone

- $\xrightarrow{H_2O}$ R-CH$_2$-COOH Carboxylic acid
- $\xrightarrow{C_2H_5OH}$ R-CH$_2$-COOR' Ester
- $\xrightarrow{NH_3}$ R-CH$_2$-CONH$_2$ Amide
- $\xrightarrow{R'-NH_2}$ R-CH$_2$-CONH-R'

$$R-\overset{\overset{O}{\|}}{C}-CHN_2 \; + \; H_2O \xrightarrow[\substack{Without \\ Catalyst}]{HCOOH} R-CO-CH_2OH$$

Diazoketone   Hydroxyketone

Mechanism

The route for synthesis of a carboxylic acid from a lower member is known as Arndt-Eistert synthesis while decomposition of the intermediate diazoketone is Wolff rearrangement. The mechanism of reaction is analogous to Curtius rearrangement.

Step 1: Nucleophilic attack on carbonyl carbon of the acid chloride molecule yields an intermediate which lose chloride ion to form diazoketone.

$$H_2C=\overset{\oplus}{N}=\overset{\ominus}{N}: \quad \longleftrightarrow \quad H_2\overset{\ominus}{C}-\overset{\oplus}{N}\equiv N:$$

$$Cl-\overset{O}{\overset{\|}{C}}-R \;+\; H_2\overset{\ominus}{C}-\overset{\oplus}{N}\equiv N: \xrightarrow[\text{attack}]{\text{Nucleophilic}} \quad \xrightarrow{-HCl} \quad \text{Diazoketone}$$

Step 2: Diazoketone spits off a molecule to form a carbene. Carbene rearrange itself to give a ketene as an intermediate.

$$R-\overset{O}{\overset{\ominus}{C}}=\overset{H}{\overset{\oplus}{C}}-N\equiv N \xrightarrow[-N\equiv N]{Ag_2O} R-\overset{\overset{O}{\|}}{C}-\overset{..}{C}H \xrightarrow{1,2\text{-shift}} R-\overset{H}{\underset{}{C}}=C=O$$

Carbene   Ketene

Step 3: Ketene undergo hydrolysis to produce higher homologue of carboxylic acid.

$$R-\overset{H}{\underset{}{C}}=C=O \xrightarrow{H_2O} R-\overset{H_2}{C}-COOH$$

Ketene   Carboxylic acid

If the second step of Arndt-Eistert synthesis is carried out in alcohol or ammonia or an amine as a solvent, the final product is an ester or an amide of the homologous carboxylic acids.

Applications

a) *Synthesis of acids and their homologues:* The reaction is applicable to the synthesis of almost all types of acyclic, alicyclic, aromatic, carbocyclic and heterocarboxylic acids. For example: Synthesis of α-naphthylacetic acid.

α-naphthoic acid →(SOCl₂, CH₂N₂)→ Diazo-α-acetonaphthone →(Ag₂O, H₂O, Δ)→ α-naphthylacetic acid

b) *Synthesis of natural products:* Arndt-Eistert synthesis has been used for synthesizing various natural products like alkaloids (Mescaline), corticoids (Deoxycorticosterone), etc. For example: Synthesis of mescaline by Arndt-Eistert synthesis.

3,4,5-trimethoxybenzoyl chloride →(CH₂N₂, AgNO₃-NH₂)→ 2-(3,4,5-trimethoxyphenyl)acetamide →(LiAlH₄, [H])→ Mezcaline

c) Peptides that contain β-amino acids feature a lower rate of metabolic degradation and are therefore of interest for pharmaceutical applications. The method is also used for preparation of β-amino acids.

Principle

The reaction was initially given by Aston and Greenburg in 1940. The transformation of α-haloketone into ester along with a migration of one of the alkyl (or aryl) groups to α-position of another moiety, when treated with alkali alkoxide (for example: $NaOC_2H_5$, $NaOCH_3$) is an example of Aston-Greenburg rearrangement. The rearrangement is useful for preparation of the esters of tertiary α-carbon.

General Reaction

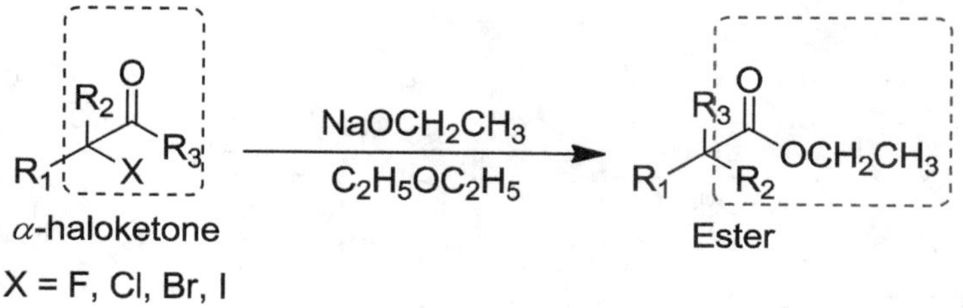

α-haloketone $\xrightarrow[\text{$C_2H_5OC_2H_5$}]{\text{$NaOCH_2CH_3$}}$ Ester

$X = F, Cl, Br, I$

Mechanism

The rearrangement involves following steps.

Step 1: The addition of an alkoxyl group to carbonyl group results in the formation of epoxide.

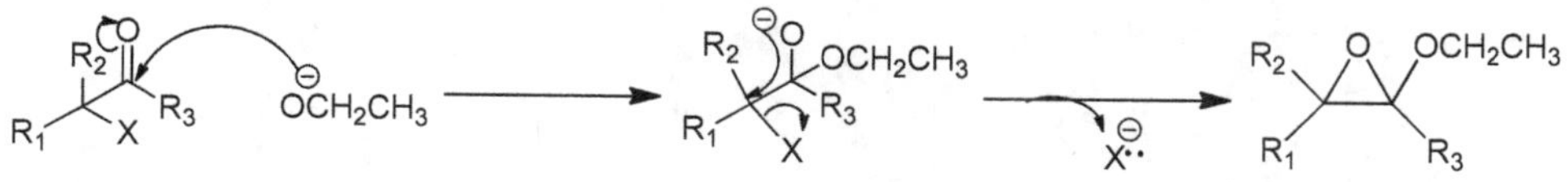

Step 2: The migration of the alkyl group in the epoxide to form ester

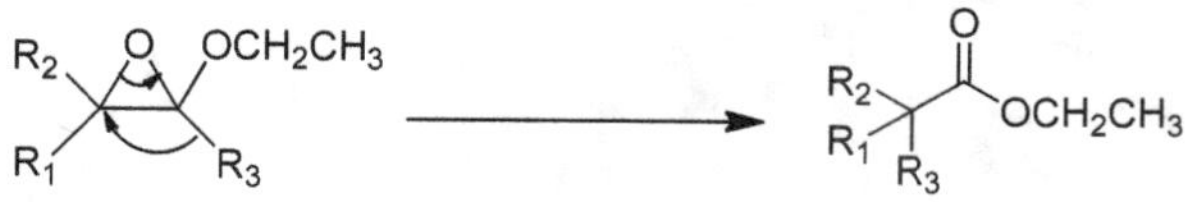

Applications

a) Synthesis of ethyl pivalate from 3-bromo-3-methylbutan-2-one.

3-bromo-3-methylbutan-2-one $\xrightarrow[C_2H_5OC_2H_5]{NaOCH_2CH_3}$ Ethyl pivalate

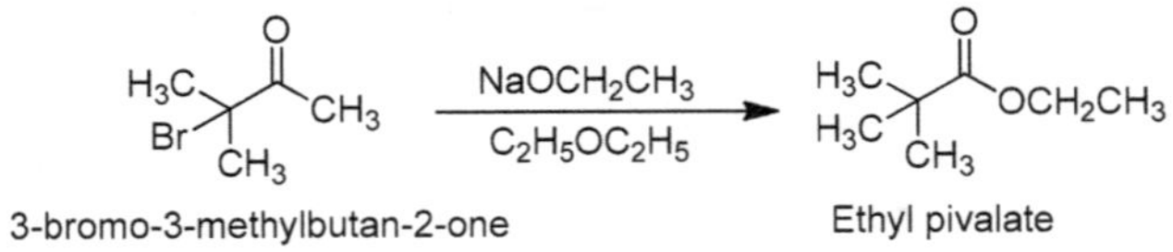

b) Conversion of chloroprogesterone to 3-keto-17-isoetienate.

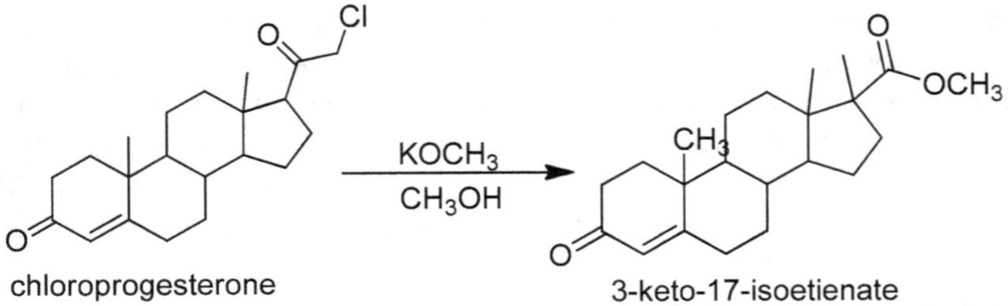

chloroprogesterone $\xrightarrow[CH_3OH]{KOCH_3}$ 3-keto-17-isoetienate

Principle

Aza-Claisen rearrangement involves simultaneous breaking of one bond and the formation of a new bond; however, in this rearrangement, one allyl carbon atom is displaced by a nitrogen atom. In general, the nitrogen atom in *aza*-Claisen rearrangement is either in a heterocycles or a quaternary amine (ammonium salt), and the extension of heterocycle ring or cleavage of ammonium salt is probably the driving force for this reaction. Although one of the earliest examples of the *aza*-Claisen rearrangement: the reaction between *N*-methylpyrrole and dimethyl acetylenedicarboxylate (DMAD) was reported by Alder in 1931, this reaction is currently receiving renewed attention.

General Reaction

1,2-divinylpyrrolidine   (1*E*,5*E*)-3,4,7,8-tetrahydro-2*H*-azonine

Mechanism

Concerted mechanism involving simultaneous bond breaking and making to form cyclic product.

Applications

   a) *Aza-Claisen rearrangement is widely applicable in synthesis of various types of nitrogen-containing compounds.*

N-allylpyrrolidine

2-phenylacetyl fluoride

E-3S,8R-1-benzyl-8-(tert-butyldimethylsilyloxy)-3-phenyl-2,3,4,7,8,9-hexahydro-1*H*-azonin-2-one

14. Baeyer Indole Synthesis

Principle

The synthesis of indole *via* alkaline reduction of *o*-nitro cinnamic acid in presence of iron was reported by Baeyer and Emmerling in 1869. Fischer Indole synthesis was known to be the most versatile and famous approach for synthesis of indoles, due to convinced advantages over Baeyer Indole synthesis.

General Reaction

o-nitro-cinnamic acid   Indole

Mechanism

Step 1:

Step 2: Acidic Workup

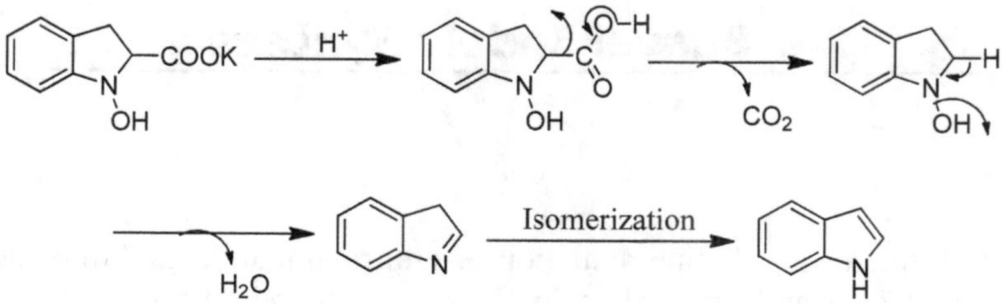

$$\xrightarrow{\text{Isomerization}}$$

Applications

a) *Synthesis of 1H-indole from 2-ethylaniline.*

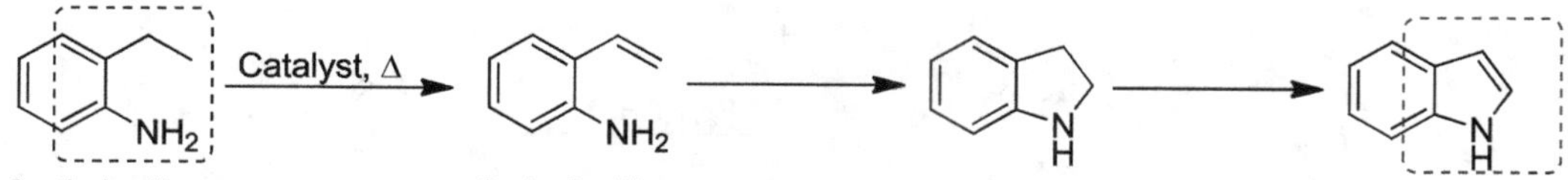

2-ethylaniline $\xrightarrow{\text{Catalyst, }\Delta}$ 2-vinylaniline $\longrightarrow$ indoline $\longrightarrow$ 1*H*-indole

b) Preparation of diethyl 1H-indole-3,6-dicarboxylate from ethyl 4-(1-cyano-2-ethoxy-2-oxoethyl)benzoate.

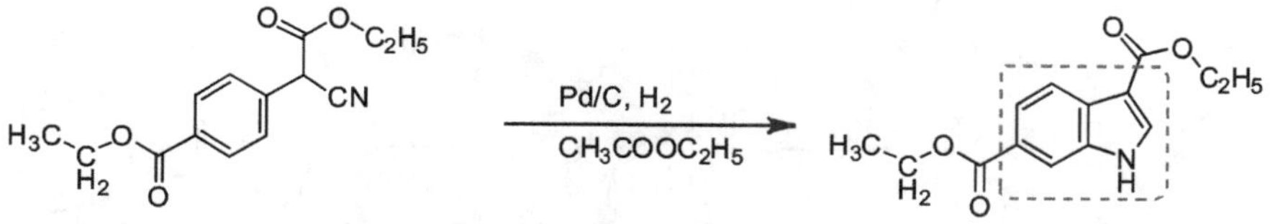

Ethyl 4-(1-cyano-2-ethoxy-2-oxoethyl)benzoate $\xrightarrow[\text{CH}_3\text{COOC}_2\text{H}_5]{\text{Pd/C, H}_2}$ Diethyl 1*H*-indole-3,6-dicarboxylate

15. Baeyer Oxindole Synthesis

Principle

A synthesis of oxindole *via* acidic reduction of *o*-nitrophenylacetic acid by tin and the subsequent cyclization of the resulting reduced intermediate is known as Baeyer oxindole synthesis. Von Baeyer in 1878 reported Baeyer oxindole synthesis. Different acidic reducing agents have been employed in thesis of substituted oxindoles with better yield. For example: $Zn + H_2SO_4$, $Zn + CH_3COOH$ $Fe + HCOOH$, $SnCl_2 + NH_4Cl$.

General Reaction

o-nitrophenylacetic acid

Oxindoles

R = H, Br, NH₂, Alkyl

Mechanism

The detailed mechanism for the cyclization of *o*-nitrophenylacetic acid into oxindoles in presence of tin and hydrochloric acid is illustrated below:

Applications

a) *The reaction is applicable for synthesis of different substituted oxindoles. For example: Synthesis of 5-bromoindolin-2-one from 2-(5-bromo-2-nitrophenyl)acetic acid.*

2-(5-bromo-2-nitrophenyl)acetic acid

5-bromoindolin-2-one

Principle

Baeyer pyridine synthesis was reported and explored by von Baeyer in 1910. The conversion of γ-pyrones or pyrone analogues into pyridines *via* formation of pyronium salt intermediates, followed by reaction with ammonium salt (ammonium carbonate) is known as Baeyer pyridine synthesis. Reaction is useful in the synthesis of different nitrogen heterocyclic compounds. Many other methods have been reported for the synthesis of pyridine and its derivatives such as distillation of allyl ethylamine over heated lead oxide, passing a mixture of acetylene and hydrocyanic acid through a red-hot tube, heating pyrrole with sodium methylate and methylene iodide to 200°C, heating isoamyl nitrate with phosphorus pentoxide, and heating piperidine in acetic acid with silver acetate. All these methods gave fewer yields compared to Baeyer's method of pyridine synthesis. Hence this approach is more famous and versatile for the synthesis of pyridines.

General Reaction

Pyronium salt

Pyridines

For example: Conversion of γ-pyrones into 4-methoxypyridine

γ-pyrones

4-methoxypyridine

Mechanism

Step 1:

Step 2:

Applications

a) Different substituted pyridines can be easily prepared from pyrones and its analogues by Baeyer pyridine synthesis. For example: Synthesis of 4,5-dimethoxy-2-methylpyridine

5-methoxy-2-methyl-4*H*-pyran-4-one

ClO_4

4,5-dimethoxy-2-methylpyridine

Principle

Rearrangement of electron deficient oxygen atom

Baeyer-Villiger reaction is the oxidation of ketones into esters by means of peracids. With an insertion of oxygen in carbonyl compounds by Baeyer-Villiger oxidation, a new way of synthesis of esters was discovered. Bayer-Villiger oxidation is particularly an oxidation of aldehydes into carboxylic acids or ketones into esters. The oxidizing agent used in a reaction is peracid. The best oxidizing is peroxytrifluro acetic acid; other peroxy derivatives can also be used such as permonosulphuric acid (For examples: peroxymonosulfuric acid, perbenzoic acid, peracetic acid BF_3-H_2O_2). Inert organic solvents are used in reaction. The choice of solvent depends upon the solubility of the reactants. Commonly used solvents are glacial acetic acid and chloroform. The reaction was given by Baeyer and Villiger in 1899.

The most electron-rich alkyl group (more substituted carbon atom) migrates first. The migration order of alkyl group: tertiary alkyl > cyclohexyl > secondary alkyl > benzyl > phenyl > primary alkyl > methyl >> H. For aromatic substitutions: p-CH_3O-Ph > p- CH_3-Ph > p-Cl-Ph > p-Br-Ph > p-NO_2-Ph.

General Reaction

For Ketone

For aldehyde

Mechanism

Step 1: Formation of carbonium ion

Step 2: Nucleophilic attack by peroxy acid

Step 3: Migration of alkyl or aryl group to form ester in ketones; for aldehydes, it is the hydrogen atom of carboxy group that migrates.

Applications

a) *For the synthesis of esters and acids.*

$$R\text{--}\underset{\text{Ketones}}{\underset{R'}{C}}\text{=O} \;+\; \underset{\text{Peracids}}{RCOOOH} \longrightarrow \underset{\text{Esters}}{R\text{--}C(\text{=O})\text{--}O\text{--}R'} \xrightarrow{\text{Hydrolysis}} \underset{\text{Acids}}{R'\text{--}COOH} \;+\; \underset{\text{Alcohols}}{R\text{--}OH}$$

b) *For the synthesis of anhydrides. For example: When α-diketones or o-quinones are reacted with peracids gives anhydrides.*

α-naphthaquinone $\xrightarrow{RCOOOH}$ Benzo[*c*]oxepine-1,3-dione

c) *For the synthesis of lactones. For example: When cyclic ketones reacted with peracids undergo ring expansion to give lactones. Cyclohexanone yields ε-caprolactone.*

Cyclohexanone $\xrightarrow{RCOOOH}$ ε-caprolactone

Principle

The reaction of benzenediazonium chloride with hydrofluoroboric acid followed by heating to give fluorobenzene is known as Baltz-Schiemann reaction. It is well known method to insert a fluorine atom into aromatics. In this reaction, when diazonium salt is treated with a hydrofluoroboric acid, diazonium fluoroborate precipitates out. The dry diazonium flururoborate on heating affords fluorobenzene. The reaction was reported by Balz and Schiemann in 1927. The general procedure can be directly extended for preparing other aromatic halides.

Note: Diazonium flurororoborates are relatively more stable than other diazonium salts.

General Reaction

Mechanism

The reaction proceeds by S_N1 mechanism.

Step 1: Formation of arenium ion

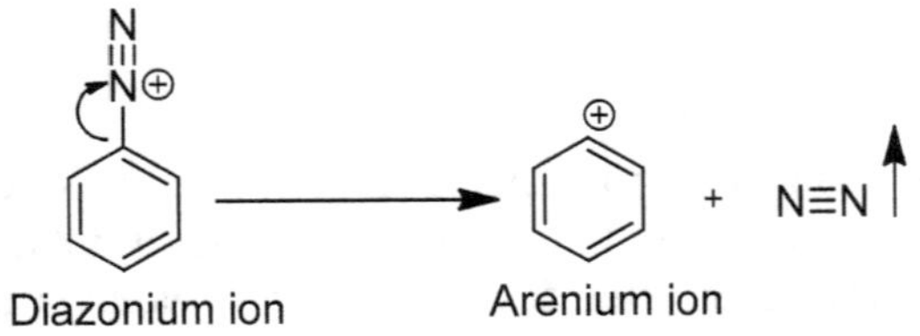

Diazonium ion   Arenium ion

Step 2: Formation of fluorobenzene

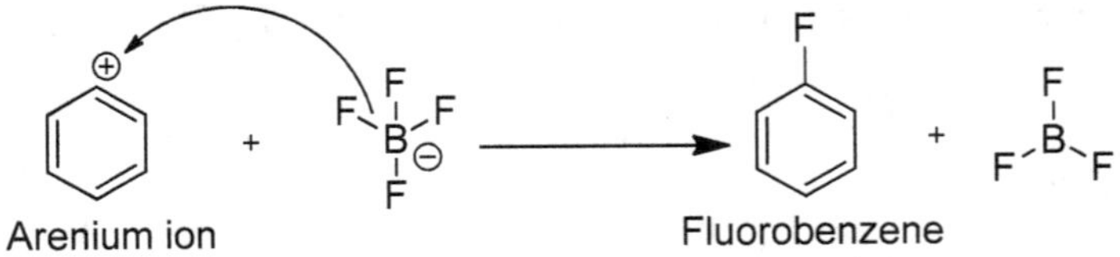

Arenium ion   Fluorobenzene

Applications

   a) Formation of substituted fluorobenzenes.

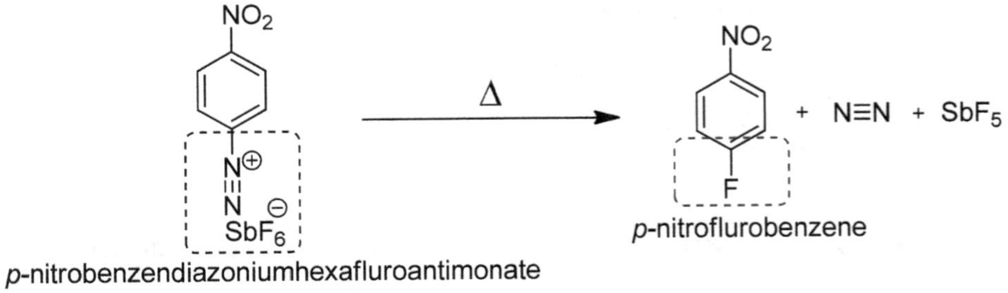

p-nitrobenzendiazoniumhexafluroantimonate

p-nitroflurobenzene

19. Baker-Venkataraman Rearrangement

Principle

Base catalyzed acyl transfer reaction that converts α-acyloxyketones to β-diketones (1,3-diketones) under basic conditionsis known as Baker-Venkataraman rearrangement. In this reaction, the migration of ester acyl group of an *o*-acylated phenolic ester takes place to form 1,-3-diketones. The reaction is analogous to the Claisen condensation, and proceeds through the formation of an enolate, followed by intramolecular acyl transfer. This intramolecular acyl transfer reaction initially reported by Baker and Venkataraman in 1933.

General Reaction

Mechanism

Step 1: Abstraction of α-hydrogen from *o*-acylated phenolic ester by base to form enolate ion.

Step 2: Migration of acyl group (Intramolecular rearrangement)

Acyl group migrates within a molecule to form a more stable intermediate through intramolecular attack of Π electrons at carbonyl carbon atom of ester.

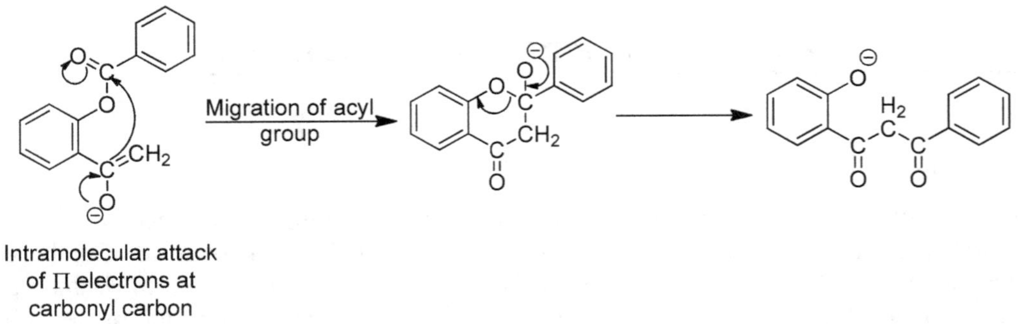

Intramolecular attack
of Π electrons at
carbonyl carbon

Step 3: Acidic workup (hydrolysis reaction)

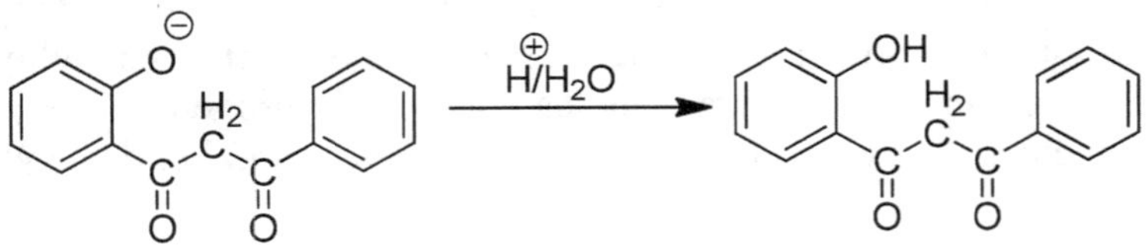

Applications

a) Baker-Venkataraman rearrangement has become a major reaction in flavone chemistry.

In this reaction migration of acyl group has been confined to aromatic or heteroaromatic acyl group. The resulting molecules can be applied to the synthesis of chromones and flavones. For example:

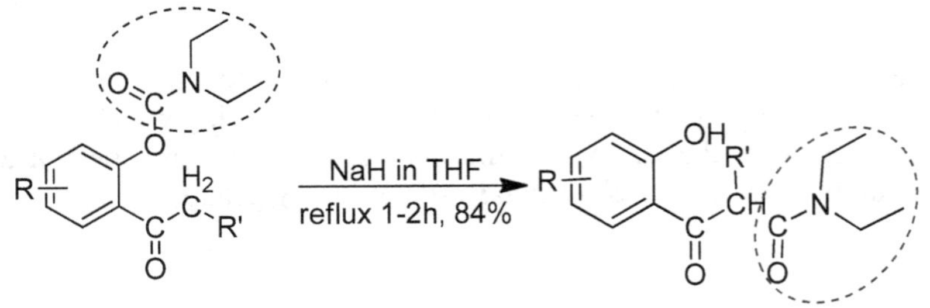

b) Transformation of 4-methoxy-5-methyl-2-propionylphenyl diethylcarbamate into 4-hydroxy-6-methoxy-3,7-dimethyl-2H-chromen-2-one.

20.　Bamberger Rearrangement

Principle

An intermolecular rearrangement of *N*-phenylhydroxylamines in presence of aqueous sulfuric acid to give the corresponding 4-aminophenols is known as Bamberger rearrangement. The reaction was reported by Bamberger in 1894. In Bamberger rearrangement, the hydroxyl group is introduced into aromatic ring through nucleophilic attack of hydrosulfate ion on an anilenium ion, which is generated by the heterolytic N-O bond cleavage of *O*-protonated *N*-phenylhydroxylamine. The alkoxy, halogen, phenoxy are incorporated into an aromatic ring when a rearrangement of *N*-phenylhydroxylamine is carried out in nucleophilic solvents, such as alcohols, hydrogen halides, and phenols. There is much mechanistic evidence to favor the heterolytic N-O bond cleavage.

General Reaction

H
N
OH
N-phenylhydroxylamine
H_2SO_4
H_2O
NH_2
HO
4-aminophenol

Mechanism

Under acidic conditions, phenylhydroxylamine is protonated and subsequently transformed into anilenium by releasing a water molecule, during which an aromatic ring carries a partial positive charge so that the nucleophilic attacking occurs at the *para* position (the amino group is an *ortho* and *para*-determining group).

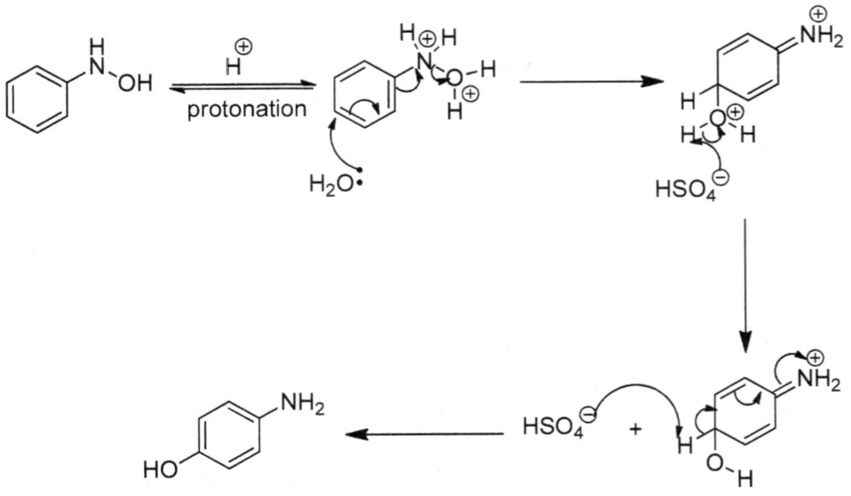

Applications

a) Bamberger rearrangement has been extended to pyridines and isoquinolones. It is reported that the nitro group (NO_2) will migrate to the ortho position when it is connected to nitrogen atom of aniline. However, in the case of O-phenylhydroxylamine, the amino group will migrate to both the ortho and the para positions.

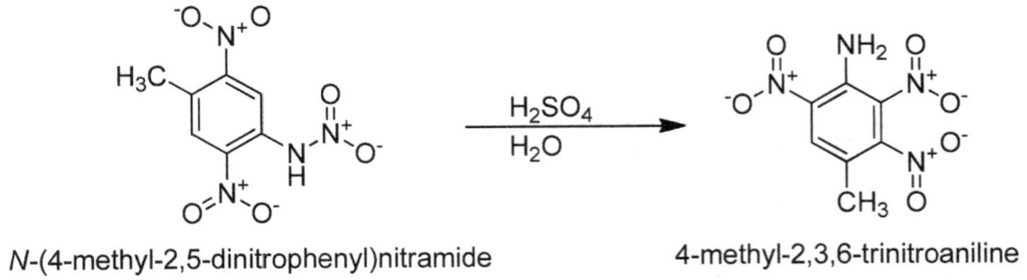

N-(4-methyl-2,5-dinitrophenyl)nitramide

4-methyl-2,3,6-trinitroaniline

b) Synthesis of substituted 4-aminophenols (Conversion of N-(2,6-dimethylphenyl)hydroxylamine into 4-amino-3,5-dimethylphenol).

N-(2,6-dimethylphenyl)hydroxylamine

4-amino-3,5-dimethylphenol

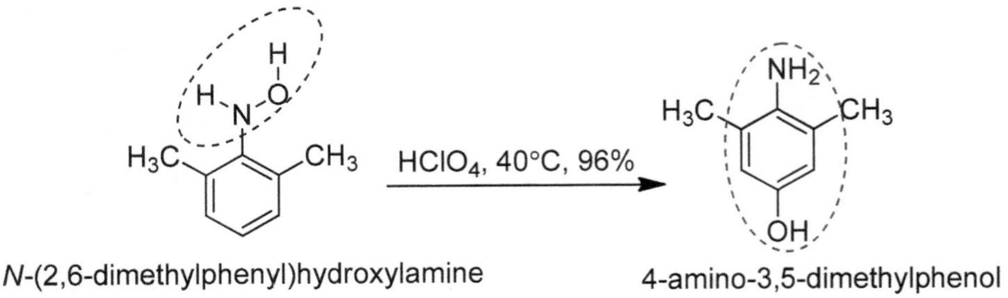

21. Barbier Reaction

Principle

Barbier reaction was reported in 1898. The reaction was closely related to Grignard reaction. In Grignard reaction, haloalkane react initially with magnesium to form Grignard reagent, followed by attack on the carbonyl group to form carbon-carbon bond; on the other hand, in Barbier reaction, both haloalkanes and carbonyl compounds are concurrently mixed with magnesium and form carbon-carbon bonds in single step. Barbier reaction is suitable for halides, such as allyl and benzyl bromides, and reaction is efficiently performed through single-electron transfer. Zero-valent metals like lithium offers exceptional yields in Barbier reaction. Barbier reaction goes through water; many metals have been reported to be effective in aqueous Barbier reaction, including aluminum, antimony, bismuth, cadmium, magnesium, manganese, samarium, gallium, indium, lead, and zinc. Nonpolar solvents favor regioselectivity, whereas polar solvents (For example: DMSO) which is strongly coordinated with the tin of an allylic tin, lead to *γ-syn* selection.

General Reaction

$$R-X \quad + \quad R_1\text{-}CO\text{-}R_2 \quad \xrightarrow{\text{Metal}} \quad R_2\text{-}C(OH)(R)\text{-}R_1$$

Alkyl halides   Ketone   Alcohol

X = Br, Cl

Mechanism

Step 1: Generation of Grignard reagent

$$H_2C{=}CH{-}CH_2{-}X \quad + \quad Mg$$

$$\downarrow$$

$$H_2C{=}CH{-}CH_2^{\ominus} \quad + \quad {}^{\oplus}MgX \; \rightleftharpoons \; \left[H_2C{=}CH{-}CH_2{-}MgX \right] \leftarrow H_2C{=}CH{-}\overset{\bullet}{C}H_2 \quad + \quad \overset{\bullet}{M}gX$$

Ionic mechanism

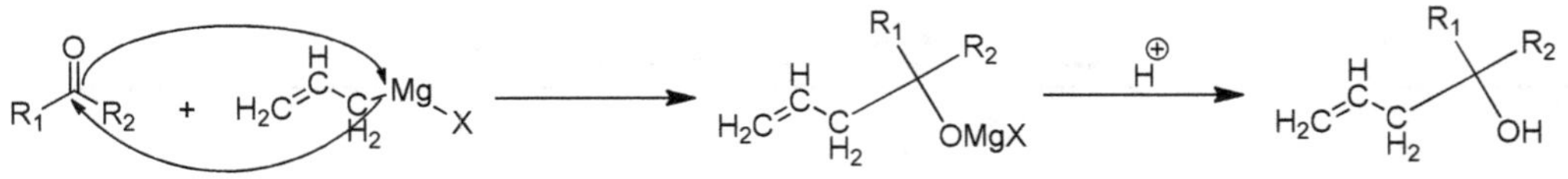

Applications

a) The reaction is widely applied in organic synthesis. For example:

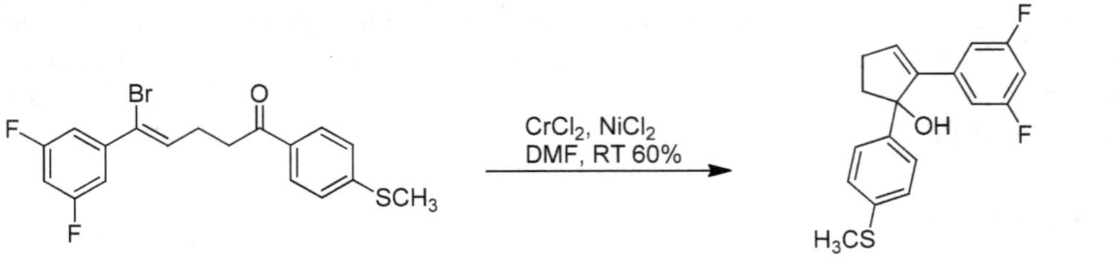

(Z)-5-bromo-5-(3,5-difluorophenyl)-1-(4-
(methylthio)phenyl)pent-4-en-1-one

2-(3,5-difluorophenyl)-1-(4-
(methylthio)phenyl)cyclopent-2-enol

22.   Barton Reaction

Principle

The reaction was first given by Barton in 1960. The conversion of a nitrite ester into a nitroso alcohol in photolytic conditions through homolytic cleavage of a nitrogen-oxygen bond followed by an intramolecular δ-hydrogen abstraction through resulting oxygen-centered radical is known as the Barton reaction. The formed nitroso compound can tautomerize to an oxime derivative that further hydrolyzes to an aldehyde or can be oxidized to nitrile.

General Reaction

butyl nitrite → 4-nitrosobutan-1-ol → (*E*)-4-hydroxybutanal oxime → 4-hydroxybutanal → tetrahydrofuran-2-ol

Mechanism

The homolytic cleavage of the nitrogen-oxygen bond is reversible, which initially conforms to a geminate recombination. However, in the subsequent stages, the partners are separated; therefore, the "solvent cage" effect is not shown in this reaction, the transformation of nitrite ester into nitroso compound is displayed here in detail.

Applications

a) *The reaction is applicable for transformation of an angular methyl groups (Δ-methyl groups) in steroids into carbonyl groups. For example: Activation of the methyl group three carbon atoms away from hydroxyl group, specific for methyl groups occurring in steroids.*

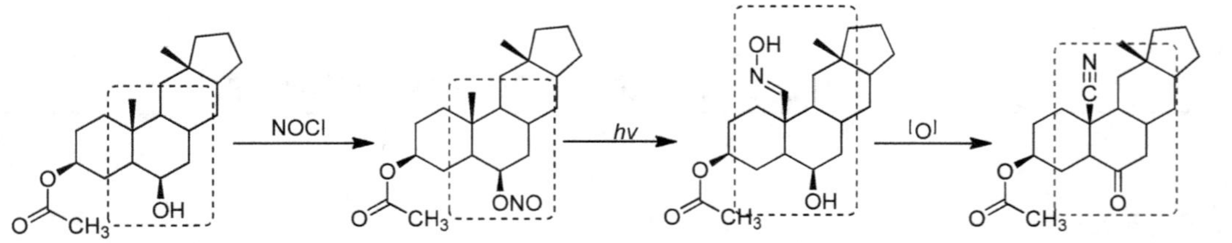

NOCl
hv
|O|
3β-acetoxy-6β-hydroxyandrostane
3β-acetoxyandrostan-6β-yl-nitrite

Principle

The reaction was discovered by Barton and Zard in 1985. The preparation of 2-substituted pyrroles (2-pyrrole-carboxylates/2-sulfonyl pyrroles) *via* alkaline condensation between alkyl isocyanoacetate (tosylmethyl isocyanide) and α,β-unsaturated nitroalkenes (β-nitroacetates) is known as Barton-Zard pyrrole reaction. The reaction is useful for the synthesis of pyrroles with various substituents at the β positions but not appropriate for the synthesis of pyrroles without substituent at position 2. A nonionic strong base (DBU and guanidine) is used in this reaction. Barton-Zard reaction is effectively used to synthesize polypyrroles and porphyrins clubbed with various aromatics or bicyclic frameworks, starting from aromatic nitro compounds and ethyl isocyanoacetate. These polypyrroles and porphyrins can be used as functional dyes.

General Reaction

Nitroalkenes   Alkyl α-isocyanoacetates   Substituted pyrroles

R_1 = H, alkyl, aryl; R_2 = H, alkyl;

R_3 = CH_3, C_2H_5; **Base** = DBU, guanidine

Nitroethene   Alkyl α-isocyanoacetates   1*H*-pyrrol-2-yl acetate

Mechanism

Step 1: Abstraction of proton by base to generate carbanion

Step 2: Attack of carbanion at nitroalkenes

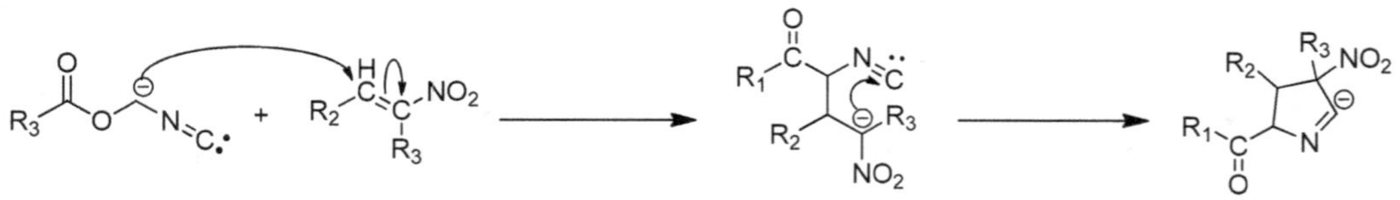

Step 3:

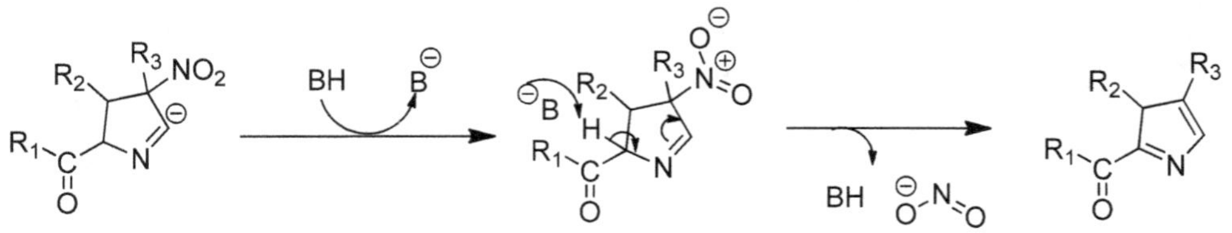

Step 4: [1,5]-migration to form substituted pyroles

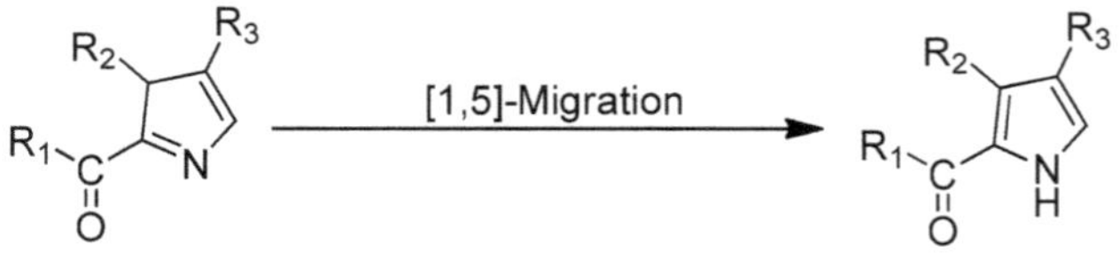

Applications

a) *The reaction is useful for synthesis of polypyrroles and porphyrins.*

For example: Synthesis of (E)-ethyl 3-acetoxy-4-(3,7-dimethylocta-2,6-dien-1-yl)-1H-pyrrole-2-carboxylate.

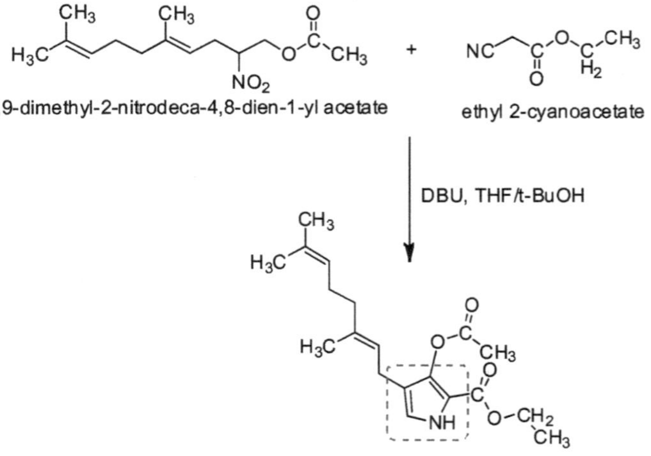

(*E*)-5,9-dimethyl-2-nitrodeca-4,8-dien-1-yl acetate   ethyl 2-cyanoacetate

DBU, THF/t-BuOH

(*E*)-ethyl 3-acetoxy-4-(3,7-dimethylocta-2,6-dien-1-yl)-1*H*-pyrrole-2-carboxylate

b) *(Z)-ethyl 3-(4-methyl-1H-pyrrol-3-yl)acrylate.*

(1*E*,3*E*)-penta-1,3-dien-1-yl propionate   +   1-((isocyanomethyl)sulfonyl)-4-methylbenzene

NaH; DMSO/$(C_2H_5)_2O$

(*Z*)-ethyl 3-(4-methyl-1*H*-pyrrol-3-yl)acrylate

24. Bartoli Indole Synthesis

Principle

Giuseppe Bartoli reported synthesis of indole in 1989. The reaction of an *ortho*-substituted nitroarenes with 3 moles of vinylmagnesium bromide to yield 7-substituted indole is known as Bartoli indole synthesis. This reaction proved to be an efficient method for the synthesis of 7-substituted indoles. In Bartoli's earlier work, he had extensively studied the reaction between nitroarenes and a lesser amount of Grignard reagents and found that aromatic nitroso compounds can be obtained by using 2 moles of vinylmagnesium bromide and nitroarenes.

General Reaction

Substituted nitrobenzene

Vinylmagnesium bromide

Substituted indole

Mechanism

The mechanism for Bartoli indole synthesis is proposed as given below, on the basis of fact that nitroso compounds may be generated from nitroarenes and 3 moles of Grignard reagents. The reaction has been modified to take place on solid-state supporting material.

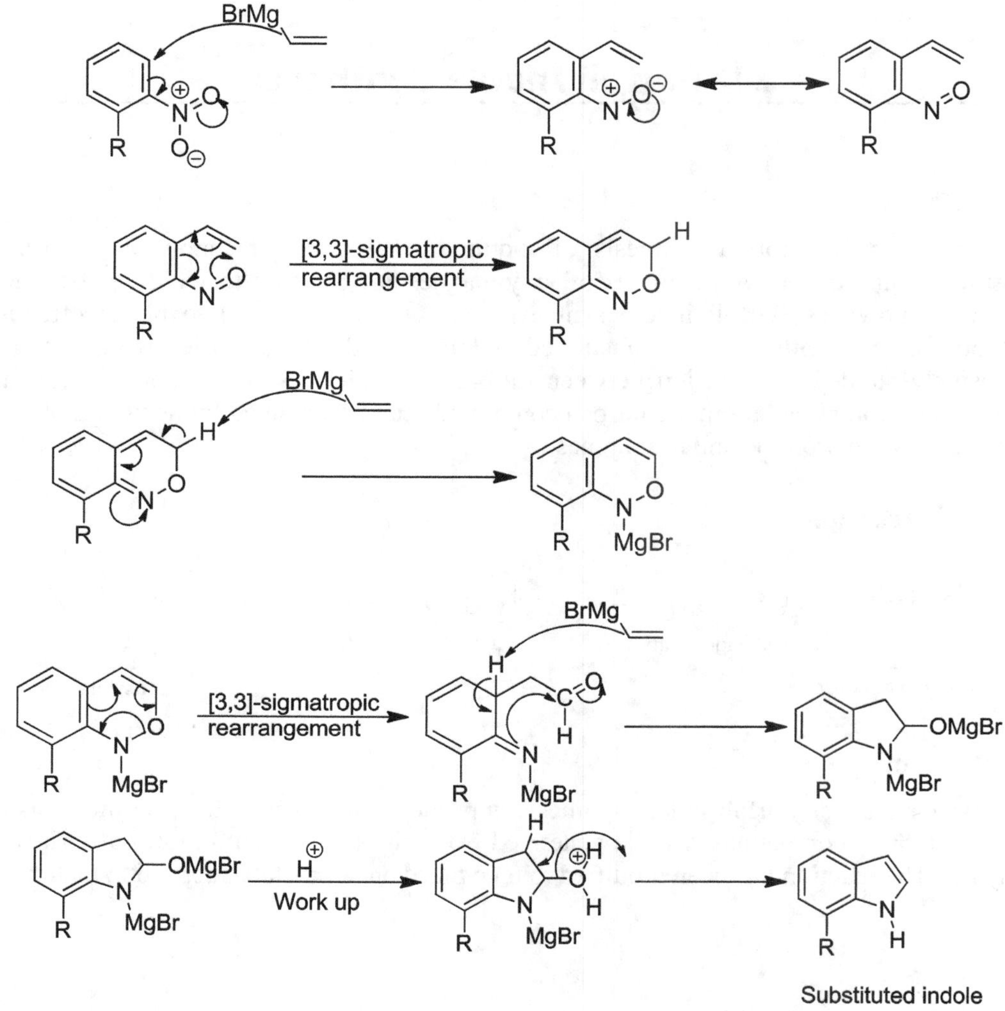

[3,3]-sigmatropic rearrangement

[3,3]-sigmatropic rearrangement

Work up

Substituted indole

Applications

a) The reaction has common application in the synthesis of 7-substituted indole analogues. Bartoli indole synthesis has also useful in synthesis of different benzofused nitrogen heterocyclic compounds such as synthesis of azaindoles and optically pure 7-alkoxyltryptophan. For example: Synthesis of 7-bromo-2,4-dimethyl-1H-indole from 1-bromo-4-methyl-2-nitrobenzene and 3 moles of vinylmagnesium bromide.

THF, -40°C, 67%

1-bromo-4-methyl-2-nitrobenzene

Vinylmagnesium bromide

7-bromo-2,4-dimethyl-1H-indole

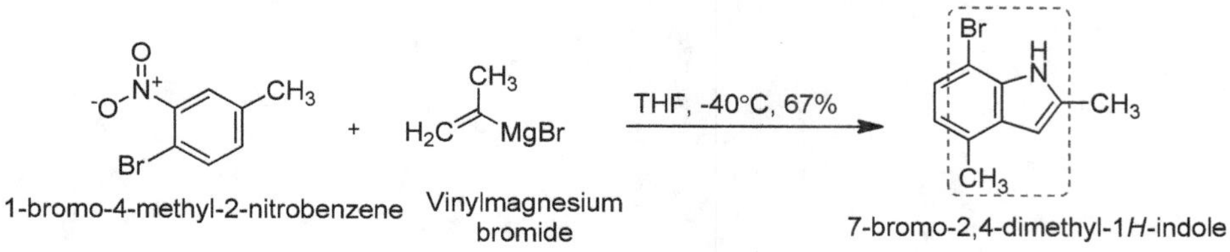

25. Baumann-Fromm Thiophene Synthesis

Principle

Baumann and Fromm reported synthesis of thiophene in 1891. The reaction between substituted styrenes and sulfur to afford 2,4-diarylthiophenes is commonly known as Baumann-Fromm thiophene synthesis. The major product formed in reaction is 2,4-diarylthiophenes along with minor product 2,5-diarylthiophenes. Similarly, butadiene's can also react with hydrogen sulfide to yield thiophenes. However, at 600°C in the presence of a ferrous sulfide-alumina catalyst, styrene reacted with hydrogen sulfide to give 60% benzothiophene.

General Reaction

Styrene

2,4-diphenylthiophene

Mechanism

It is believed that this reaction proceeds through a radical mechanism.

Step 1:

Step 2:

Step 3: Formation of 2,4-diarylthiophenes through sequence of steps with the thermal elimination of hydrogen sulphide gas.

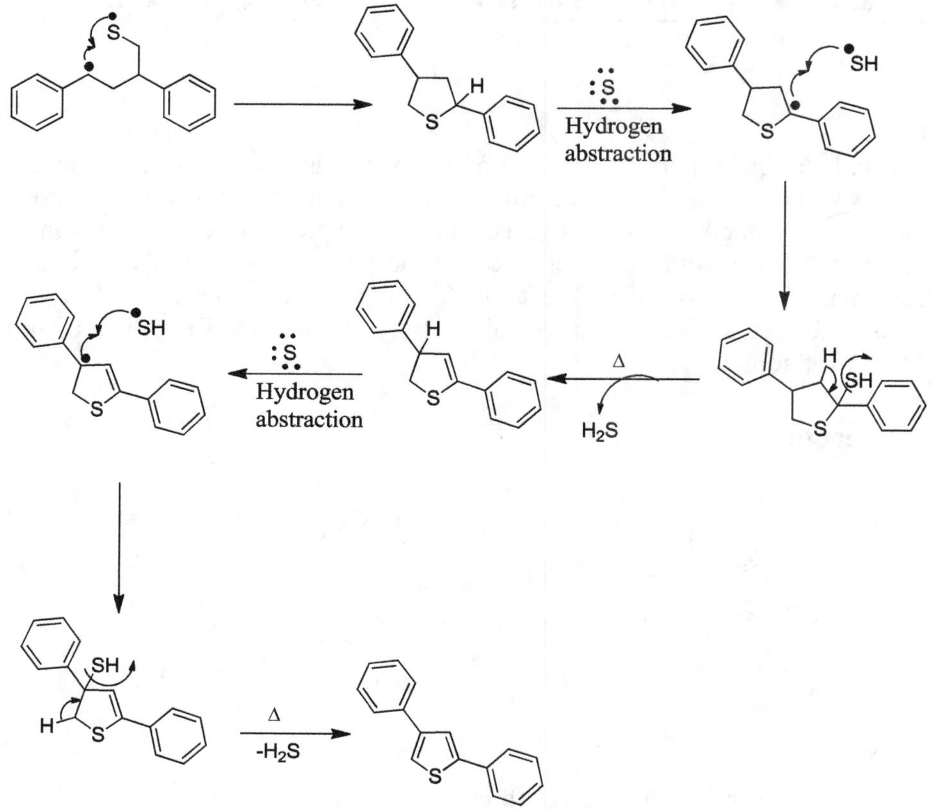

Applications

a) *The reaction is useful in synthesis of 2,4-diarylthiophene. For example: Reaction of acetophenone with hydrogen sulphide in presence of copper chromite to afford 2,4-diphenylthiophene.*

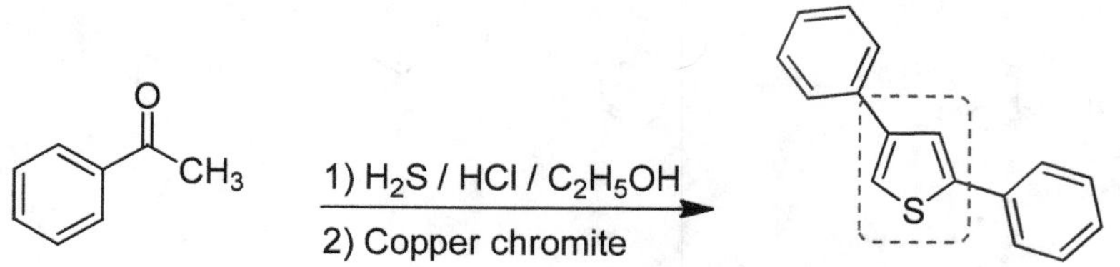

Acetophenone

1) H_2S / HCl / C_2H_5OH

2) Copper chromite

2,4-diphenylthiophene

26. Beckmann Rearrangement

Principle

The reaction of ketones with hydroxylamines results in the formation of ketoximes. Acid catalyzed rearrangement involving the conversion of ketoximes into *N*-substituted amides is known as Beckmann rearrangement. Reaction is catalyzed by using acidic reagents like H_3PO_2, H_2SO_4, P_2O_5, BF_3, SO_3, $SOCl_2$, and PCl_5 in ether. It is an example of rearrangement involving electron deficient nitrogen atom.

General Reaction

Ketoxime → *N*-substituted amide

For example: Conversion of Benzophenone oxime into Benzanilide

Benzophenone oxime → (PCl₅/Ether) → Benzanilide

Mechanism

Step 1: Generation of electron deficient nitrogen atom

Step 2: Intramolecular rearrangement reaction (*1,2* shift)

Migration of alkyl group towards electron deficient nitrogen atom results in the formation of carbocation which on acidic hydrolysis followed by tautomerism gives *N*-substituted amides.

Applications

a) *The method is useful for identification and determination of configuration of ketoximes.*

For example: Two isomeric ketoximes (syn and anti) gives different amides after Beckmann rearrangement. The amides obtained in reaction are characterized by their hydrolyzed products.

b) *Synthesis of ε-caprolactum from cyclohexanone oxime and its conversion into perlon polymers.*

27. Benzidine Rearrangement

Principle

Hofmann reported Benzidine rearrangement in 1863. An acid-catalyzed transformation of hydrazobenzenes (*N,N*-diaryhydrazines) to 4,4'-diaminobiphenyls (benzidines) is known as Benzidine rearrangement. Major product obtained in this reaction is benzidine, hence the reaction is referred as benzidine rearrangement. In this reaction, hydrobenzene (1,2-diphenylhydrazine) is warmed with dilute mineral acids (HCl or H_2SO_4) to give benzidine as a major product (70%) and isomeric diphenylene ([1,1'-biphenyl]-2,4'-diamine). Different acids are used as catalysts to synthesize benzidine by using this rearrangement.

General Reaction

Benzidine (70%)

1,2-diphenylhydrazine
(Hydrazobenzene)

Dil. HCl or H_2SO_4

+

(30%)
[1,1'-biphenyl]-2,4'-diamine (Diphenylene)

In addition to these two compounds, three other side products are also obtained in a reaction with lesser amount, namely *o*-benzidine, *o*-semidine, and *p*-semidine. Reaction of solid hydrazobenzene with dry HCl by benzidine rearrangement affords all these products in different concentration as follows:

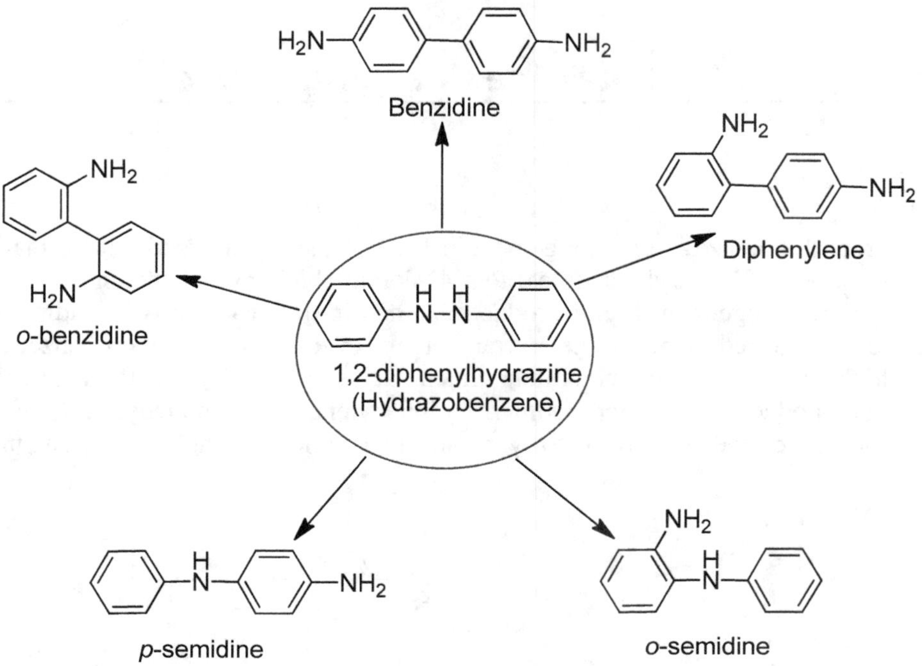

Mechanism

The benzidine rearrangement is proven to be intramolecular in literature.

When a mixture of *o,o'*-dimethoxy hydrazobenzene and *o,o'*-diethoxy benzidine undergo rearrangement, they gives *m,m'*-diethoxy benzidine. However no cross product, ethoxymethoxy benzidine is detected.

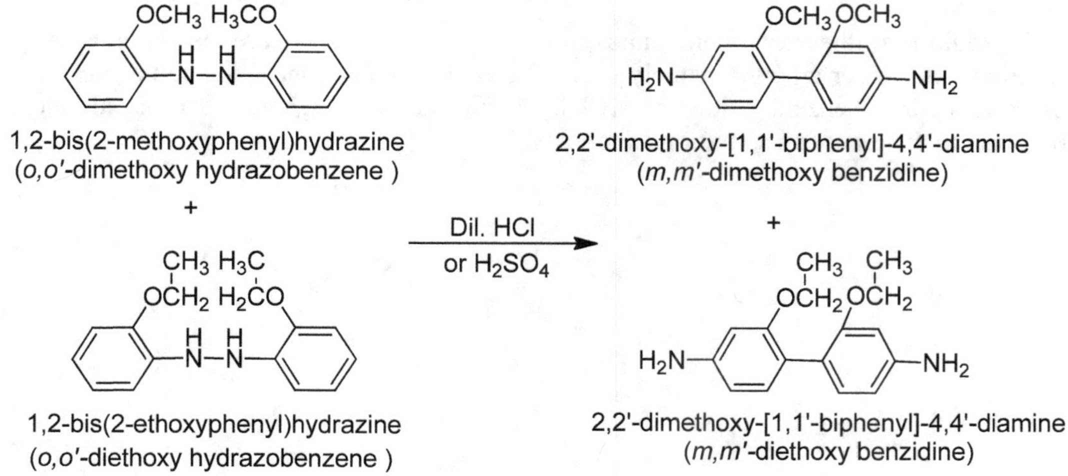

Based on extensive literature, it is believed that the rearrangement occurs *via* cationic intermediates, as given here:

1. Formation of 4,4'-diaminobiphenyl (benzidine)

2. Formation of [1,1'-biphenyl]-2,4'-diamine

[1,1'-biphenyl]-2,4'-diamine

Applications

a) The reaction has general application in the synthesis of biphenyls with or without amino groups.

b) The 4,4'-diaminobiphenyl (benzidine) and its analogues are used in the synthesis of dyes. For example: **Benzopurin 4B** *an important red dye and is obtained from o-toluidine as a starting material. The rectangle in a structure of* **Benzopurin 4B** *indicates part of molecule formed from benzidine rearrangement.*

Benzopurpurin 4B

Principle

Rearrangement reaction of 1,2-diketones (For example: Benzil) in presence of base like potassium hydroxide results in the formation of α-hydroxy carboxylic acid known as Benzilic acid rearrangement. It is an example of intramolecular rearrangement reaction. The condensation of benzaldehyde in the presence of cyanide ion (benzoin condensation) gives benzoin. Oxidation of benzoin results in the formation of benzil. Benzil undergoes intramolecular rearrangement in presence of base to form Benzilic acid.

Rearrangement Reactions: These are the reactions where reshuffling of the sequence of the atoms takes place to form a new structure.

Intramolecular rearrangements: These are the reactions which occur within the same molecule. In this type of rearrangement, the migrating group does not get completely detached from the system in which rearrangement taking place.

Intermolecular rearrangements: These are the reactions which occur between two molecules. In this type of rearrangement, the migrating group is first removed from one molecule and then attached at another site

General Reaction

Mechanism

Step 1: Attack of nucleophile (base ⁻OH) on carbonyl carbon atom of benzil.

The carbonyl carbon makes a room for attack of nucleophile by transfer of electron towards more electronegative oxygen atom.

Step 2: 1,2-migration of phenyl ring to second carbonyl carbon of benzil (intramolecular carbanion addition).

The phenyl ring along with its shared electron pair migrates to the carbonyl carbon. Phenyl group acting as a carbanion and its addition occurs on second carbonyl carbon to form α-hydroxy acids.

Migration of phenyl ring

Rapid proton transfer

α-hydroxy acid
(Benzilic acid salt)

Applications

a) *Benzil on treatment with sodium alkoxide is converted into alkyl benzilate. Bezilic acid rearrangement is utilized for the formation of corresponding esters by using alkoxides in place of base.*

Benzilic acid rearrangement

CH_3ONa

Benzil

Methyl benzilate

b) *Aliphatic ketones like ketopinic acid also undergo benzilic acid rearrangement to give citric acid.*

$HOOC-H_2C-\overset{\overset{O}{\|}}{C}-\overset{\overset{O}{\|}}{C}-CH_2-COOH$ $\xrightarrow[\text{ii) H}^{\oplus}]{\text{i) KOH}}$ $HOOC-H_2C-\overset{\overset{OH}{|}}{\underset{\underset{COOH}{|}}{C}}-CH_2-COOH$

Ketopinic acid

Citric acid

29. Benzoin Condensation

Principle

Condensation: A chemical reaction in which two molecules combine to form a larger molecule together with the loss of small molecule like H_2O, HCl, KBr, etc.

Benzoin condensation is the condensation reaction between two molecules of aromatic aldehydes (identical or different) under the catalytic influence of cyanide ion (nucleophile) to give α-hydroxy ketone (benzoin).

General Reaction

In case of benzaldehyde, the condensation in presence of cyanide ion as nucleophile results in the formation of benzoin (α-hydroxy ketone).

Benzaldehyde + Benzaldehyde $\xrightarrow{\ \ominus CN\ }$ Benzoin

Mechanism

Step 1: Attack of nucleophile on the aromatic aldehydes.

CN^- ion behaves as a good nucleophile, which attacks over the carbonyl carbon of first molecule of benzaldehyde. When CN^- ion attached to carbonyl carbon atom, it results in increased acidity of C-H bond; because CN^- ion is good electron withdrawing group. As the consequence, the removal of acidic proton takes place, which is rapidly transferred to adjacent oxygen atom. The cyanide ion helps in stabilization of carbanion thus formed.

Benzaldehyde + Nucleophile $\rightleftharpoons$ (Rapid proton transfer) Carbanion

Step 2: The addition of carbanion to second molecule of benzaldehyde

Carbanion + Benzaldehyde $\rightleftharpoons$ (Rapid proton transfer)

Step 3: Loss of cyanide ion.

The **CN⁻** ion is good nucleophile as well as a good leaving group. Rapid loss of **CN⁻** ion results in the formation of benzoin, a 2-hydroxy ketone (α-hydroxy ketone).

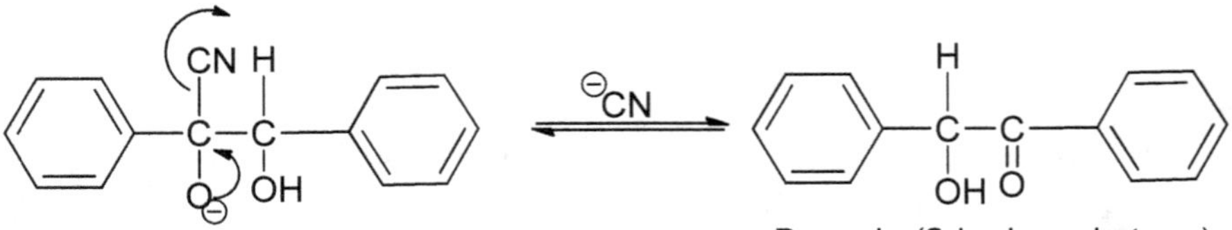

Loss of cyanide ion   Benzoin (2-hydroxy ketone)

Applications

a) Heterocyclic aldehydes also undergo benzoin condensation.

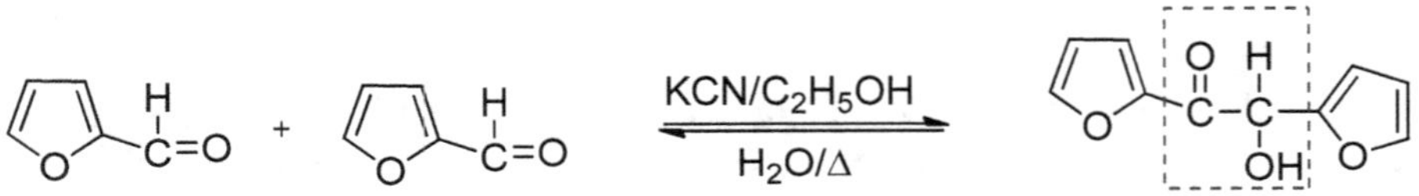

furan-2-carbaldehyde furan-2-carbaldehyde   1,2-di(furan-2-yl)-2-hydroxyethanone

Reagents: KCN/C_2H_5OH, H_2O/Δ

b) Benzaldehyde on benzoin condensation yields benzoin that can readily be converted into variety of compounds.

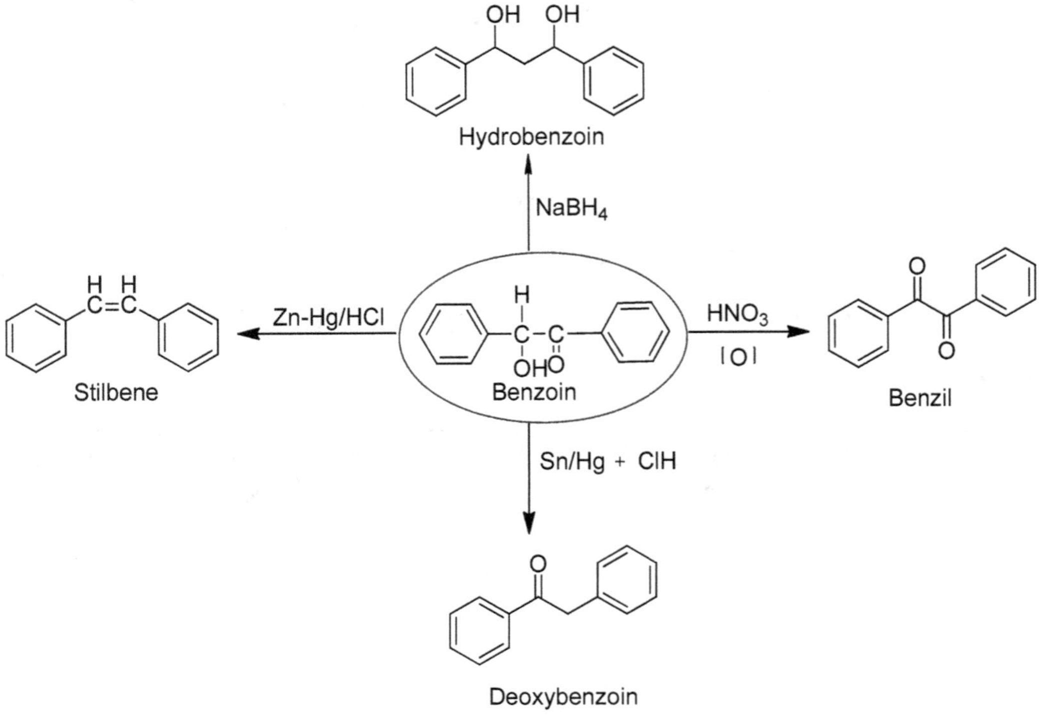

Hydrobenzoin

$NaBH_4$

Stilbene   Zn-Hg/HCl   Benzoin HNO_3 $[O]$ Benzil

Sn/Hg + ClH

Deoxybenzoin

30. Birch Reduction (Metal-Ammonia Reduction)

Principle

Wooster in 1937 reported the Birch reduction. In this reaction the reduction of aromatic compounds takes place with the help of sodium in liquid ammonia and water. Birch extended the Wooster's reported reaction in 1944 and extensively explored the reduction of benzene and aromatic derivatives with alkali metal (For example: Li, Na, K) in liquid ammonia in presence of an alcohol (as a proton donor) to produce corresponding cyclohexa-1,4-diene compounds. This partial reduction of aromatics by alkali metal in liquid ammonia in presence of alcohol is commonly known as Birch reduction or metal ammonia reduction.

General Reaction

Benzene affords 1,4-dihydrocyclohexadiene and naphthalene gives 1,4-dihydronapthalene. Liquid ammonia serves as solvent. Primary amines can also be used as solvent with advantage as it allows higher temperature of reaction (Boiling point of ethanamine: 19°C; Boiling point of liquid ammonia: -33°C).

Benzene   Li, Liq. NH₃ / C₂H₅OH   Cyclohexa-1,4-diene

Naphthalene   Li, Liq. NH₃ / C₂H₅OH   1,4-dihydronaphthalene

Mechanism

The metal ammonia solution contains both the solvated electron and alkali cation, in which the solvated electron has extremely high single-electron redox potential, so that the solvated electron can even reduce electron enriched aromatic systems.

Step 1: The mechanism of Birch reduction starts with the addition of an electron from a metal to form resonance stabilized anion radical.

Step 2: Resonance stabilized anion radical accepts a proton from an alcohol to form a neutral radical.

Step 3: Addition of one more electron to the neutral radical from metal (Li) produces an anion which on protonation by alcohol gives dihydro product.

cyclohexa-1,4-diene

The double bonds which are isolated in the 1,4-dihydro structure are not readily reduced like aromatic ring, so the reduction terminates at the dihydro stage.

As the repulsion between the anion and radical centre is minimum in resonance stabilized anion radical (*Step 1*), therefore formation of 1,4-dihydro product takes place but not 1,2-dihydro compound. At high temperature (50-120°C), formation of 1,2-dihydro compound takes place; ammonia becomes the proton source instead of alcohol.

Resonance stabilized
anion radical

1,2-dihydro
compound

The 1,2-dihydro product have a conjugated double bond and hence undergoes further metal ammonia reduction to form a tetrahydro derivative.

Electron withdrawing substituents stabilize the carbanions whereas electron donating substituents destabilize them. Hence, reduction takes place on a carbon bearing an electron withdrawing substituents and not on electron releasing substituents.

Applications

The reaction is useful for transformation of alkyl aryl ethers into 1-alkoxycyclohexa-1,4-dienes. Phenols and isolated double bonds are not reduced by this method.

For example: Reduction of anisole to form 1-methoxycyclohexa-1,4-diene (1,4-dihydro compound). Hydrolysis of 1,4-dihydro derivative (enol ethers) affords cyclohex-2-enone.

31. Bischler-Napieralski Reaction

Principle

The reaction constitutes an elegant method for the synthesis of isoquinoline derivatives. The reaction was initially given by Bischler and Napieralski in 1893. It is a two-step reaction *through* a cyclization of *N*-acyl phenylethylamines by using dehydrating agents to form 1-alkyl/1-aryl 3,4-dihydroisoquinolines, followed by dehydrogenation or oxidation to yield isoquinolines. Reaction is commonly referred as Bischler-Napieralski isoquinoline synthesis. The dehydrating agents used include phosphoric acid, sulfonic acid, phosphorus oxychloride, phosphorus pentoxide, zinc chloride, aluminum chloride, and ferric chloride. The reaction proceeds smoothly in boiling toluene or xylene in the presence of dehydrating agents, but low boiling point solvents such as acetonitrile or dichloromethane also helpful in synthesis of isoquinoline derivatives.

General Reaction

The reaction proceeds through following steps:

a) The acyl analogues of β-aryl ethylamine are heated with phosphorus pentoxide or anhydrous zinc chloride, on dehydration results into the formation of dihydroisoquinoline derivative.

b) The dihydroisoquinoline derivative on $KMnO_4$ oxidation or by dehydrogenation with S, Se or Pd-C gives corresponding isoquinoline analogues.

Mechanism

Step 1: Formation of carbonium by treatment of phosphorus oxychloride with β-phenyl ethylamide.

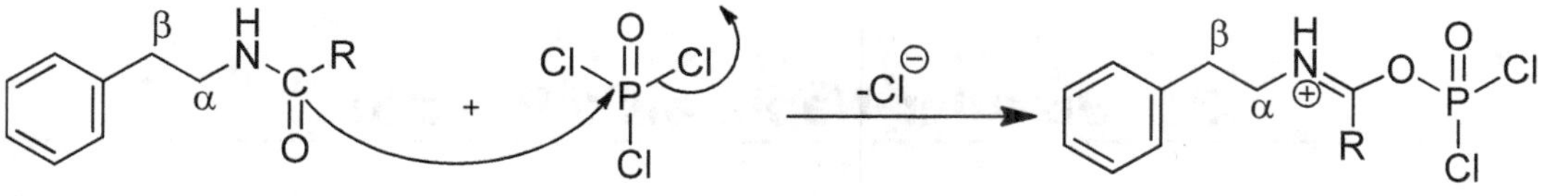

β-phenyl ethylamide

Step 2: An electrophilic intramolecular attack on the aromatic double bond to form 3,4-dihydroisoquinoline.

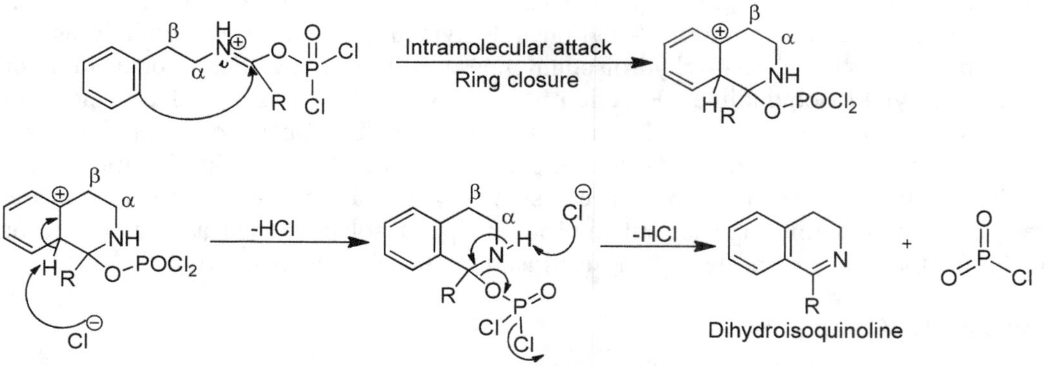

Step 3: Catalytic dehydrogenation to afford corresponding isoquinoline derivative.

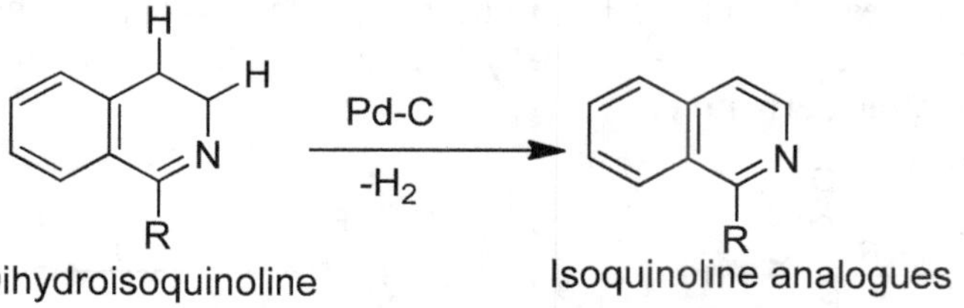

Applications

Naturally occurring alkaloids has been efficiently synthesized by suing this method; for example: laudanosine, laudanine, papaverine, cotarnine and hydrastinine; salsoline, pectenine, and corypallin; anhalonium alkaloids; and takatonine and petaline.

a) Synthesis of Papaverine from Homoveratralamine.

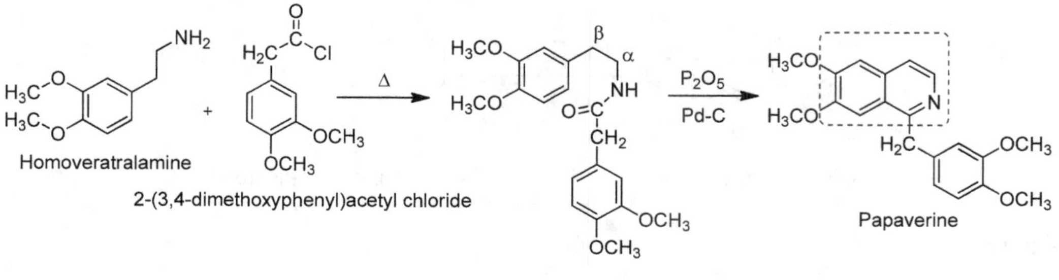

b) Synthesis of Harman from Tryptamine.

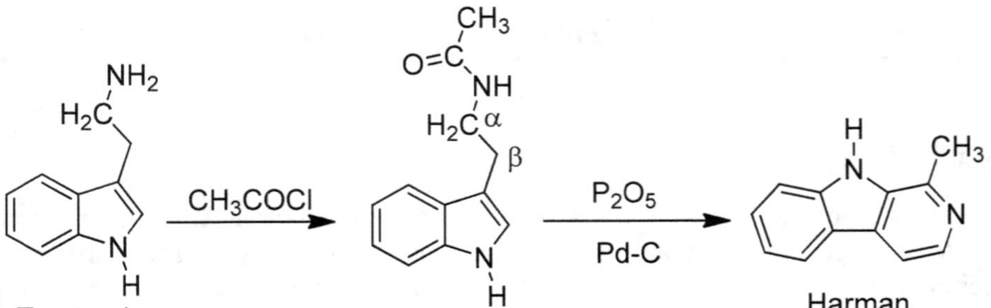

32. Blaise Reaction

Principle

Blaise in 1901 reported the reaction between a nitrile and zinc reagent from α-bromoester (Reformatsky reagent) to give an enamino ester or the further product of a β-keto ester upon acidic hydrolysis. The reaction is known as Blaise reaction. It is a example of condensation reaction. Although it is a one-step synthesis of higher aliphatic β-keto esters from a stable and accessible starting material, the easy introduction of functionalities and the straightforward nature of the conversion have been overshadowed by the problems of low yield, narrow scope, and competing side reactions.

General Reaction

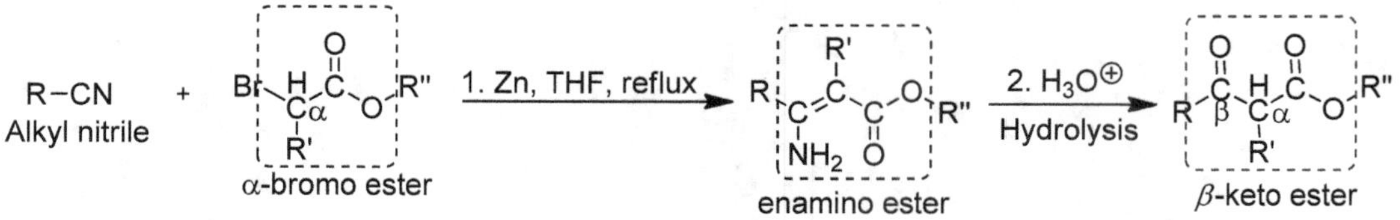

R–CN (Alkyl nitrile) + α-bromo ester → (1. Zn, THF, reflux) enamino ester → (2. $H_3O^{\oplus}$, Hydrolysis) β-keto ester

Mechanism

The reaction proceeds by a mechanism similar to that of Reformatsky reaction, involving condensation of a nitrile with α-bromoester; the intermediate is finally transformed into β-keto ester by acidic hydrolysis.

Step 1: Formation of organozinc intermediate by oxidative addition of Zinc into an α-bromoester.

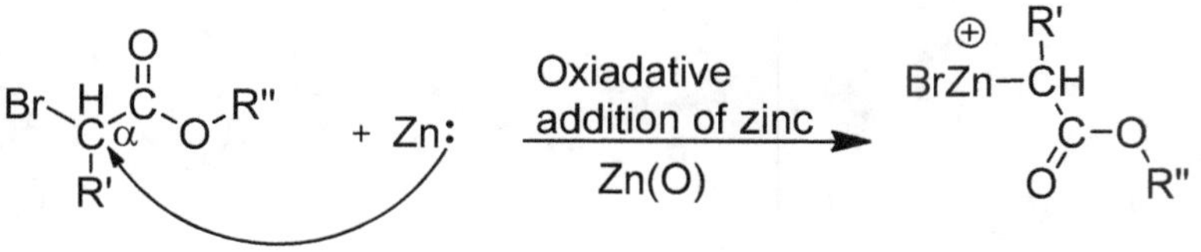

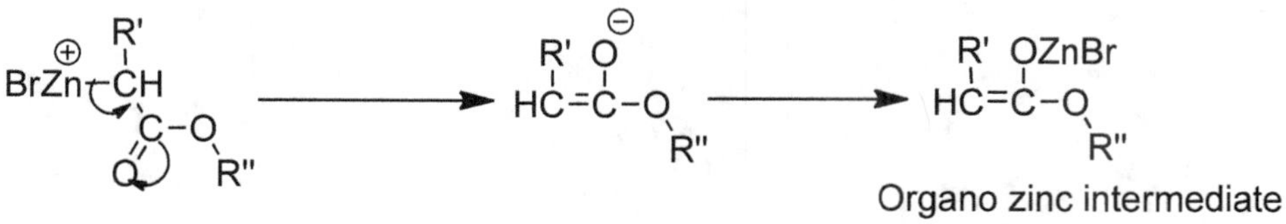

Organo zinc intermediate

Step 2: Condensation of nitrile with organozinc intermediate (α-bromoester/Reformatsky reagent) to give enamino ester which on acidic hydrolysis affords β-keto ester.

Applications

The enamino and β-keto esters are easily prepared by using this reaction, which can be asymmetrically reduced to afford chiral β-amino or hydroxyl esters and further hydrolyzed to β-amino or hydroxyl acids subsequently cyclized to produce β-lactams or lactones.

a) Preparation of β-Keto esters.

3-methylbut-2-enenitrile   +   Reformatsky reagent $\xrightarrow{\text{Zn, THF}}$ (Z)-methyl 3-amino-5-methylhexa-2,4-dienoate

33. Bohlmann-Rahtz Pyridine Synthesis

Principle

The Bohlmann-Rahtz pyridine synthesis allows the generation of substituted pyridines in two steps. Condensation of enamines with ethynylketones leads to an aminodiene intermediate that, after thermal *E/Z* isomerization undergoes cyclodehydration to yield 2,3,6-trisubstituted pyridines. The reaction is known as Bohlmann-Rahtz pyridine synthesis.

General Reaction

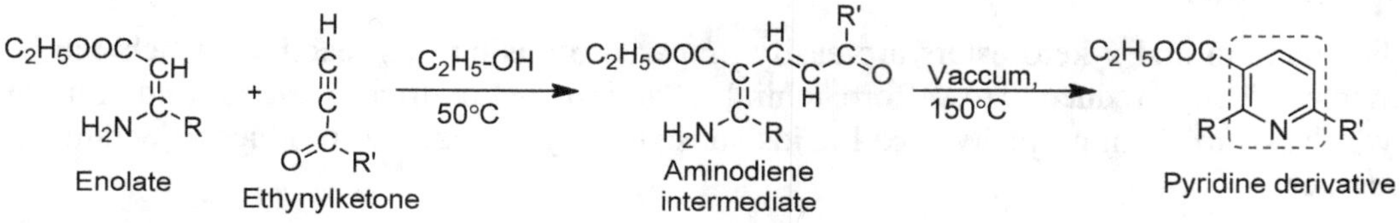

Enolate + Ethynylketone $\xrightarrow[50^\circ C]{C_2H_5\text{-}OH}$ Aminodiene intermediate $\xrightarrow[150^\circ C]{\text{Vaccum,}}$ Pyridine derivative

Mechanism

Step 1: Michael addition of enamine and ethynylketone to form an aminodiene intermediate.

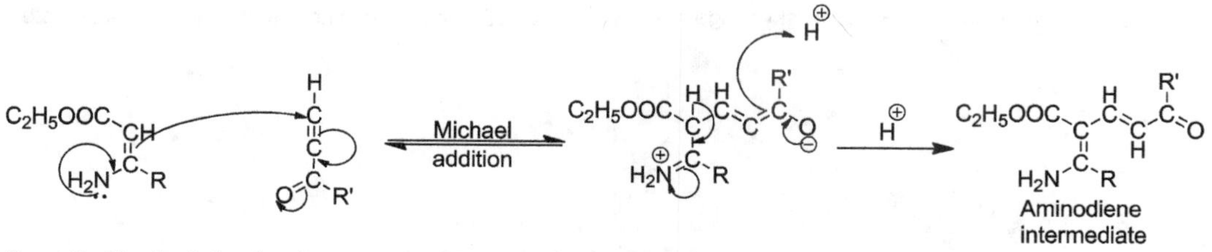

Step 2: Cyclodehydration to give 2,3,6-trisubstituted pyridines.

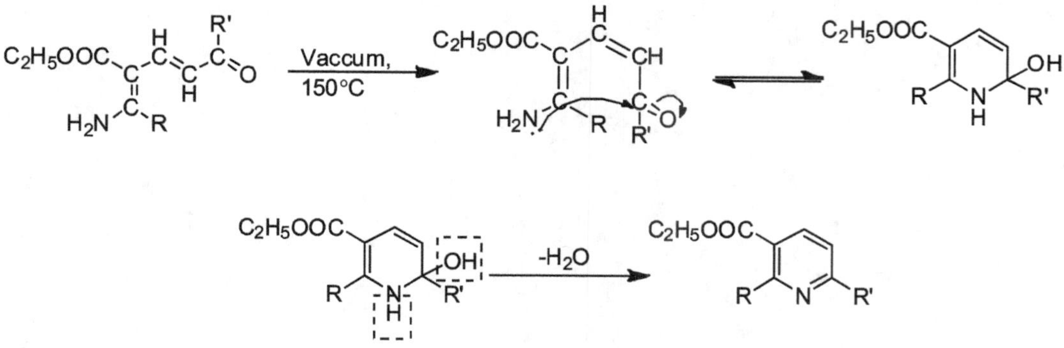

Applications

a) The reaction is useful for the synthesis of substituted pyridine analogues. For example: Synthesis of ethyl 2,6-dimethylnicotinate.

(Z)-ethyl 3-aminobut-2-enoate + 4-(trimethylsilyl)but-3-yn-2-one $\xrightarrow[50°C]{C_2H_5\text{-}OH}$ (Z)-ethyl 2-(1-aminoethylidene)-5-oxohexanoate $\xrightarrow[50°C,\ 1h]{Toluene/AcOH}$ Ethyl 2,6-dimethylnicotinate

Principle

Bodroux in 1904 reported the synthesis of substituted amide. In this reaction synthesis of substituted amide is carried out by the reaction between a simple aliphatic or aromatic ester and an aminomagnesium halide (1:2). An aminomagnesium halide is obtained by treatment of a primary or secondary amine with a Grignard reagent at room temperature. However, the direct aminolysis of carboxylic esters with primary amines requires high temperatures and the use of an autoclave. Apart from magnesium amides, other metal amides have also been developed, including the alkali, aluminum, tin, and titanium amides. The modified Bodroux amide synthesis is the reaction between esters and lithium aluminum amide obtained from the $LiAlH_4$ and amine at 25°C in acetic anhydride. More recent modifications for preparing amide include the reaction between a carboxylate, amine, and Grignard reagent in a 1:1:1.5 molar ratios.

General Reaction

Aliphatic or aromatic ester   Aminomagnesium halide   Substituted amide

R₁, R₂, R₃ = Alkyl, Aryl; **R₄** = H, Alkyl, Aryl;

X = Cl, Br, I

Mechanism

The mechanism is analogous to reaction between a Grignard reagent and ester.

Applications

a) The reaction is used to synthesize substituted amides.

Acid   Alkyl magnesium halide   Amide

For example: Synthesis of N,N-dimethylacetamide from Methyl acetate and Magnesium chloride dimethylamide

Methyl acetate   Magnesium chloride dimethylamide   N,N-dimethylacetamide

35. Bouveault Aldehydes Synthesis

Principle

Formylation of an alkyl or aryl halide to the homologous aldehyde by transformation to the corresponding organometallic reagent followed by addition of dimethylformamide (Metal: Li, Mg, Na, and K) is known as Bouveault aldehydes synthesis. The reaction was first reported by Bouveault in 1896. It is the reaction for synthesizing aldehydes by the treatment of *N,N*-disubstituted formamides with either Grignard reagent or organic lithium reagent in an ether as a solvent. The reaction works only in certain cases because the reaction is too complicated and sometimes produces tertiary amines as major product.

General Reaction

$$R-X \xrightarrow[\text{2. DMF; 3. } H^{\oplus}]{\text{1. M: Li, Mg, Na, K}} R-CHO$$

Alkyl halide → Aldehyde

Mechanism

The detailed mechanism for formation of aldehyde from alkyl halide is illustrated below:

M: Li, Mg, Na, K

Commins modification:

Applications

The reaction is very useful for the transformation of aryl/alkyl halides into corresponding aldehydes.

a) Synthesis of 2-phenylacetaldehyde from (bromomethyl)benzene.

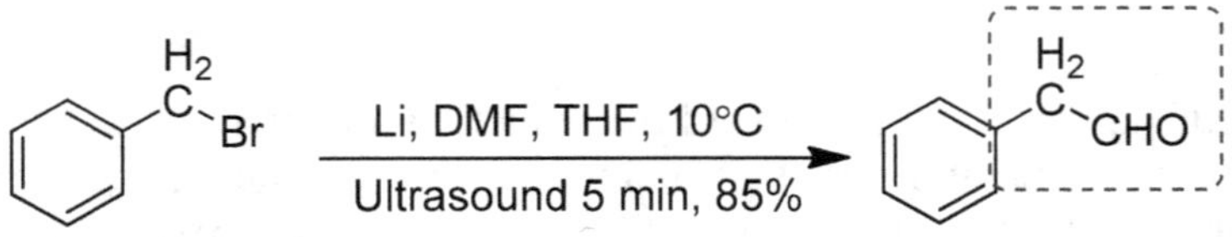

(bromomethyl)benzene → Li, DMF, THF, 10°C / Ultrasound 5 min, 85% → 2-phenylacetaldehyde

b) Synthesis of 5-methoxythiophene-2-carbaldehyde.

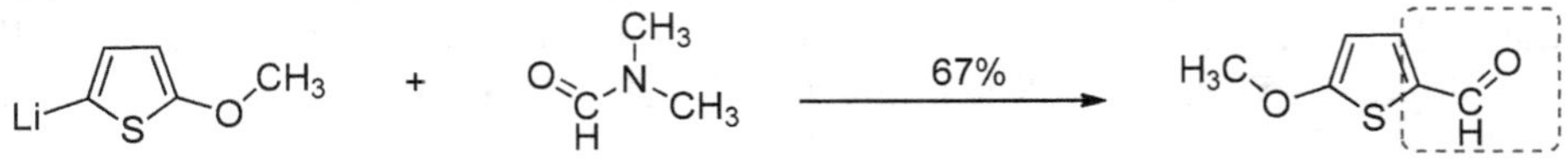

(5-methoxythiophen-2-yl)lithium   +   *N,N*-dimethylformamide   → 67% → 5-methoxythiophene-2-carbaldehyde

36. Bouveault-Blanc Reduction

Principle

Bouveault and Blanc in 1903 performed the reduction of esters into alcohols with sodium in ethanol, in which sodium serves as a single electron reducing agent and ethanol as a proton donor is known as the Bouveault-Blanc reduction. Apart from ethanol, other alcohols can also be used as proton donors. If the reaction is performed without proton donor, the reduction of esters with sodium gives acyloins (*Acyloin Condensation*). The Bouveault-Blanc reduction under appropriate conditions also effects the ring reduction of aromatics (*Birch Reduction*). The reaction is an elegant substitute to the lithium aluminium hydride reduction of esters for industrial production.

General Reaction

Ester $\xrightarrow{\text{Na, C}_2\text{H}_5\text{OH}}$ Alcohols + R'—OH

For example: Preparation of mixture of propan-1-ol and ethanol from ethyl propionate

Ethyl propionate $\xrightarrow{\text{Na, C}_2\text{H}_5\text{OH}}$ Propan-1-ol + Ethanol

Preparation of mixture of propan-1-ol and methanol from methyl propionate

Methyl propionate $\xrightarrow{\text{Na, C}_2\text{H}_5\text{OH}}$ Propan-1-ol + Methanol

Mechanism

A radical mechanism is illustrated below.

Sodium serves as single electron reducing agent and ethanol is the proton donor. In absence of proton donor dimerisation take place, like acyloin condensation.

Step 1: Single electron transfer from the sodium metal.

Step 2: Formation of aldehyde.

Step 3: Single electron transfer from the sodium metal.

Step 4: Formation of alcohol

Applications

a) The reaction is generally used to reduce esters, aldehydes, and ketones to the corresponding alcohols.

Ethyl 2-(benzylamino)-3-methoxyhexanoate   2-(benzylamino)hexan-1-ol

b) *In addition, if a double bond exists on the appropriate location of the ester, a cyclic or spirocyclic alcohol will form.*

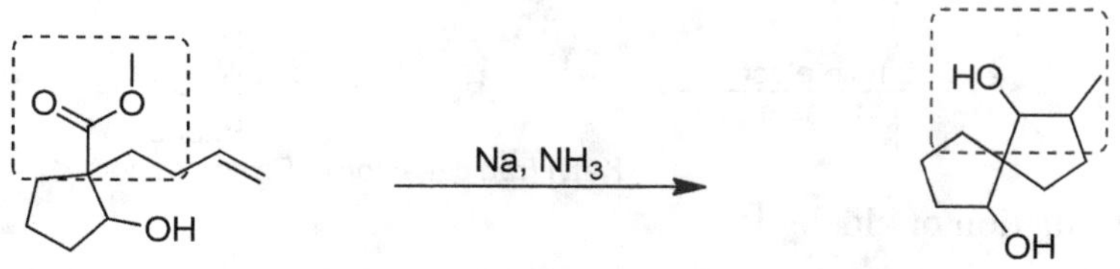

Methyl 1-(but-3-en-1-yl)-2-hydroxycyclopentanecarboxylate

$\xrightarrow{\text{Na, NH}_3}$

2-methylspiro[4.4]nonane-1,6-diol

Principle

Bruckner first reported the structure of 1,3-dimethyl-6,7-methylenedioxyisoquinoline in 1935. It is the elegant method to synthesize isoquinolines from allyl phenyl ether *via* the process of a *Claisen Rearrangement*, double-bond migration, amination, and cyclization.

General Reaction

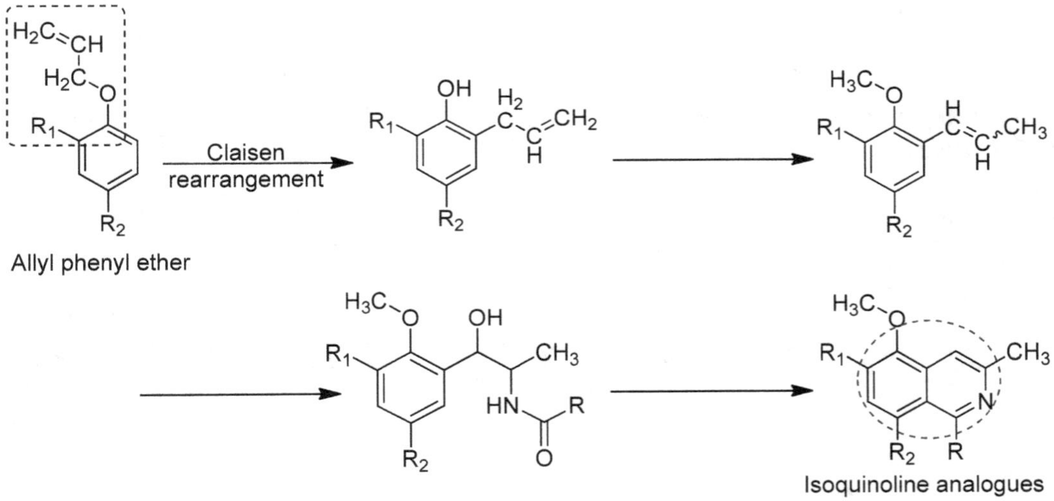

Allyl phenyl ether

Isoquinoline analogues

Mechanism

Step 1: Claisen rearrangement

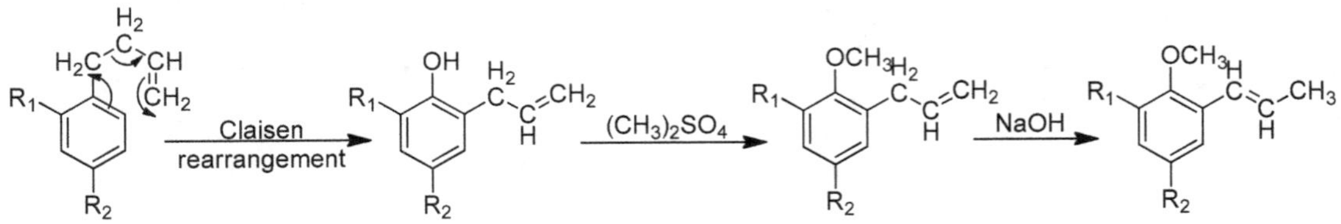

Step 2:

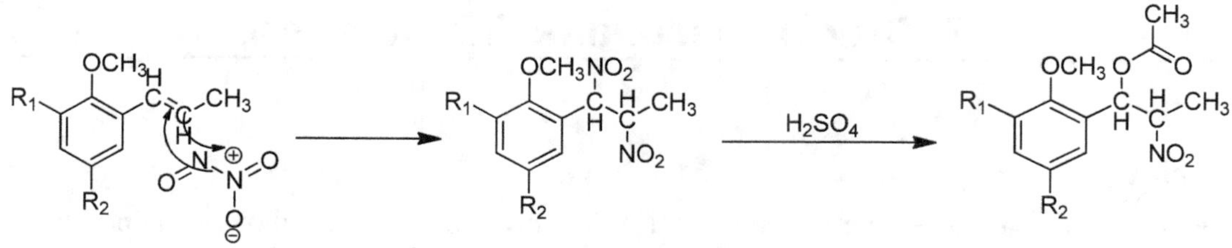

Step 3:

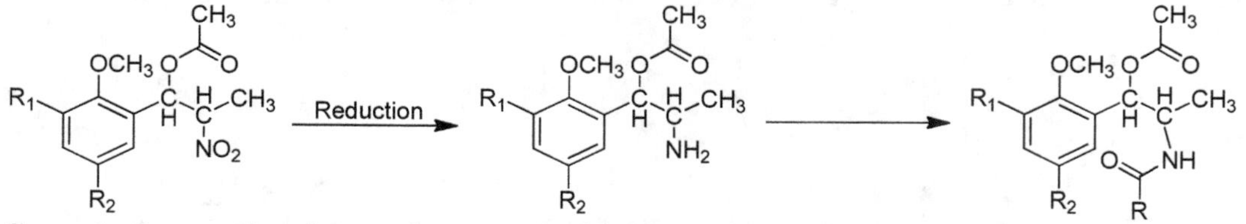

Step 4: Intramolecular cyclization with the elimination of water molecule to give isoquinoline analogues

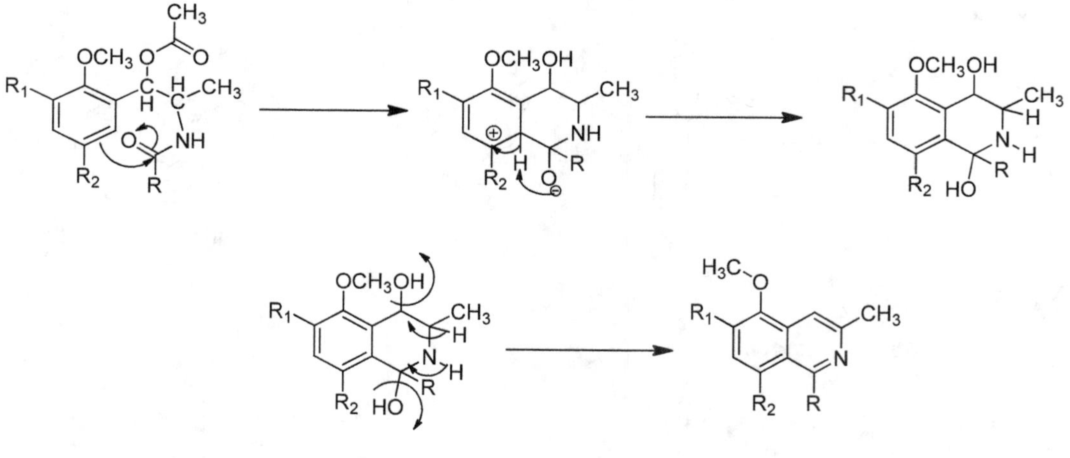

Applications

a) *Synthesize the substituted isoquinoline anlogues. For example: Synthesis of 5,8-dimethoxy-1,3-dimethylisoquinoline.*

2-allyl-1,4-dimethoxybenzene

5,8-dimethoxy-1,3-dimethylisoquinoline

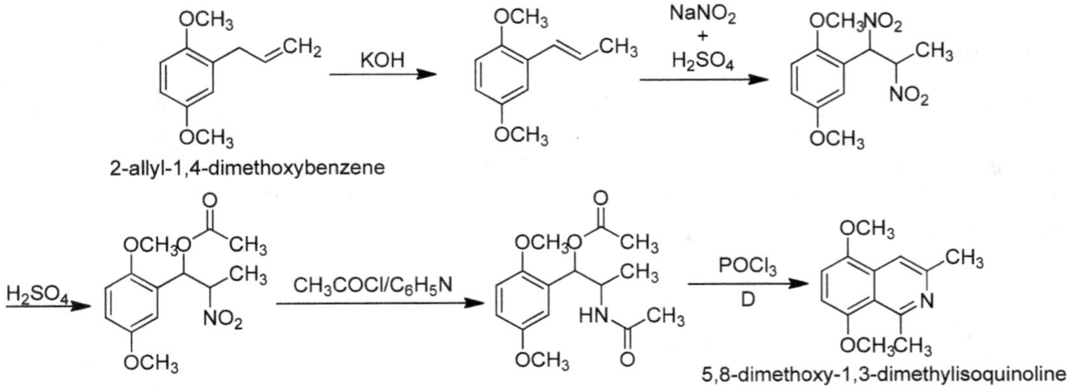

Principle

Bucherer in 1906 carried out synthesis of carbazoles from aryl hydrazines, sodium bisulfite, and 1-/2-naphthols (1- or 2-naphthylamines) is known as Bucherer carbazole synthesis. The reaction between 2-aminonaphthalene-1-sulfonic acid and phenylhydrazine also gives carbazole, indicating that sulfonation of naphthylamine might be the early step in the reaction pathway.

General Reaction

Mechanism

The mechanism for transformation of naphthalene-2-ol by reacting with phenylhydrazine into 7*H*-benzo[*c*]carbazole is illustrated below:

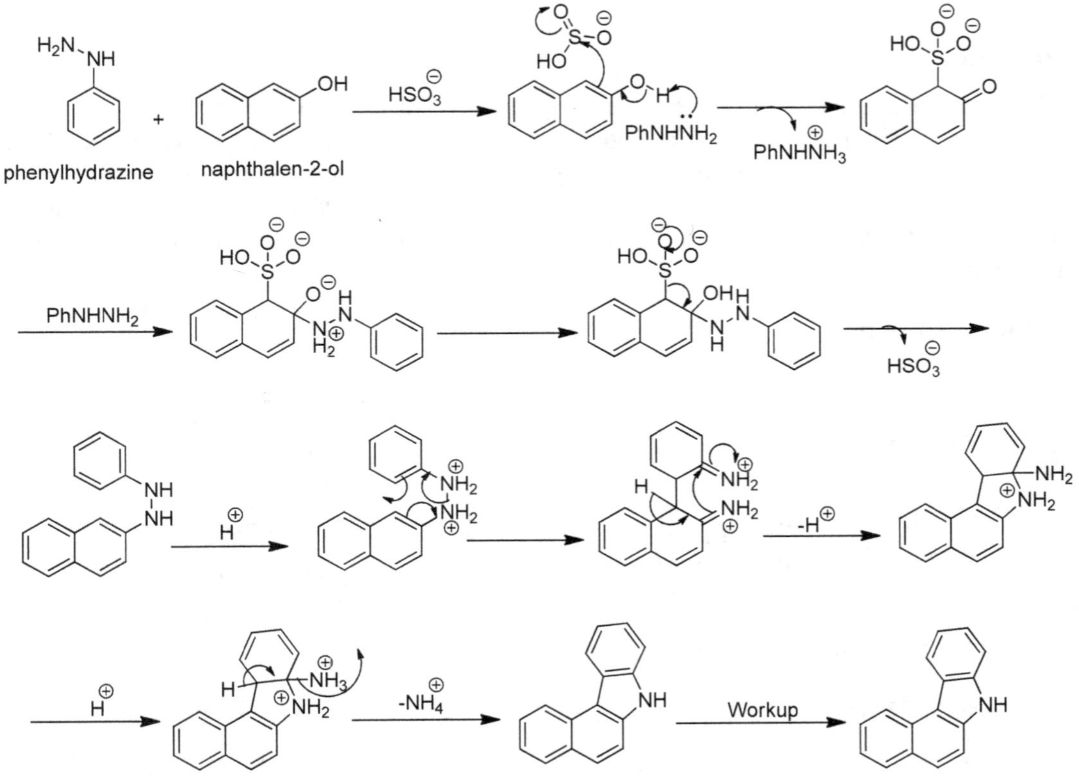

phenylhydrazine   naphthalen-2-ol

7*H*-benzo[*c*]carbazole

Applications

a) The reaction is useful for the synthesis of fused carbazole derivatives or dibenzocarboles.

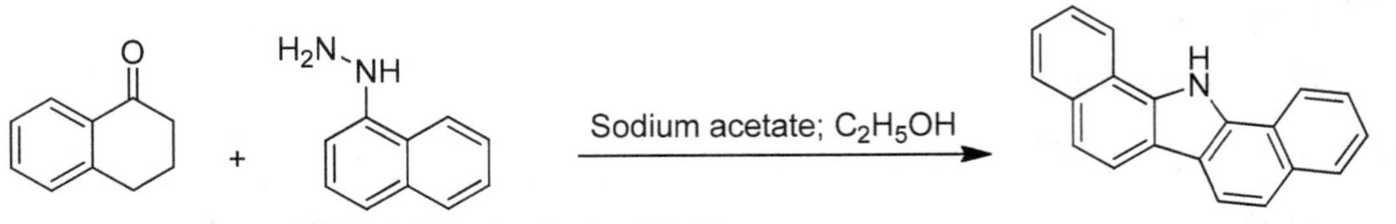

α-tetralone α-naphthylhydrazine hydrochloride   13*H*-dibenzo[*a,i*]carbazole

Principle

The reaction is undergone by terminal alkynes (except ethyne). Terminal alkynes react with cuprous chloride and base (ammonium hydroxide) to afford copper alkynides which can be coupled together by oxidation to give dialkynes. The oxidation is performed in presence of oxygen and acetic acid. This oxidative coupling reaction is commonly known as Cadiot-Chodkiewick oxidative coupling. The copper (I) catalyzed coupling of a terminal alkyne and an alkynyl halide offers access to unsymmetrical bisacetylenes.

General Reaction

$$R-C\equiv C-H \quad + \quad CuCl \quad \xrightarrow{NH_4OH} \quad R-C\equiv C-Cu$$

Alkyne (Terminal alkyne)   copper(I) chloride   Alkynyl copper reagent

$$R-C\equiv C-X \quad + \quad Cu-C\equiv C-R' \quad \xrightarrow{CH_3COOH/O_2} \quad R-C\equiv C-C\equiv C-R' \quad + \quad Cu-X$$

Alkynyl halide   Alkynyl copper reagent   Bis acetylene

For example: Preparation of hexa-2,4-diyne from prop-1-yne

$$H_3C-C\equiv C-H \quad + \quad CuCl \quad \xrightarrow{NH_4OH} \quad H_3C-C\equiv C-Cu$$

Prop-1-yne   Copper(I) chloride   Prop-1-yn-1-ylcopper

$$H_3C-C\equiv C-Cl \quad + \quad Cu-C\equiv C-CH_3 \quad \xrightarrow{CH_3COOH/O_2} \quad H_3C-C\equiv C-C\equiv C-CH_3 \quad + \quad Cu-Cl$$

1-chloroprop-1-yne   Prop-1-yn-1-ylcopper   Hexa-2,4-diyne   Copper(I) chloride
         (Bis acetylene)

Mechanism

Preparation of alkynyl copper reagent

$$R-C\equiv C-H \quad \xrightarrow[-BH^{\oplus}]{Base\ (B)} \quad R-C\equiv C^{\ominus} \quad + \quad CuX \quad \longrightarrow \quad R-C\equiv C-Cu \quad + \quad {}^{-}X\!:^{\ominus}$$

Alkyne   Alkynyl carbanion   Copper halide   Alkynyl copper reagent

Step 1: Oxidative addition to form adduct in presence of CH_3COOH/O_2

$$R-C{\equiv}C-Cu \quad + \quad X-C{\equiv}C-R' \xrightarrow[\text{addition}]{\text{Oxidative}} R-C{\equiv}C-\overset{\overset{X}{|}}{Cu}-C{\equiv}C-R'$$

Cu (III) intermediate
(Adduct)

Step 2: Reductive elimination to produce bis-acetylene (dialkyne)

$$R-C{\equiv}C-Cu-C{\equiv}C-R' \xrightarrow[\text{elimination}]{\text{Reductive}} R-C{\equiv}C-C{\equiv}C-R' \quad + \quad Cu-X$$

Adduct

Applications

a) *The reaction is useful for the synthesis of different substituted dialkynes. For example:
Preparation of octa-3,5-diyne-2,7-diol.*

Acetaldehyde   Ethyne, sodium salt   But-3-yn-2-ol   octa-3,5-diyne-2,7-diol

40. Cannizzaro's Reaction

Principle

In presence of alkali, the aldehydes with no α-hydrogens undergo intermolecular oxidation and reduction (auto oxidation-reduction or dismutation of aldehyde). In this case, one molecule of aldehyde is oxidized to sodium salt of carboxylic acid and the other is reduced to an alcohol. Cannizzaro's reaction is a characteristic of aromatic aldehydes as aromatic aldehyde lacks the α-hydrogen. This reaction was first reported by Cannizzaro in 1853.

General Reaction

a) Formation of sodium formate and methanol.

b) Preparation of benzyl alcohol and sodium benzoate.

c) Preparation of benzyl alcohol and sodium formate.

Mechanism

The reaction involves hydride shift (intermolecular hydride transfer) and it can be explained by taking example of formaldehyde.

Step 1: Attack of base (nucleophile) on carbonyl carbon of methanal

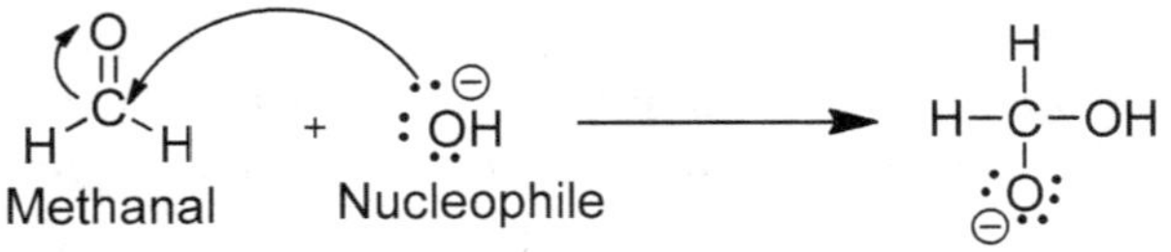

Methanal   Nucleophile

Step 2: Intermolecular hydride shift to second molecule of aldehyde

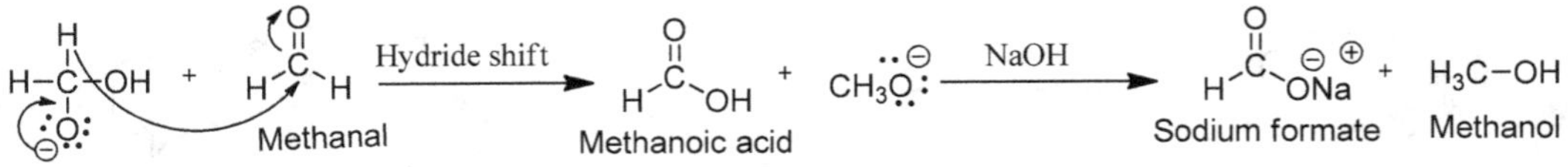

Methanal   Methanoic acid   Sodium formate   Methanol

In case of dialdehydes (dials) such as glyoxal, the hydride transfer takes place within the molecule to the adjacent aldehyde group, which results in the formation of salts of hydroxy acid. This is known as **internal Cannizzaro's reaction**.

Mechanism of Cannizzaro reaction:

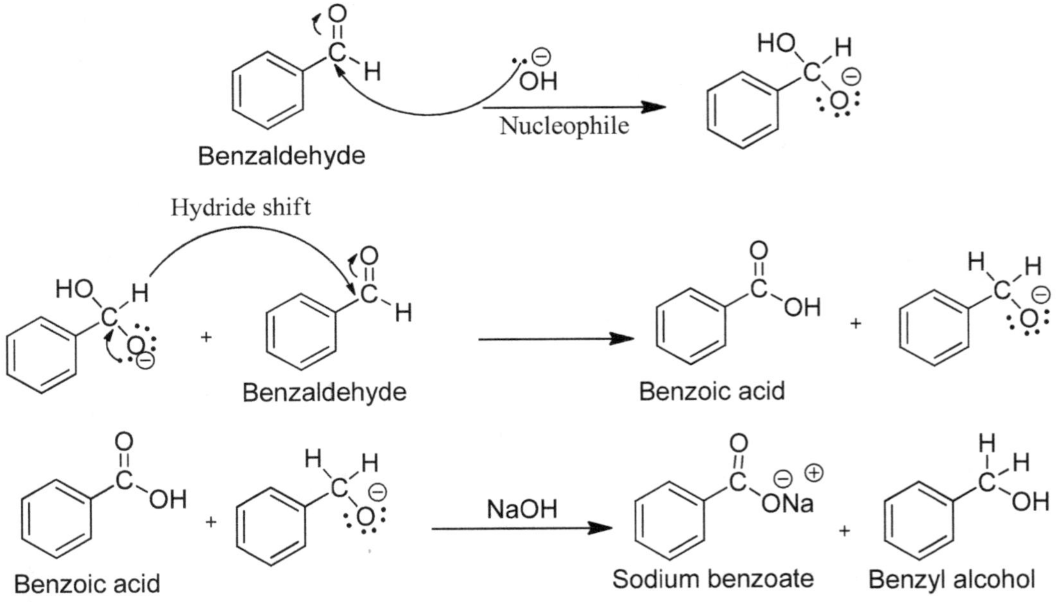

Benzaldehyde   Nucleophile

Hydride shift

Benzaldehyde   Benzoic acid

Benzoic acid   NaOH   Sodium benzoate   Benzyl alcohol

Crossed Cannizzaro's reaction uses formaldehyde as reducing agent in presence of 30% sodium hydroxide.

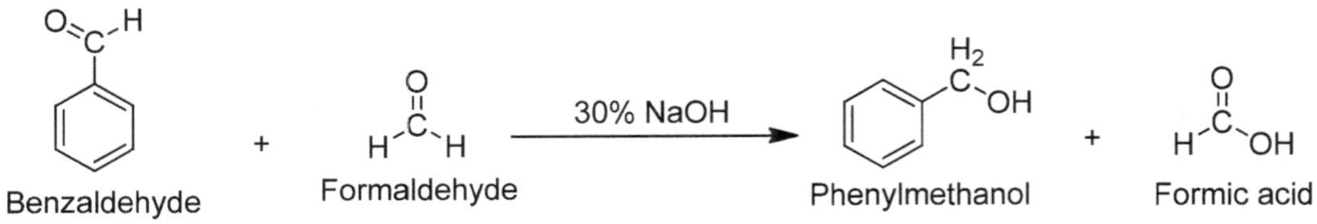

Benzaldehyde   Formaldehyde   30% NaOH   Phenylmethanol   Formic acid

Applications

a) *Cannizzaro's reaction has general application in synthesis of alcohols from aldehydes without α-hydrogens. For example: Quinoline-4-carbaldehyde undergoes Cannizzaro's reaction to give quinoline-4-carboxylic acid and quinolin-4-ylmethanol.*

Quinoline-4-carbaldehyde   Quinoline-4-carboxylic acid   Quinolin-4-ylmethanol

41. Carbylamine Reaction

Principle

Alcoholic solutions of primary amines (aliphatic and aromatic) react with chloroform in the presence of alkali solution to yield isocyanide (isonitrile) or carbylamine is known as Carbylamine reaction. The reaction was first reported by Hoffmann in 1868, hence also referred as Hoffmann isonitrile synthesis.

The formation of isocyanides is readily noted by their characteristic nauseating odor. Due to this reason, this reaction has been used as a qualitative test for primary amines.

General Reaction

$$R-NH_2 \xrightarrow[\Delta]{CHCl_3/KOH} R-\overset{\oplus}{N}\equiv\overset{\ominus}{C} \quad + \quad 3KCl \quad + \quad 3H_2O$$

Amine → Alkyl isocyanide or carbylamine

Mechanism

Step 1: Formation of dichlorocarbene by α-elimination of chloroform.

$$HCCl_3 \xrightarrow[\alpha\text{-elimination}]{\overset{\ominus}{O}H \quad \Delta} :CCl_2$$

Step 2: Reaction of primary amine with dichlorocarbene to form isocyanides

Amine   Dichlorocarbene

$R-N\equiv C$ + H_2O

Applications

a) *The reaction is used to convert primary amines into isonitriles. The isonitriles have gained much attention in organic synthesis; because of their application as versatile building blocks for heterocycles and in multicomponent reactions to generate combinatorial libraries.*

For example: conversion of aromatic primary amines into isonitriles

Aniline + $CHBr_3$ $\xrightarrow[45\%]{\text{NaH, THF}}$ Isocyanobenzene

b) *Conversion of aliphatic primary amines into isonitriles; For example: N,N'-diethylpropane-1,3-diamine is converted into N,N'-diethyl-3-isocyanopropan-1-amine.*

$+$ $CHBr_3$ $\xrightarrow[\Delta]{\text{NaOH, Benzene}}$

N,N'-diethylpropane-1,3-diamine   N,N'-diethyl-3-isocyanopropan-1-amine

42. Carroll Rearrangement

Principle

Carroll in 1940 reported the synthesis of γ,δ-unsaturated ketones. Thermal transformation of allyl acetoacetates into γ,δ-unsaturated ketones or base catalyzed reaction between allyl alcohols and β-ketoesters to form γ,δ-unsaturated ketones is known as Carroll rearrangement. The reaction does not get popularity in organic synthesis due to certain disadvantages; a) harsh conditions (temperatures of 130-220°C after *in situ* preparation of the β-ketoester) and lack of an efficient method for synthesis of β-ketoesters. The adsorption of tertiary allylic acetoacetate on neutral alumina will also facilitate the Carroll rearrangement. Recently, it was found that this rearrangement can be catalyzed by a ruthenium complex with high regioselectivity, but a lower regiospecificity compared to the without catalyst Carroll rearrangement.

General Reaction

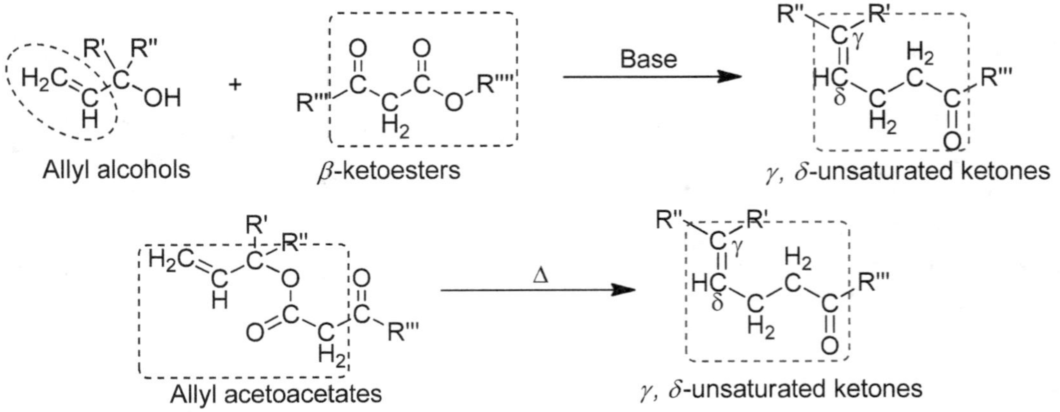

Allyl alcohols + β-ketoesters $\xrightarrow{\text{Base}}$ γ,δ-unsaturated ketones

Allyl acetoacetates $\xrightarrow{\Delta}$ γ,δ-unsaturated ketones

Mechanism

Step 1:

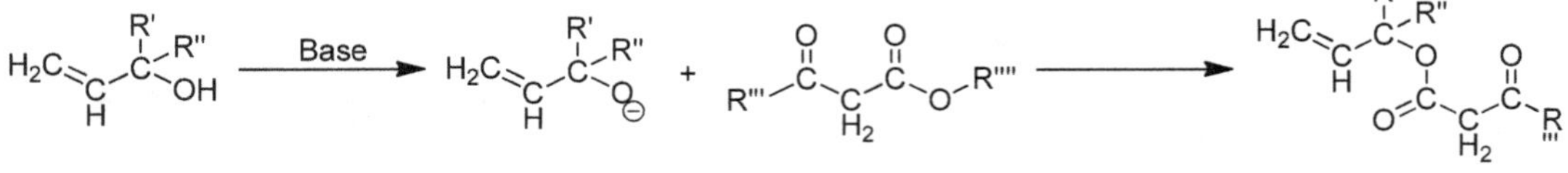

Step 2: Keto-enol tautomerism

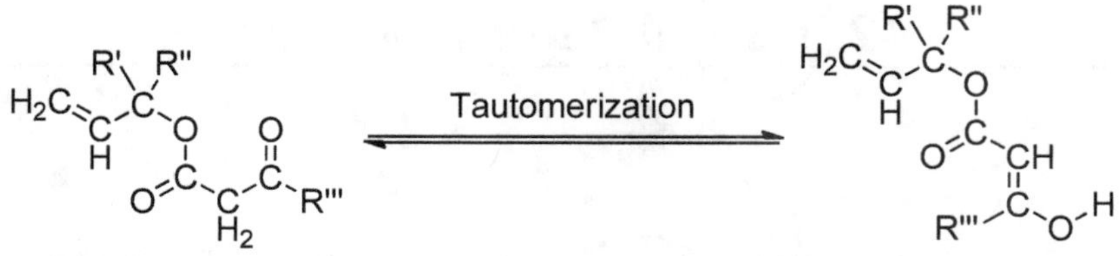

Step 3:

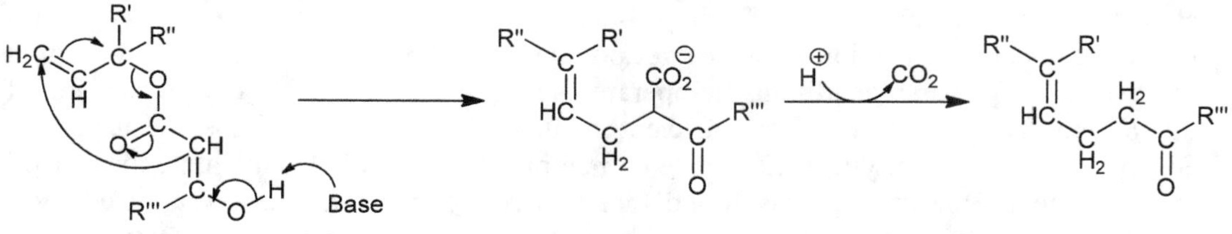

Applications

a) *The reaction is useful in the synthesis of γ,δ-unsaturated ketones.*

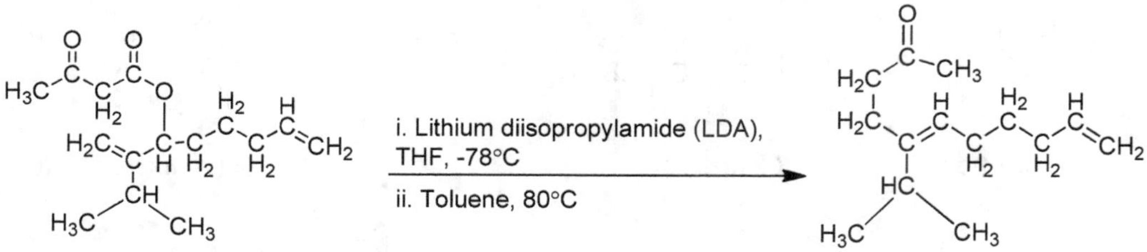

i. Lithium diisopropylamide (LDA), THF, -78°C

ii. Toluene, 80°C

2-methyl-3-methylenenon-8-en-4-yl 3-oxobutanoate

(Z)-5-isopropylundeca-5,10-dien-2-one

Principle

The amination of pyridine and other heterocyclic compounds by amide ion is known as Chichibabin amination reaction. This reaction is aromatic nucleophilic substitution reaction which takes place at ortho position of pyridine ring. If *o*-position is blocked, then the reaction takes place at the *p*-position. The reaction was reported by Chichibabin in 1914. It is the synthesis of 2-aminopyridine from pyridine and sodium amide in boiling inert solvent (For example: xylene, toluene, and benzene). It is a simple method for synthesis of 2-aminopyridine compared to traditional procedure for aromatic amines *via* nitration and subsequent reduction works poorly for pyridines. The impurities in sodium amide may have catalytic effect. In this reaction, if the amount of sodamide is increased, then a second amino group might be introduced to the aromatic ring. However, when both position **2** and **6** are blocked by substituents, the amino group will be introduced into position **4**, though with a low yield. The source of amine in reaction is sodium amide ($NaNH_2$). In addition, if organic amine is connected to a pyridine ring, then cyclic amine will form under these conditions. This reaction also work for other nitrogen-containing heterocycles, including purines, pyrazines, naphthyridines, triazine, and pyrimidines. Reaction takes place with $NaNH_2$ or KNH_2 in liquid ammonia at high temperature (100-200°C). If excess of sodamide is used, 2, 6-diaminopyridine is obtained.

General Reaction

a) Preparation of 2-amino pyridine.

b) Preparation of 2-amino quinoline from quinoline by reaction with sodium amide.

Mechanism

Resonance stabilization of pyridinium ion

Step 1: Nucleophilic attack of amide anion at α-position

Step 2: Chichibabin reaction involves addition-elimination mechanism. The leaving group has been the hydride ion which undergoes combination with a proton from the initially formed aminopyridine to evolve hydrogen. On hydrolysis of the sodium salt of 2-aminopyrdine, it gives 2-aminopyridine and sodium hydroxide.

Sodium salt of 2-amino pyridine   2-amino pyridine

Applications

a) The reaction is useful in synthesis of nitrogen-containing aromatic amines. For example: Synthesis of 1,5-naphthyridin-2-amine.

1,5-naphthyridine   1,5-naphthyridin-2-amine

b) Synthesis of 3-(5,6,7,8-tetrahydro-1,8-naphthyridin-2-yl)propan-1-amine.

3,3'-pyridine-2,5-diyldipropan-1-amine

3-(5,6,7,8-tetrahydro-1,8-naphthyridin-2-yl)propan-1-amine

c) 2-aminopyridine finds use in the preparation of drug sulfapyridine.

4-amino-*N*-(pyridin-2-yl)benzenesulfonamide
(Sulfapyridine)

Principle

Chichibabin reported synthesis of substituted pyridines in 1906. The thermal cyclo-condensation of aldehydes and ammonia by passing aldehydes and ammonia over a contact catalyst such as alumina to give substituted pyridines is known as Chichibabin pyridine synthesis. Besides aldehydes, ammonia gas also reacts with acetylene or acetonitrile over a heated contact catalyst to give pyridine derivatives. The Chichibabin pyridine synthesis also works with aliphatic and aromatic ketones, α,β-unsaturated aldehyde, and keto acids.

General Reaction

Mechanism

The reaction mechanism is illustrated by the reaction between acetaldehyde and ammonia.

Step 1: Formation of imine (I) intermediate.

Step 2: Aldol addition followed by dehydration to give α,β-unsaturated aldehyde (II).

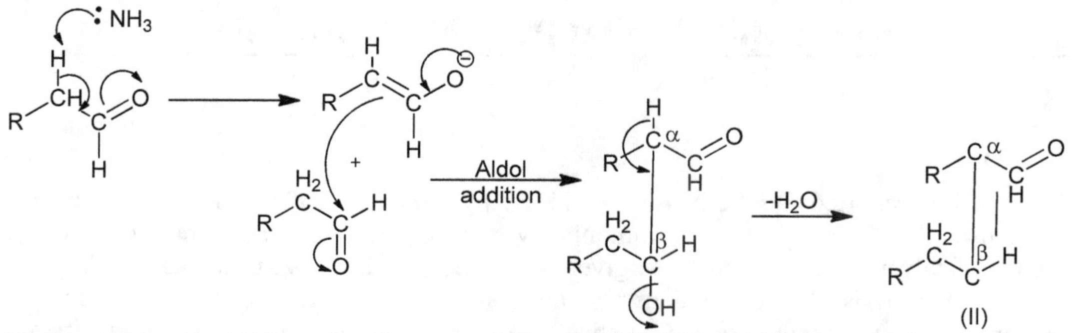

Step 3: Michael addition of α,β-unsaturated aldehyde (II) and imine (I) intermediate to give dihydro product.

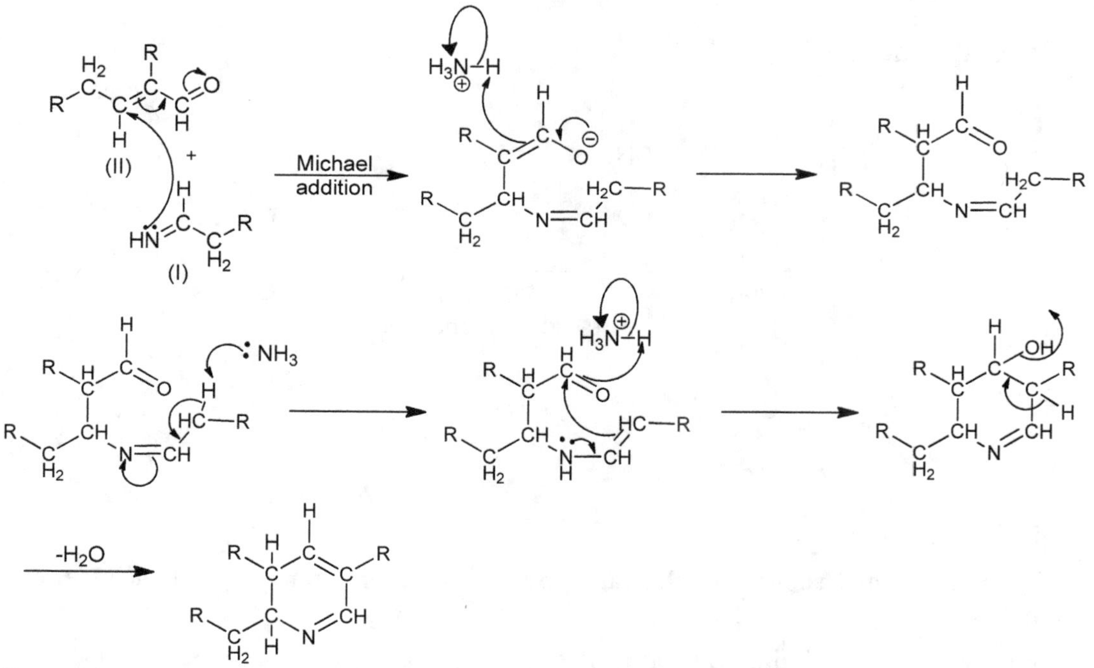

Step 4: Auto-oxidation to afford substituted pyridine.

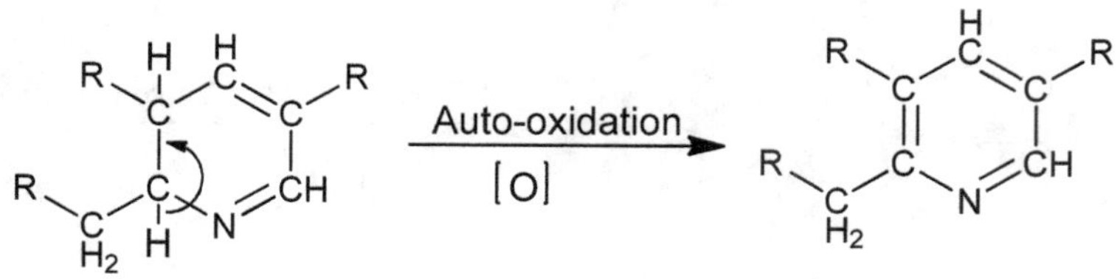

Applications

The reaction has been used for the synthesis of pyridine and collidine derivatives.

a) Synthesis of mixture of substituted pyridines.

2-methyl-5-phenylpyridine

2-methyl-3,5-diphenylpyridine

2-phenylacetaldehyde   +   OHC—CH₃ $\xrightarrow{\text{NH}_3,\ \text{C}_2\text{H}_5\text{OH},\ 230°C,\ 1000PSI}$

acetaldehyde

2-methyl-3,5-diphenylpyridine

b) Synthesis of 3,5-bis(3,4-dimethoxybenzyl)pyridine.

2-(3,4-dimethoxyphenyl)acetaldehyde $\xrightarrow{\text{NH}_3,\ \text{C}_2\text{H}_5\text{OH},}$ 3,5-bis(3,4-dimethoxybenzyl)pyridine

45. Chugaev Elimination

Principle

The transformation of alcohols into olefins by the pyrolysis of corresponding xanthate esters is Chugaev reaction. The reaction was reported by Chugaev in 1899. Pyrolysis in this reaction proceeds *via* an intramolecular *cis*-elimination without rearrangement of a carbon skeleton. The reaction is also applicable for the conversion of secondary and tertiary alcohols into olefins; because xanthate esters of primary alcohols have higher thermal stability and thus are relatively difficult to decompose during heating.

General Reaction

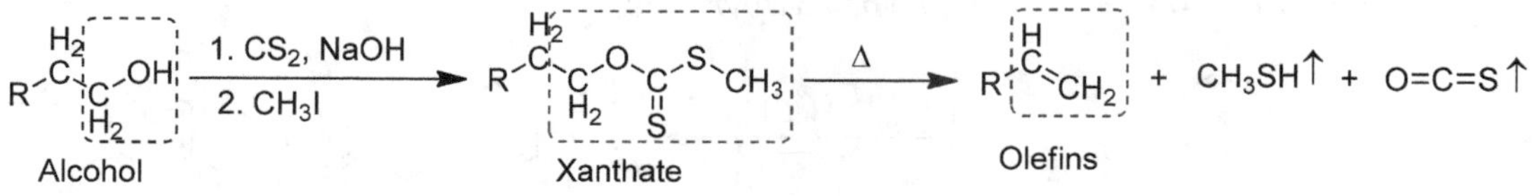

Mechanism

The reaction takes place *via* an intramolecular cyclic concerted process, as given below:

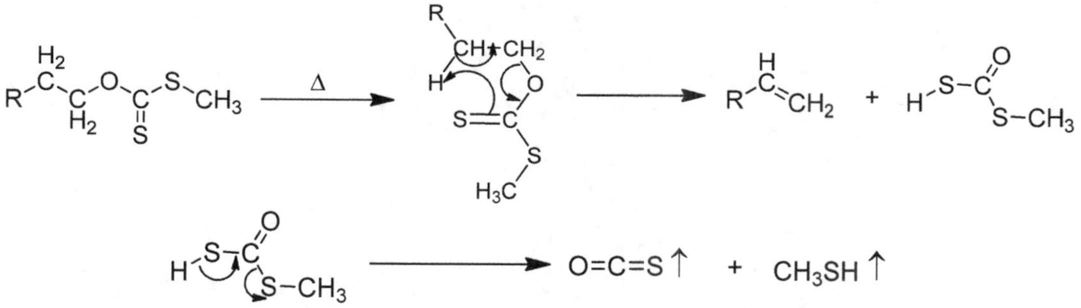

Applications

a) The reaction is particularly useful in synthesis of olefins. For example: Preparation of prop-1-ene from 3-methylbutan-1-ol.

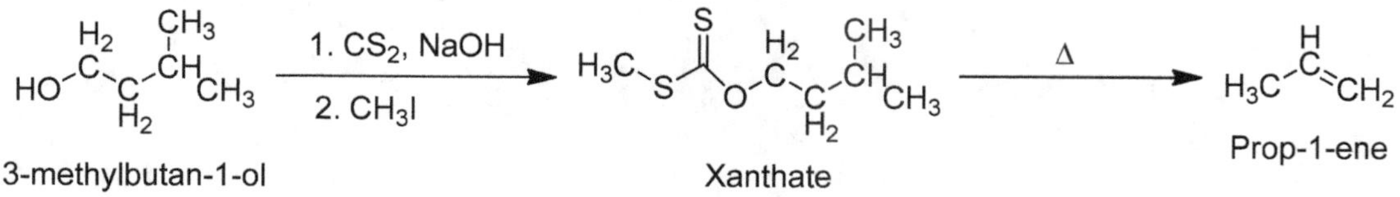

b) Synthesis of 1,2,3,4-tetrahydro-1,1'-biphenyl.

S-methyl O-((1R,2R)-2-phenylcyclohexyl)carbonodithioate

$\xrightarrow{\Delta}$

1,2,3,4-tetrahydro-1,1'-biphenyl

46. Claisen Condensation

Principle

The reaction was first reported by Geuther in 1863. Claisen explored this self condensation reaction. It is a self-condensation of ester in the presence of alkali alkoxide in alcohol to form β-keto esters (For example: Ethyl acetoacetate from Ethyl acetate) and is generally known as acetoacetic ester condensation. The condensation is generally performed under basic conditions (For example: Sodium acetate) to generate β-keto-esters from aliphatic carboxylic acid esters.

General Reaction

For example: Preparation of ethylacetoacetate from self condensation of two molecules of ethyl acetate

Mechanism

Claisen condensation is a nucleophilic substitution reaction. Sodium ethoxide acts as a catalyst and provides conditions necessary for the reaction. The reaction proceeds through following steps:

Step 1: Abstraction of α-hydrogen of ester by base (formation of carbanion).

Step 2: Attack of carbanion on carbonyl carbon of second ester molecule.

$$H_3C-\overset{O}{\underset{}{C}}-O-\overset{H_2}{C}-CH_3 \;+\; \overset{\ominus}{H_2}C-\overset{O}{\underset{}{C}}-O-\overset{H_2}{C}-CH_3 \longrightarrow H_3C-\overset{O^{\ominus}}{\underset{OC_2H_5}{C}}-\overset{H_2}{C}-\overset{O}{\underset{}{C}}-O-\overset{H_2}{C}-CH_3$$

Carbanion

Step 3: Removal of ethoxide ion (Formation of ethyl acetoacetate).

$$H_3C-\overset{O^{\ominus}}{\underset{OC_2H_5}{C}}-\overset{H_2}{C}-\overset{O}{\underset{}{C}}-O-\overset{H_2}{C}-CH_3 \longrightarrow H_3C-\overset{O}{\underset{}{C}}-\overset{H_2}{C}-\overset{O}{\underset{}{C}}-O-\overset{H_2}{C}-CH_3 \;+\; \overset{\ominus}{O}-C_2H_5$$

ethyl 3-oxobutanoate

Applications

a) Synthesis of quinine from ethyl 6-methoxyquinoline-4-carboxylate and N-benzoyl homomeroquinone.

Ethyl 6-methoxyquinoline-4-carboxylate + N-benzoyl homomeroquinone $\xrightarrow{C_2H_5ONa}$

Quinine

b) Synthesis of thiamine.

Ethyl 3-ethoxypropanoate + H-COOC$_2$H$_5$ (Ethyl formate) $\xrightarrow{Na}$ Thiamine

c) *Synthesis of isoflavone.*

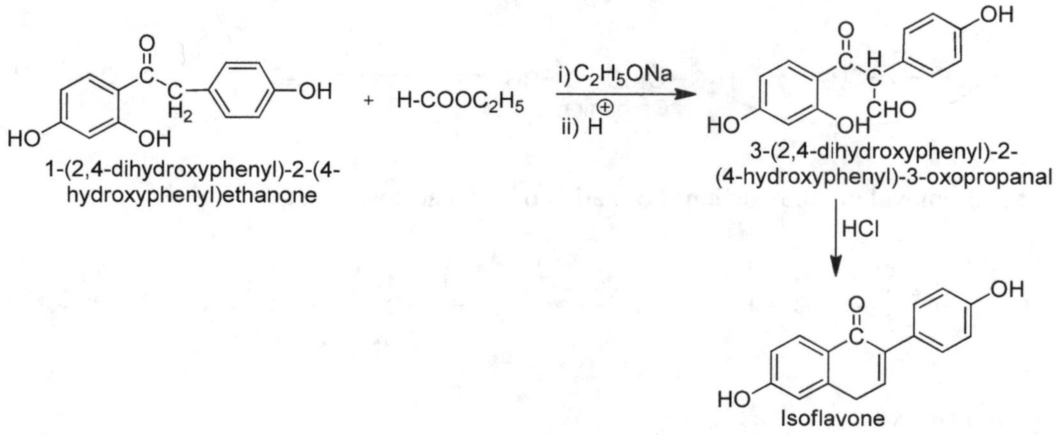

1-(2,4-dihydroxyphenyl)-2-(4-hydroxyphenyl)ethanone

$+$ H-COOC$_2$H$_5$

i) C$_2$H$_5$ONa
ii) H$^{\oplus}$

3-(2,4-dihydroxyphenyl)-2-(4-hydroxyphenyl)-3-oxopropanal

HCl

Isoflavone

47. Claisen Rearrangement

Principle

Ludwig Claisen in 1912 reported this rearrangement reaction. The thermal isomerization of allyl vinyl ether or of its nitrogen or sulfur containing analogous to produce a bifunctionalized molecule is known as Claisen rearrangement. It is highly stereoselective [3,3]-sigmatropic rearrangement of allyl vinyl or allyl aryl ethers to afford γ,δ-unsaturated carbonyl compounds or *o*-allyl substituted phenols, respectively. The reaction can be chemoselective, regioselective, diastereoselective, and entantioselective; give varieties of multifunctionalized molecules. When allyl aryl ethers undergo Claisen rearrangement, the γ,δ-unsaturated carbonyl compound will convert into *o*-allyl phenols *via* re-aromatization; however, if the *ortho* position is blocked by other substituents other than hydrogen the allyl group will migrate to the *para* position of the aromatic ring followed by enolization (*para* Claisen rearrangement). In addition, rearrangement works faster in a polar solvent, and the relative rate of the reaction for allyl vinyl ether decreases in order of water>trifluoroaceticacid>methanol>ethanol>2propanol>acetonitrile>acetone≈benzene >cyclohexane.

General Reaction

a) Preparation of 2-allyl phenol (*o*-allyl phenol) from Allylphenyl ether.

Allylphenyl ether → 200°C → 2-allyl phenol (*o*-allyl phenol)

b) Preparation of 4-allyl-2,6-dimethylphenol from Allylphenyl ether.

Allylphenyl ether → 200°C → 4-allyl-2,6-dimethylphenol

c) Preparation of 2-allylnaphthalen-1-ol from 1-(allyloxy)naphthalene (α-Naphthyl allyl ether).

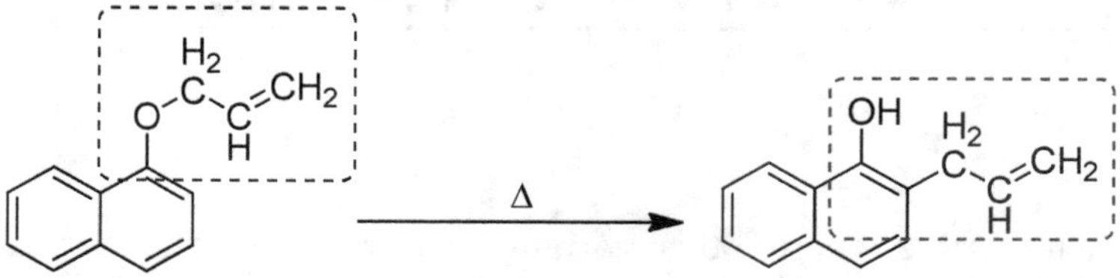

1-(allyloxy)naphthalene
(α-Naphthyl allyl ether)

Δ

2-allylnaphthalen-1-ol

d) Preparation of phenyl allyl phenol.

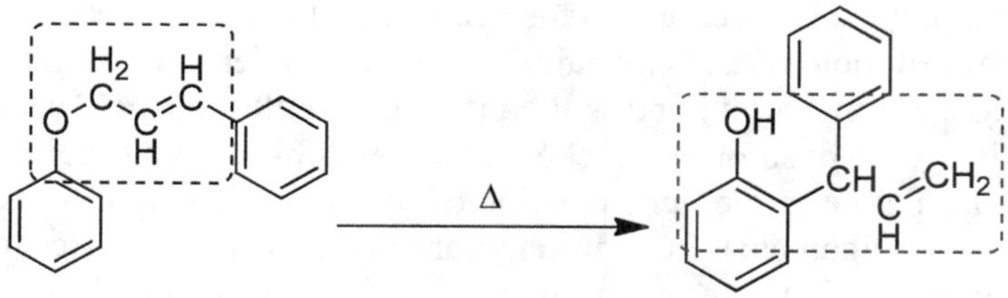

Phenyl cinnamyl ether

Δ

Phenyl allyl phenol

Mechanism

a) In the reaction, the migration of allyl group occurs at the ortho position to give the product. In this reaction C-1 and oxygen bond cleavage and C-3 and ortho carbon (or aromatic ring) bond formation takes place. Formation of dienone intermediate; tautomerism to afford 2-allyl phenol.

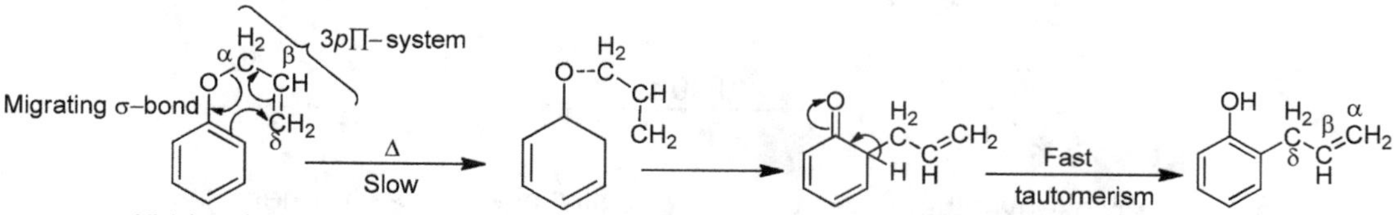

Migrating σ–bond $3p\Pi$–system

Δ
Slow

Transition state

Dienone intermediate

Fast
tautomerism

Allylphenyl ether
[3,3]-sigmatropic
rearrangement

2-allyl phenol
(o-allyl phenol)

b) Formation of 4-allyl-2,6-dimethylphenol

Allylphenyl ether

4-allyl-2,6-dimethylphenol

Applications

a) The Claisen rearrangement is frequently useful in the synthesis of o-eugenol.

Guaiacol allyl ether

o-eugenol

b) Claisen rearrangement is also useful in synthesis of polymer.

(*E*)-1-((*E*)-dec-1-en-1-yloxy)undec-2-ene

(2*R*,3*S*)-2-octyl-3-vinylundecanal

Principle

The condensation reaction between an aromatic aldehyde and an aliphatic aldehyde or ketone in the presence of a base or an acid to form an α,β-unsaturated aldehyde or ketone with high chemoselectivity is known as the Claisen-Schmidt condensation. Claisen and Schmidt reported this condensation reaction in 1881. In a reaction, asymmetric ketones react with the aromatic aldehyde through a less substituted position (a methyl group) under basic conditions or through more substituted position under acidic catalysis. The reaction between arylaldehydes and cycloalkanones specifically gives α,α-diarylidenecycloalkanones instead of α-arylidenecycloalkanones.

General Reaction

Benzaldehyde   Acetophenone β-hydroxy keone

Benzylidenacetophenone
α,β-unsaturated aryl ketone

Mechanism

This is base catalyzed reaction. The step wise mechanism of the reaction is as follows:

Step 1: Abstraction of α-hydrogen (Formation of carbanion)

In alkaline medium, carbonyl compound containing α-hydrogen looses a proton and forms an enolate ion (carbanion)

Acetophenone   Enolate ion

Step 2: Attack of nucleophile (carbanion) on second molecule, *i.e.*, aromatic aldehyde

The aromatic aldehyde devoid of α-hydrogen and hence it behaves as a carbanion acceptor.

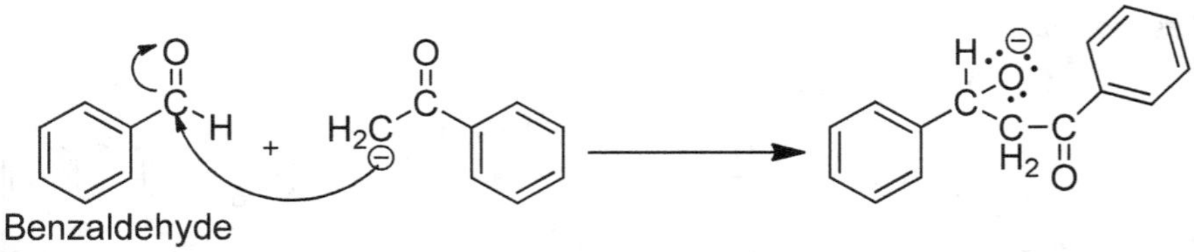
Benzaldehyde

Step 3: Proton abstraction

Proton is abstracted from water to form β-hydroxyketone which readily undergoes dehydration to form α,β-unsaturated carbonyl compound.

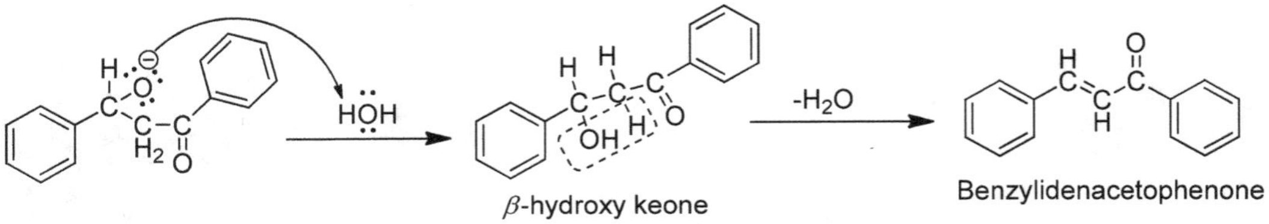
HOH
-H2O
β-hydroxy keone
Benzylidenacetophenone

Applications

The reaction widely useful in synthesis of chalcones, flavanones, 1,3-diarypropanes, etc.

a) Synthesis of flavones.

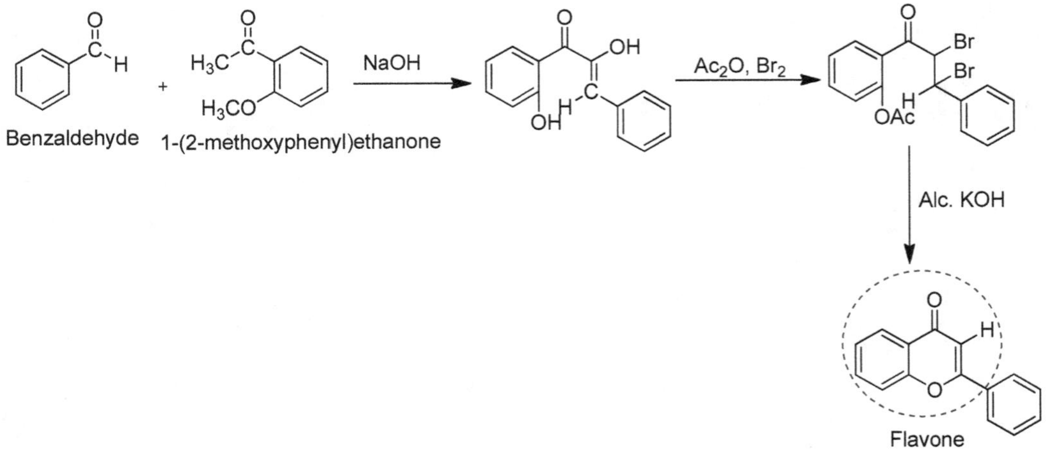
NaOH
Ac2O, Br2
Alc. KOH
Benzaldehyde
1-(2-methoxyphenyl)ethanone
Flavone

b) Synthesis of flavonol.

49. Clemmensen Reduction

Principle

The reduction of carbonyl groups of aldehydes or ketones to methylene groups in presence of Zinc-amalgam (Zn-Hg) and hydrochloric acid is known as Clemmensen reduction. This reduction reaction was first reported by Clemmensen in 1913. The actual reduction takes place on the surface of Zinc. The reaction also works out in acetic acid with Zinc and HCl. Clemmensen reduction transforms carbonyl group into methylene group, these are some problems associated with this reduction reaction:

i)   The hydroxyl aromatics are not reduced, For example: isomer of cyclic vinyl ketone;

ii) α-keto acids or α-keto esters are partially reduced to α-hydroxyl acids or esters, respectively;

iii) The reduction of 1,3-diketones involves the rearrangement *via* pinacol-type intermediate;

iv) The reduction of β-keto esters affords pure hydrocarbon,

v)   The reduction of benzophenone is not successful, giving a resinous product;

vi) The reduction of acetophenone in dilute hydrochloric acid gives styrene instead of ethylbenzene.

General Reaction

Aromatic ketone → Aromatic hydrocarbon (Zn-Hg, Conc. HCl)

a) Reduction of acetophenone to ethylbenzene.

Acetophenone → Ethylbenzene (Zn-Hg, Conc. HCl)

b) Reduction of cyclohexanone to cyclohexane.

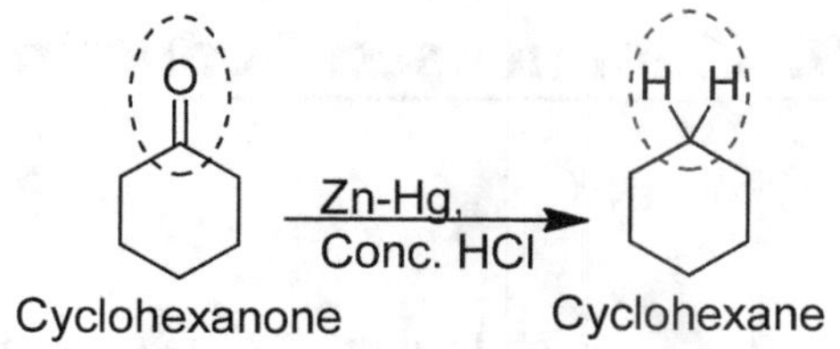

Cyclohexanone   Zn-Hg, Conc. HCl   Cyclohexane

c) Reduction of propanal to propane.

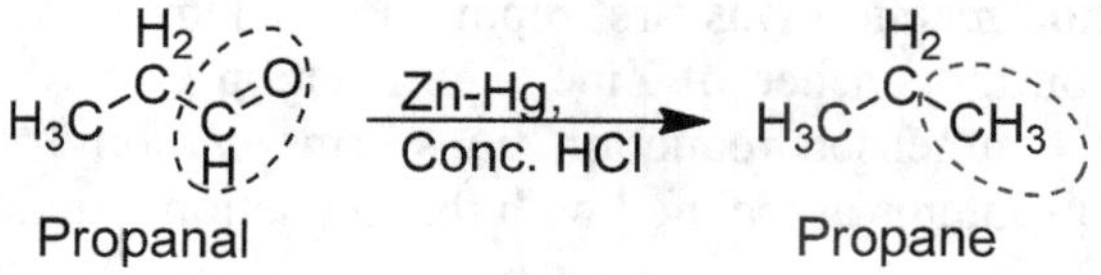

Propanal   Zn-Hg, Conc. HCl   Propane

d) Reduction of butan-2-one to butane.

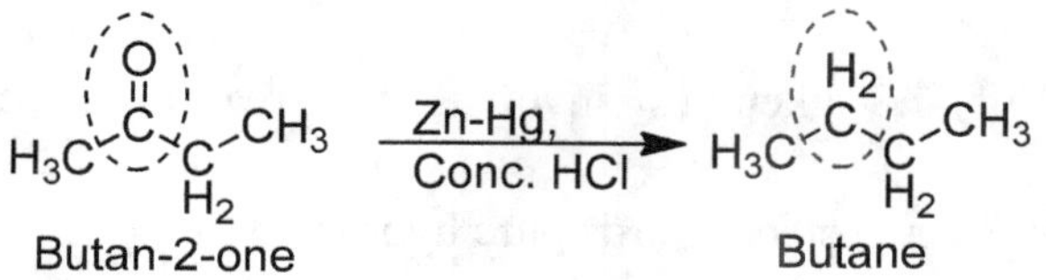

Butan-2-one   Zn-Hg, Conc. HCl   Butane

Mechanism

The detailed mechanism for the reduction of ketone into saturated hydrocarbon is illustrated below:

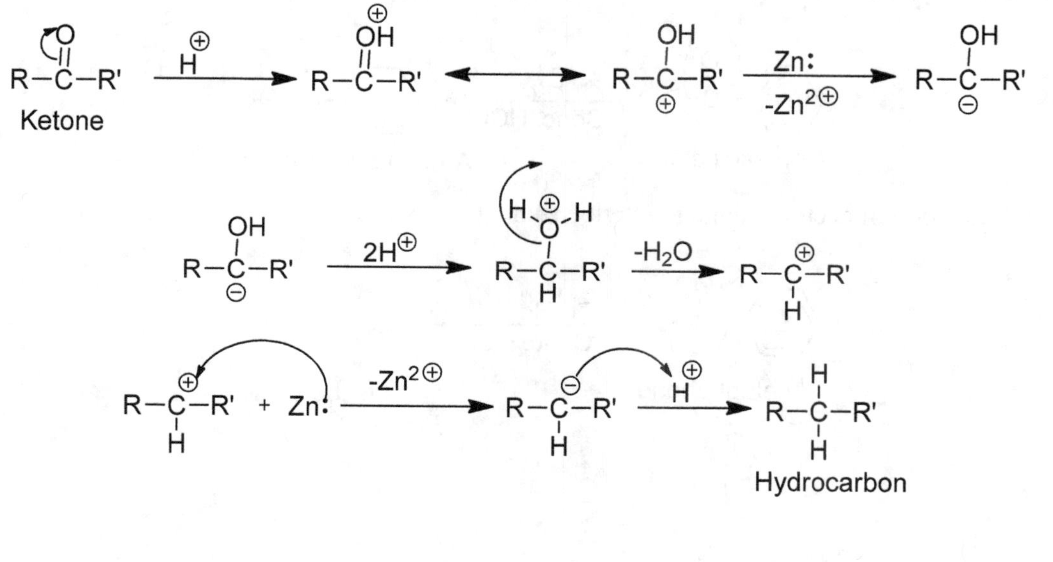

Applications

The reaction has been successfully utilized to transform carbonyl group into a methylene group, with important applications in synthesis of polycyclic aromatics and aromatics containing un-branched side hydrocarbon chains.

a) Reduction of γ-napthopyrone to 3,4-dihydro-2,2-dimethylnaphthopyran.

γ-napthopyrone → (Zn-Hg, Conc. HCl) → 3,4-dihydro-2,2-dimethylnaphthopyran

b) Conversion of Friedel-Crafts products by Clemmensen reduction.

Benzene + Propionyl chloride → (AlCl$_3$, H$_2$O) → Propiophenone → (Zn-Hg, Conc. HCl) → Propylbenzene

c) Reduction of phenolic carbonyl compounds.

Salicylaldehyde → (Zn-Hg, Conc. HCl) → o-cresol

d) Reduction of cyclic ketones.

α-hydrindone → (Zn-Hg, Conc. HCl) → Indane

50. Combes Quinoline Synthesis

Principle

Combes in 1888 reported synthesis of quinoline. The condensation of primary aromatic amines with acetoacetone or other β-diketones followed by cyclization in presence of sulfuric acid to afford quinoline is known as Combes quinoline synthesis. The reaction is useful for synthesis of 2,4-substituted quinoline skeleton; but, with limitation of low regioselectivity, for example in case of *m*-substituted anilines with two different *ortho* positions for ring closure. It was found that polyphosphoric acid is a better catalyst when compared with sulfuric acid for the cyclization. The reaction has been modified to use polyphosphoric acid as acidic catalyst to give better yield.

General Reaction

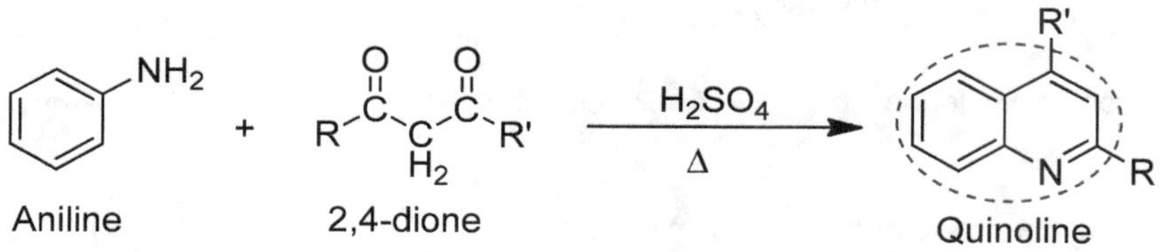

For example: Synthesis of 2,4,7-trimethylquinoline

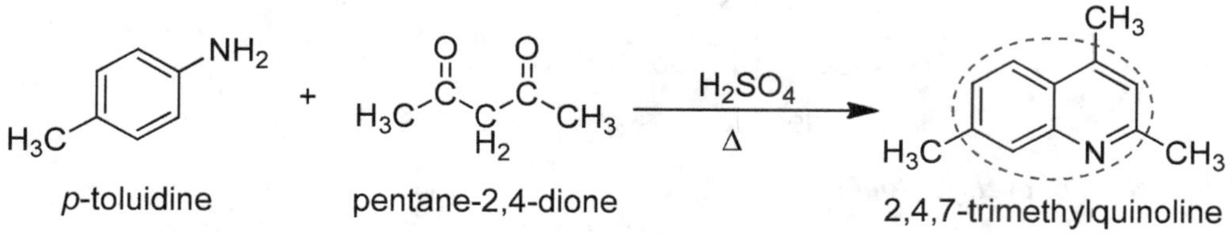

Mechanism

The reaction involves a condensation between aniline and β-diketone followed by an acid-catalyzed cyclization is given below.

Step 1: A condensation reaction between aniline (nucleophile) and β-diketone to give imine intermediate which is rapidly converted into enamine by tautomerism.

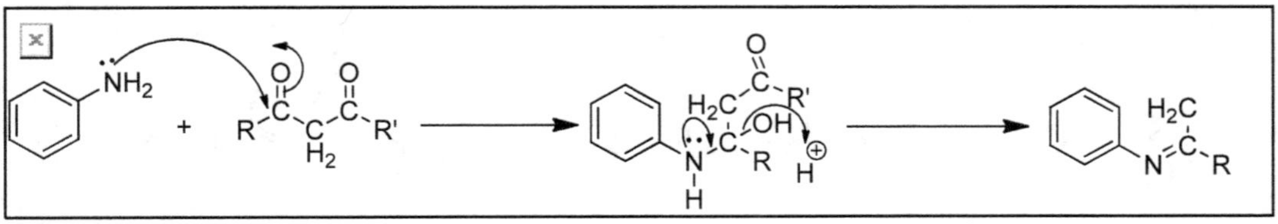

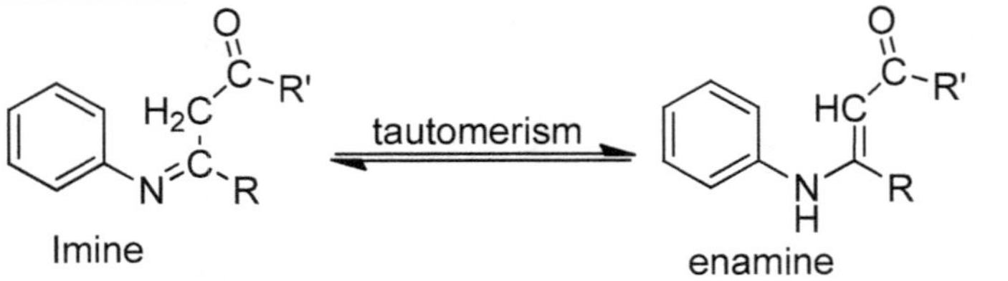

Imine

tautomerism

enamine

Step 2: Acid catalyzed cyclization (ring closure reaction) to afford substituted quinoline derivative.

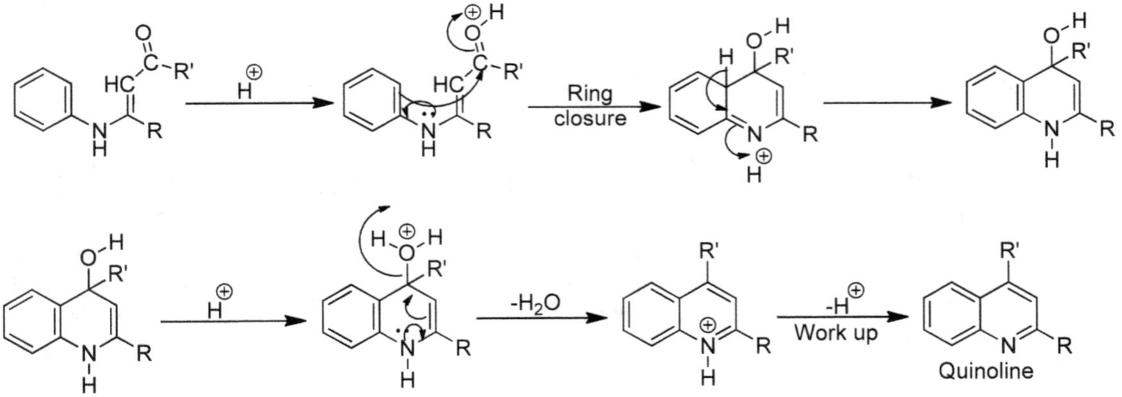

Ring closure

$-H_2O$

Work up

Quinoline

Applications

Combes quinoline synthesis is applicable to synthesize different substituted quinolines particularly at position 2 and/or 5.

a) Synthesis of (7-methoxy-2-(methylthio)quinolin-3-yl)(phenyl)methanone.

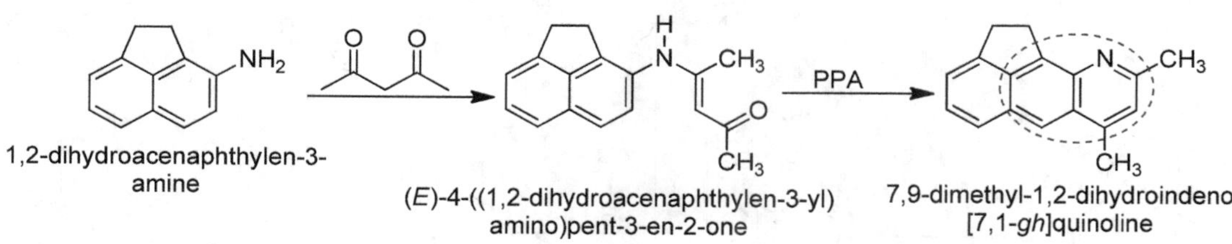

3-methoxyaniline

3,3-bis(methylthio)-1-phenylprop-2-en-1-one

n-BuLi

(*Z*)-3-((3-methoxyphenyl)amino)-3-(methylthio)-1-phenylprop-2-en-1-one

POCl₃/DMF

(7-methoxy-2-(methylthio)quinolin-3-yl)(phenyl)methanone

b) Synthesis of 7,9-dimethyl-1,2-dihydroindeno[7,1-gh]quinoline.

1,2-dihydroacenaphthylen-3-amine

(*E*)-4-((1,2-dihydroacenaphthylen-3-yl)amino)pent-3-en-2-one

PPA

7,9-dimethyl-1,2-dihydroindeno[7,1-*gh*]quinoline

51. Cope Elimination

Principle

Hoffmann's elimination involving a pyrolysis of an amine oxide prepared by the action of hydrogen peroxide on a tertiary amine is known as Cope elimination. The reaction was reported by Wolffenstein in 1898. In the reaction, thermal decomposition of *N*-oxides of tertiary amines affords olefins. The reversal of Cope elimination (addition of hydroxylamine to double bond) is the reverse Cope elimination or *retro*-Cope elimination. The temperature required for decomposition of tertiary amine oxides is 120-150°C without presence of any solvent; the reaction proceeds even at room temperature if dimethyl sulfoxide is used as a solvent. The order of the eliminating alkyl group for number of available β-protons is *p*-phenylethyl>>*t*-butyl>isobutyl·isopropyl·*n*-decyl>*n*-butyl>isoamyl>ethyl>*n*-propyl.

General Reaction

R_2N–C–C–H · 2° amine · H_2O_2 · |O| · Amine oxide · Δ · Hydroxyl amine · Ethene

N,N-dimethylpropan-2-amine · H_2O_2 · |O| · N,N-dimethylpropan-2-amine oxide · Δ · N,N-dimethylhydroxylamine · Prop-1-ene

Mechanism

The reaction is example first order elimination. In this reaction, formation of intramolecular five membered cyclic synchronous transitions state takes place consisting of hydrogen and nitrogen in coplanar condition.

Amine oxide

Five-membered cyclic
transition state

Δ

Ethene

N-ethyl-N-methylhydroxylamine

Ag_2O, Δ

Transition state

Δ

N,N-dimethylhydroxylamine   Ethene

Applications

The reaction is useful in preparation of olefins.

a) Synthesis of styrene.

N,N-dimethyl-1-phenylethanamine

H_2O_2 / $|O|$

Δ

Styrene N,N-dimethylhydroxylamine

b) Synthesis of menthene.

2-isopropyl-N,N,5-trimethylcyclohexanamine

H_2O_2 / $|O|$

Δ

Menthene

52. Cope, Oxy-Cope, and Anionic Oxy-Cope Rearrangements

Principle

Cope and Hardy in 1940 reported thermal isomerization of 1,5-diene *via* highly stereoselective [*3,3*] sigmatropic rearrangement to produce more stable regioisomeric 1,5-diene is known as Cope rearrangement. The reaction later is highly explored for different variants such as *amino*-Cope, *aza*-Cope, *diaza*-Cope, *oxaza*-Cope, *oxonia*-Cope, *oxy*-Cope, *sila*-Cope, and *thio*-Cope rearrangements. The rate of reaction is enhanced significantly; if hydroxyl substituent is present at third position of 1,5-diene and reaction is performed in presence of strong base such as potassium hydride. For example: *oxy*-Cope rearrangement. In *oxy*-Cope rearrangement, hydroxyl group is present on sp^3 hybridized carbon the starting isomer 1,5-diene. When two hydroxyl groups are present at third and fourth positions, the rearrangement occurs at a temperature as low as -78°C. The driving force for the neutral or anionic *oxy*-Cope rearrangement is to form an enol or enolate, which can irreversibly tautomerize to corresponding carbonyl compound. It has been reported that fast rate for the *oxy*-Cope rearrangement is due to bond-weakening effect of the anionic alkoxy group on adjacent C3-C4 bond; as a result, the reaction runs faster when the alkoxy anion is more exposed.

General Reaction

oxy-Cope rearrangement

Mechanism

Cope rearrangement: In Cope rearrangement through concerted mechanism, the chair-like transition state is favored, which is not polarized; as a result, no apparent solvent effect has been observed.

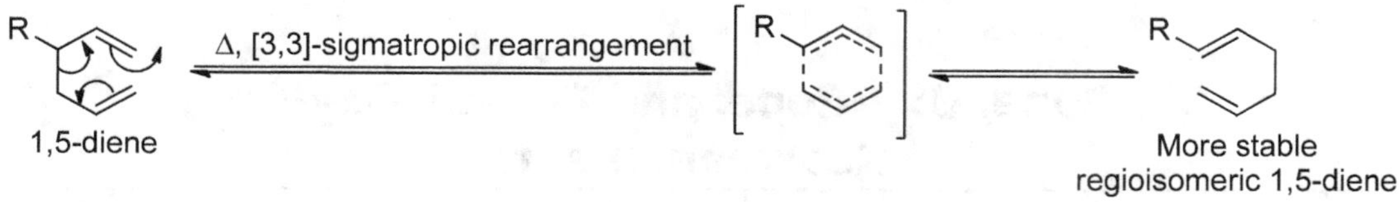

1,5-diene ⟶ Δ, [3,3]-sigmatropic rearrangement ⟶ More stable regioisomeric 1,5-diene

oxy-Cope rearrangement

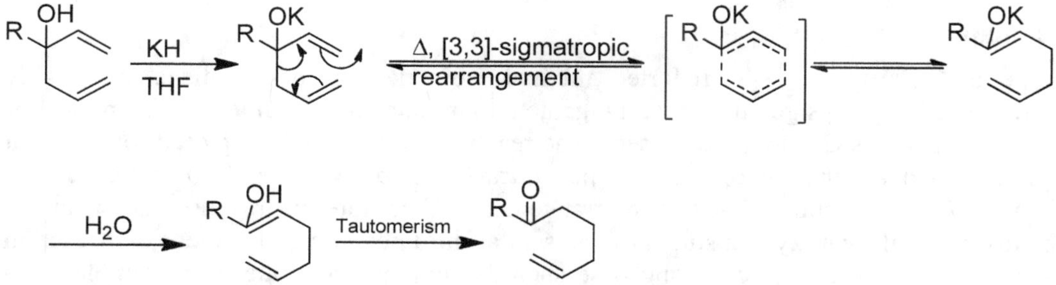

$\xrightarrow{\text{KH, THF}}$ ⟶ Δ, [3,3]-sigmatropic rearrangement

$\xrightarrow{H_2O}$ $\xrightarrow{\text{Tautomerism}}$

Applications

The reaction is well explored for the synthesis of different nucleus useful in organic chemistry.

a) Reverse aromatic cope rearrangement.

Olefination ⟶ Reverse Aromatic Cope rearrangement

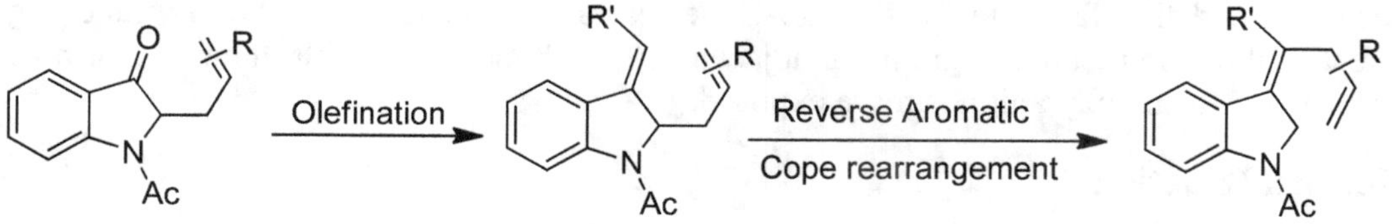

b) Formation of more stable regioisomer.

Toluene, Δ

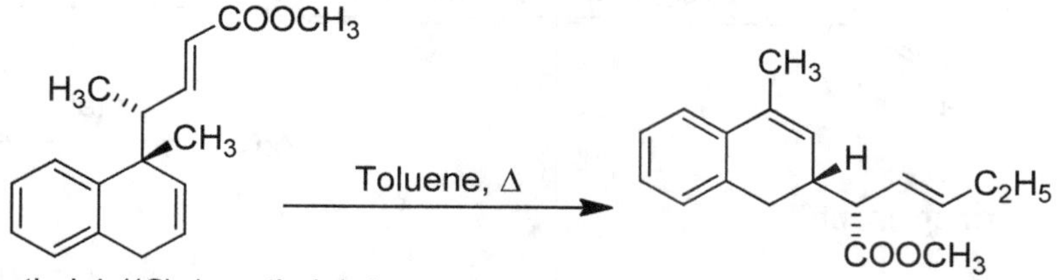

(R,E)-methyl 4-((S)-1-methyl-1,4 -dihydronaphthalen-1-yl)pent-2-enoate

(S,E)-methyl 2-((R)-4-methyl-1,2 -dihydronaphthalen-2-yl)hex-3-enoate

53. Corey-Bakshi-Shibata Reduction

Principle

Corey and his associates in 1987 reported the reduction of ketone by borane in presence of a chiral oxazaborolidine catalyst, with excellent enantioselectivity, better yield, high turnover of catalyst, and increased rate of reaction. The reaction is popularly known as Corey-Bakshi-Shibata reduction. A chiral oxazaborolidine catalyst is also known as Corey-Bakshi-Shibata catalyst or the CBS catalyst. The endocyclic CBS catalyst is highly stable at open air and to moisture. Basically, the CBS reduction occurs in four main steps as follows: a) coordination of the nitrogen atom of the Lewis base to borane, b) complexation of the ketone to the endocyclic boron (functioning as Lewis acid) *via* the Lewis acid–base interaction, c) hydride transfer from borane to the carbonyl carbon, and d) dissociation of the alkoxyborane moiety and regeneration of the catalyst. It is reported that up to two of the three hydrides of borane can be transferred for reduction.

General Reaction

$R < R'$, $R'' = H, CH_3$

Mechanism

Step 1: Preparation of CBS catalyst

Step 2: Reaction between CBS catalyst and ketone to yield secondary alcohol with regeneration of catalyst

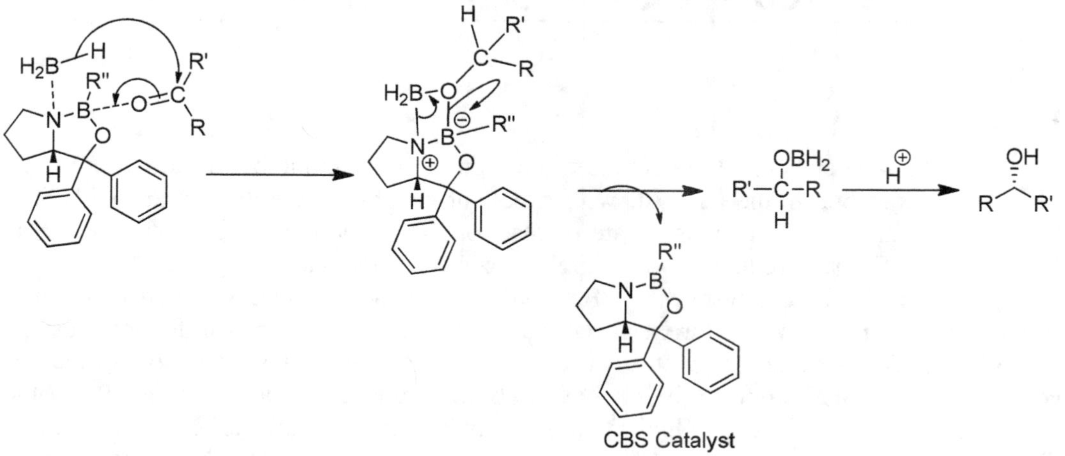

CBS Catalyst

Applications

The reaction has been modified to have a methyl group attach to the endocyclic boron atom of the CBS catalyst.

a) Generation of (S)-1,3,3-triphenylhexahydropyrrolo[1,2-c][1,3,2]oxazaborole.

Phenylboronic acid

Toluene, Δ

(S)-diphenyl(pyrrolidin-2-yl)methanol

(S)-1,3,3-triphenylhexahydropyrrolo[1,2-c][1,3,2]oxazaborole

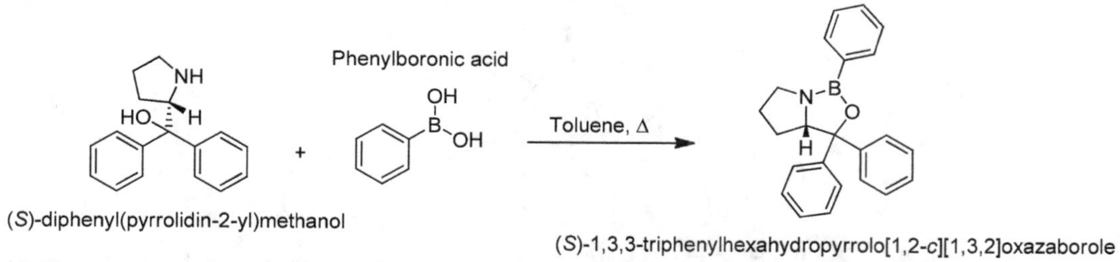

b) Conversion of acetophenone to 1-phenylethanol in presence of CBS catalyst.

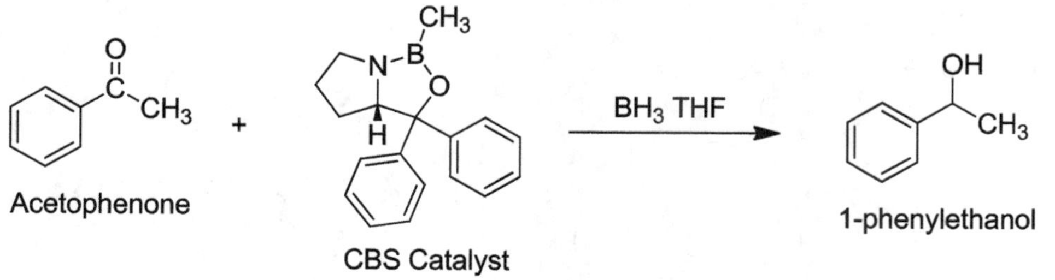

Acetophenone

CBS Catalyst

BH_3 THF

1-phenylethanol

54. Corey-House Synthesis

Principle

The reaction of haloalkanes with lithium dialkylcuprate to form alkanes in high yield is known as Corey-House synthesis. This is a versatile method for the synthesis of all types of alkanes (symmetrical or unsymmetrical) in high yields. Methane cannot be synthesized by this reaction. For high yield of alkane, the haloalkane used for the preparation of Gilman reagent, may contain methyl, 1°, 2°, or 3° alkyl group. However, haloalkane used for further coupling to produce alkane should preferably be either methyl or 1° alkyl group.

General Reaction

a) Haloalkanes on reaction with lithium metal in inert solvent (dry ether) gives alkyllithium, which on reaction with cuprous iodide yields lithium dialkylcuprate also known as Gilman reagent.

$$R-X \ + \ 2Li \ \xrightarrow{\text{Dry ether}} \ R-Li \ + \ LiX$$

Haloalkane → Alkyl lithium

$$2R-Li \ + \ CuI \ \xrightarrow{\text{Dry ether}} \ R_2Cu-Li \ + \ LiI$$

Alkyl lithium → Lithium dialkyl cuprate (Gilman reagent)

b) The Gilman reagent undergoes coupling reaction with an alkyl halide, where the alkyl group of Gilman reagent couples with the alkyl group of haloalkane to give an alkane.

$$R_2Cu-Li \ + \ R'-X \ \xrightarrow{\text{Dry ether}} \ R-R' \ + \ RCu \ + \ LiX$$

Lithium dialkyl cuprate + Haloalkane → Alkane + Lithium halide

***For example*:**

$$CH_3\text{-}Br \quad + \quad 2Li \quad \xrightarrow{\text{Dry ether}} \quad CH_3\text{-}Li \quad + \quad LiBr$$

Bromomethane   Methyllithium   Lithium bromide

$$2\ CH_3\text{-}Li \quad + \quad CuI \quad \xrightarrow{\text{Dry ether}} \quad \underset{H_3C}{\overset{H_3C}{>}}Cu\text{—}Li \quad + \quad LiI$$

Methyllithium   Lithium iodide

Lithium dimethyl cuprate
(Gilman reagent)

$$\underset{H_3C}{\overset{H_3C}{>}}Cu\text{—}Li \quad + \quad CH_3\text{-}Br \quad \xrightarrow{\text{Dry ether}} \quad CH_3\text{-}CH_3 \quad + \quad CH_3\text{-}Cu \quad + \quad LiBr$$

Lithium dimethyl cuprate   Bromomethane   Ethane   Methylcopper Lithium bromide
(Symmetrical alkane)

$$\underset{H_3C}{\overset{H_3C}{>}}Cu\text{—}Li \quad + \quad CH_3\text{-}CH_2\text{-}Br \quad \xrightarrow{\text{Dry ether}} \quad CH_3\text{-}CH_2\text{-}CH_3 \quad + \quad CH_3\text{-}Cu \quad + \quad LiBr$$

Lithium dimethyl cuprate   Bromoethane   Propane   Methylcopper Lithium bromide
(Unsymmetrical alkane)

Mechanism

Organometallic compound gives an alkyl carbanion (nucleophile) that attacks the haloalkane to give alkane with the displacement of halide ion.

It is an example of nucleophilic substitution reaction as depicted below:

Step 1: Loss of lithium from Gilman reagent

$$\underset{R}{\overset{R}{>}}Cu\text{—}Li \quad \rightleftharpoons \quad \underset{R}{\overset{R}{>}}Cu^{\ominus} \quad + \quad Li^{\oplus}$$

Gilman reagent

Step 2: Generation of nucleophile

$$\underset{R}{\overset{R}{>}}Cu^{\ominus} \quad \rightleftharpoons \quad 2\overset{\ominus}{R}\!: \quad + \quad Cu^{\oplus}$$

Nucleophile

Step 3: Nucleophilic substitution reaction to give alkane

$$:\!\overset{\ominus}{R} \quad + \quad \overset{+\delta\ \ -\delta}{R'\text{—}X} \quad \xrightarrow[\text{substitution}]{\text{Nucleophilic}} \quad R\text{—}R' \quad + \quad :\!\overset{\ominus}{X}$$

Nucleophile   Haloalkane   Alkane

Applications

a) Reaction is very useful in synthesis of different substituted alkanes. For example: Synthesis of ethylcyclohexane from bromocyclohexane.

55. Criegee Mechanism of Ozonolysis

Principle

Criegee in 1950 proposed the mechanism of ozonolysis after extensive study of reaction by Schonbein. Criegee was known to be father of modern ozone organic chemistry. It is the reaction of olefins and ozone to produce carbonyl compounds, carboxylic acids, etc. and commonly known as Criegee ozonolysis reaction.

General Reaction

Resonating structures of ozone

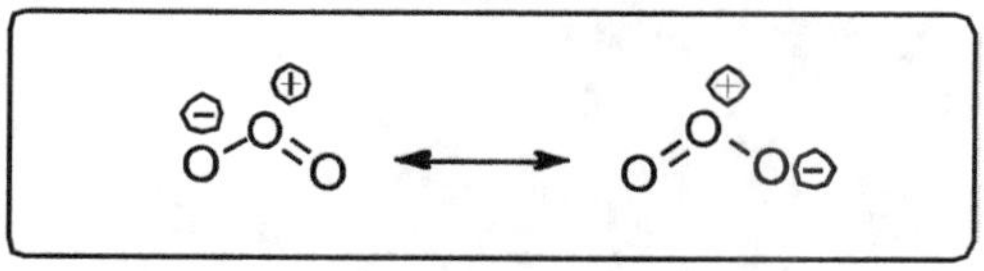

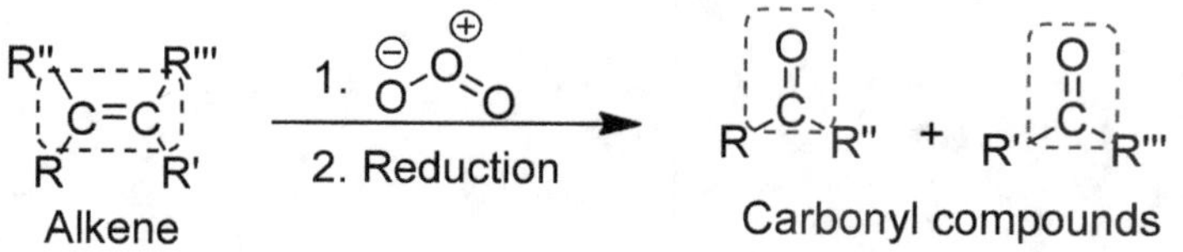

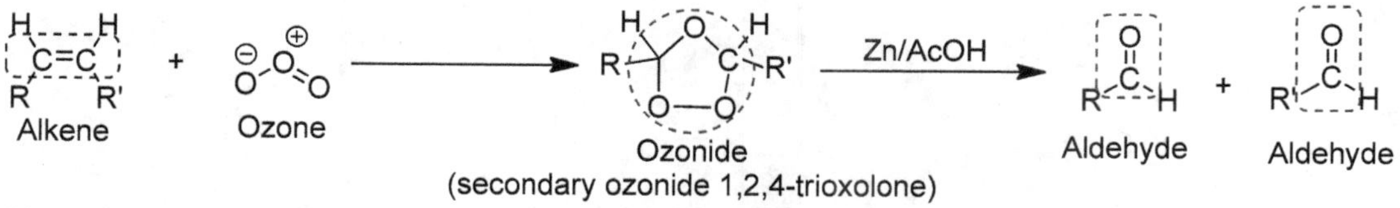

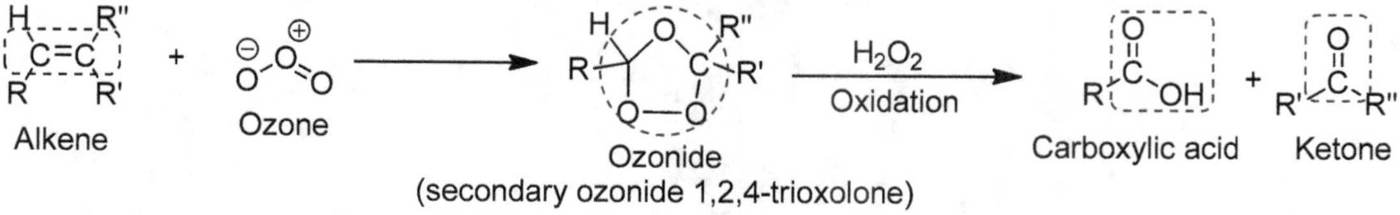

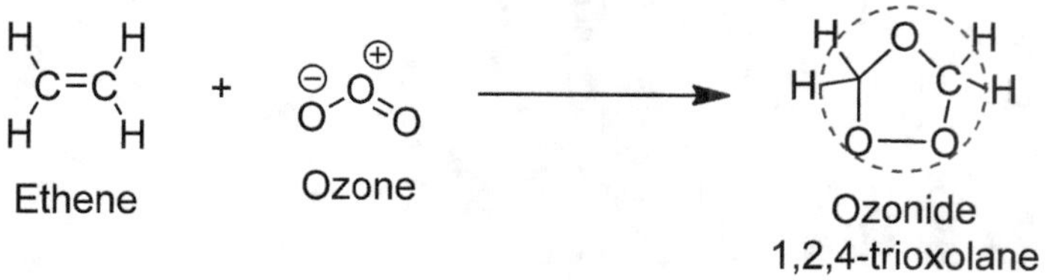

H₃C, CH₃ structures — (Z)-but-2-ene + Ozone → 3,5-dimethyl-1,2,4-trioxolane

Mechanism

Ozonolysis occurs in three steps:

a) Formation of primary ozonides; b) Splitting of primary ozonides; c) Formation of secondary ozonides.

In this reaction, ozone as an electrophile, adds to the C=C double bond *via* a concerted 1,3-dipolar cycloaddition, a symmetry allowed (*4s* + *2s*) cycloaddition, to form vibrationally excited primary ozonide (1,2,3-trioxolane, or molozonide) in a syn- or an anti-form, which quickly splits to an excited 'carbonyl oxide' (Criegee intermediate) and a carbonyl compound, which in turn couple stereospecifically into the secondary ozonide (1,2,4-trioxolane, or true ozonide) by reverse cycloaddition. Then the secondary ozonide is reduced to aldehyde or ketone *via* varieties of reducing reagents, including Zn/CH_3COOH, $(CH_3)_3P$ and $(C_6H_5)_3P$. This reaction is usually carried out in solvent such as CH_3COOH, $CHCl_3$, CCl_4, CH_2Cl_2, etc. at low temperature by passing a stream of dry ozonized oxygen until the disappearance of the olefins.

Molozonide
Primary ozonide
(1,2,3-trioxolane)

Criegee zwiterion
(Zwitterion peroxide/carbonyl oxide)

1,3-dipolar cycloaddition

Aldehyde or Ketone ← Reduction — Ozonide
Secondary ozonide
(1,2,4-trioxolane)

Applications

The reaction is very useful in determination of structures of molecules containing C=C double bonds as well as in preparation of carbonyls.

a) Reaction of cyclohex-4-ene-1,2-diol with ozone.

cyclohex-4-ene-1,2-diol

Ac₂O, Et₃N, DMAP, 75%

(2*S*,3a*S*,6a*S*)-5-oxohexahydrofuro[3,2-*b*]furan-2-yl acetate

b) Reaction of (1R,4S)-1,5,5-trimethyl-6,8-dimethylenebicyclo[2.2.2]octan-2-one with ozone.

(1*R*,4*S*)-1,5,5-trimethyl-6,8-dimethylenebicyclo[2.2.2]octan-2-one

Ac₂O, Et₃N, DMAP, 75%

(*R*)-methyl 2-(1,5,5-trimethyl-6-methylene-2-oxocyclohex-3-en-1-yl)acetate

56. Cumene Hydroperoxide Rearrangement

Principle

Acid catalyzed rearrangement of cumene hydroperoxide to phenol is referred as Cumene hydroperoxide rearrangement. The phenyl group joined to a carbon in hydroperoxide; later joined to oxygen in phenol results in formation of phenol and acetone. Cumene hydroperoxide is relatively stable organic peroxide. The decomposition products of cumene hydroperoxide are methylstyrene, acetophenone, and cumyl alcohol. Pure cumene hydroperoxide can be stored at room temperature, but possiblity for an uncontrolled reaction and explosion is high.

General Reaction

Cumene hydroperoxide
or
(2-hydroperoxypropan-2-yl)benzene

Phenol

Acetone

Mechanism

The rearrangement involves *1,2*-shift to electron deficient oxygen.

Step 1: Acid converts peroxide to protonated peroxide, which loses a molecule of water to form an intermediate in which oxygen bears only six electrons. *1,2*-shift of the phenyl group from carbon to electron deficient oxygen yields the carbocation intermediate.

Protonated
peroxide

Migration of
phenyl ring
1,2-shift

Carbocation

Step 2: Carbocation reacts with water to yield hydroxyl compound (hemiketal).

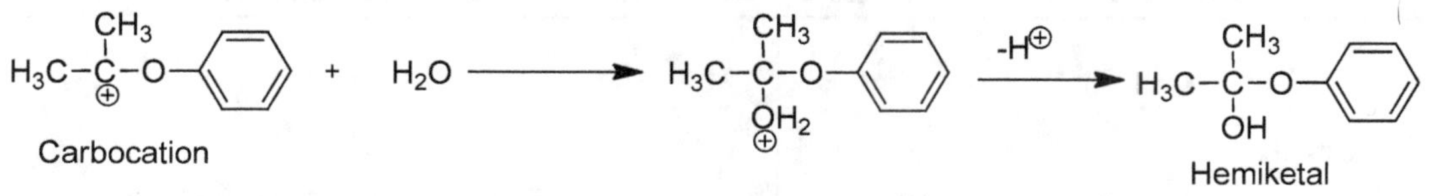

Carbocation + H_2O $\longrightarrow$ $-H^{\oplus}$ $\longrightarrow$ Hemiketal

Step 2: Hemiketal undergoes breakdown to give phenol and acetone.

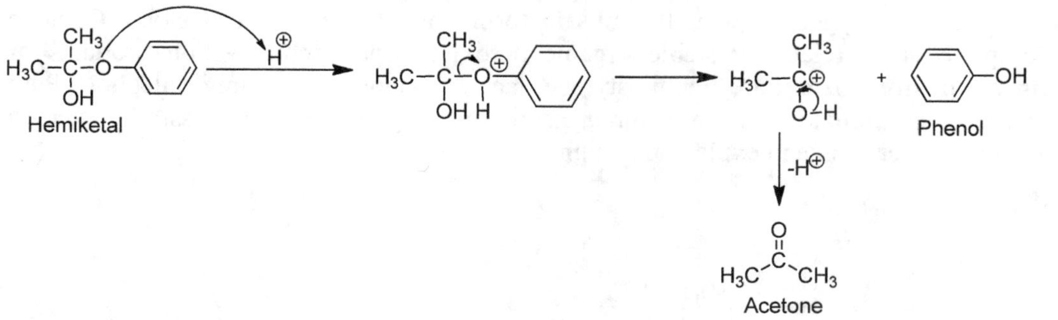

Hemiketal $\longrightarrow$ $\longrightarrow$ + Phenol $-H^{\oplus}$ Acetone

Applications

The reaction is commercially useful for the industrial synthesis of phenol, because cumene hydroperoxide is synthesized economically from benzene and propene (Friedel Crafts reaction).

a) Conversion of benzene to cumene which of autooxidation gives cumene hydroperoxide.

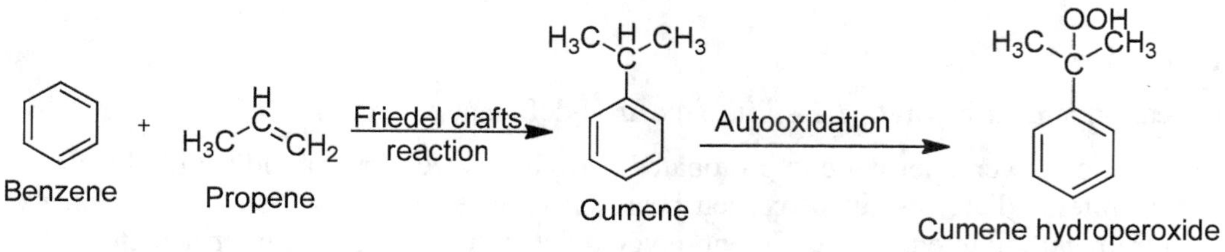

Benzene + Propene $\xrightarrow[\text{reaction}]{\text{Friedel crafts}}$ Cumene $\xrightarrow{\text{Autooxidation}}$ Cumene hydroperoxide

57. Curtius Rearrangement

Principle

Curtius in 1890 reported the formation of isocyanate by thermal decomposition of acyl azides prepared from acyl halide and sodium azide is known as Curtius rearrangement. Curtius rearrangement proceeds with complete retention of configuration at migrating carbon, and the migration from a carbon to a nitrogen atom is an irreversible intramolecular reaction in first order kinetics. The reaction is an example of intramolecular rearrangement, a solvent or salt effect is not evident. Recently, the Curtius rearrangement was performed under microwave irradiation and was modified using phase-transfer catalysis; however, the most important modification of the Curtius rearrangement is to use the Shiori reagent [diphenylphosphoryl azide (DPPA)].

General Reaction

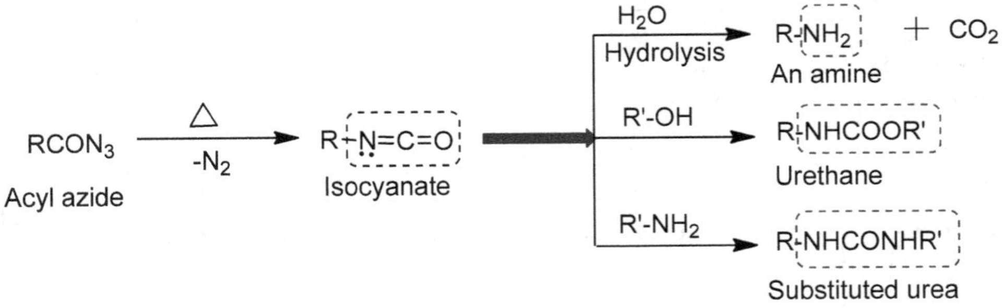

$$RCON_3 \xrightarrow[-N_2]{\triangle} R-N=C=O$$

Acyl azide — Isocyanate

$$R-N=C=O \xrightarrow[\text{Hydrolysis}]{H_2O} R-NH_2 + CO_2 \quad \text{(An amine)}$$

$$\xrightarrow{R'-OH} R-NHCOOR' \quad \text{(Urethane)}$$

$$\xrightarrow{R'-NH_2} R-NHCONHR' \quad \text{(Substituted urea)}$$

Mechanism

Step 1: Acyl azide through elimination of N≡N after thermal decomposition and produce electron deficient nitrogen atom; migration of alkyl group towards electron deficient nitrogen atom gives isocyanate intermediate.

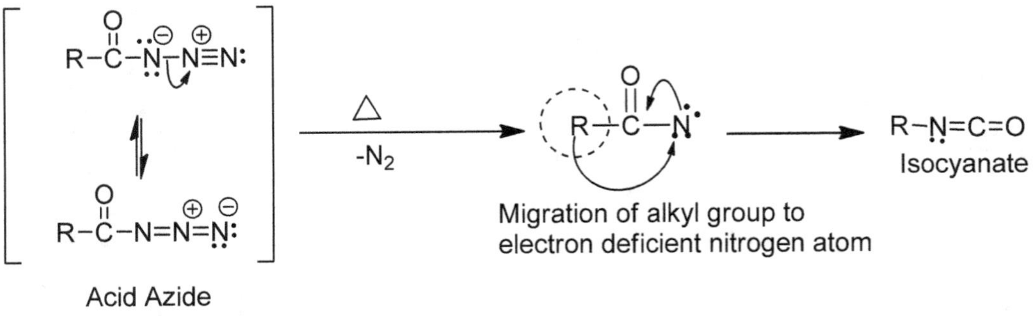

Acid Azide

$$\xrightarrow[-N_2]{\triangle}$$

Migration of alkyl group to electron deficient nitrogen atom

$$R-N=C=O \quad \text{Isocyanate}$$

Step 2: Hydrolysis of isocyanate to form an amine.

$$R-\underset{\cdot\cdot}{N}=C=O \xrightarrow[\text{H}_2\text{O}]{\text{Hydrolysis}} R-NH_2 + CO_2$$

Isocyanate   An amine

Applications

The different substituted isocyanates and respective amines has been successfully synthesized by using Curtius rearrangement.

a) Preparation of phenylmethanamine.

$$\text{C}_6\text{H}_5-\text{CH}_2\text{COOC}_2\text{H}_5 \xrightarrow[\text{ii) HNO}_2]{\text{i) NH}_2\text{NH}_2} \text{C}_6\text{H}_5-\text{CH}_2\text{CON}_3 \xrightarrow[\text{ii) H}_2\text{O}]{\text{i) }\Delta,\ \text{C}_6\text{H}_6} \text{C}_6\text{H}_5-\overset{\text{H}_2}{\text{C}}-\text{NH}_2$$

Ethyl 2-phenylacetate   2-phenylacetyl azide   Phenylmethanamine

b) Preparation of 4-methoxyaniline.

$$H_3CO-C_6H_4-COOC_2H_5 \xrightarrow[\text{ii) HNO}_2]{\text{i) NH}_2\text{NH}_2} H_3CO-C_6H_4-CON_3 \xrightarrow[\text{ii) H}_2\text{O}]{\text{i) }\Delta,\ \text{C}_6\text{H}_6} H_3CO-C_6H_4-NH_2$$

Ethyl 4-methoxybenzoate   4-methoxybenzoyl azide   4-methoxyaniline

c) Preparation of 2-aminoacetic acid.

$$H_2C\begin{smallmatrix}\text{COOC}_2\text{H}_5\\\text{COOC}_2\text{H}_5\end{smallmatrix} \xrightarrow[\text{ii) HNO}_2]{\text{i) NH}_2\text{NH}_2} H_2C\begin{smallmatrix}\text{COOH}\\\text{CON}_3\end{smallmatrix} \xrightarrow{\Delta,\ \text{C}_2\text{H}_5\text{OH}} H_2C\begin{smallmatrix}\text{COOH}\\\text{NHCOOC}_2\text{H}_5\end{smallmatrix} \xrightarrow{\text{HCl}} H_2C\begin{smallmatrix}\text{COOH}\\\text{NH}_2\end{smallmatrix}$$

Diethyl malonate   3-azido-3-oxopropanoic acid   2-((ethoxycarbonyl)amino)acetic acid   2-aminoacetic acid

58. Dakin Reaction

Principle

The reaction was initially reported by Dakin in 1909. The preparation of phenols from aryl aldehydes or aryl ketones involves the oxidation of corresponding aromatic compounds by hydrogen peroxide in presence of a base and subsequent hydrolysis of aryl formate or alkylcarboxylate intermediates is commonly referred as Dakin reaction. It has been reported that *para* or *ortho* substituents (For example: OH, NH_2) on aryl aldehydes or aryl ketones will increase the rate of reaction. Specifically, an ortho OH group accelerates Dakin reaction *via* intramolecular hydrogen bonding.

General Reaction

Aryl aldehydes or aryl ketones → Phenols + Acids

$R= NH_2, OH;$ **R'**= H, Alkyl

α-hydroxybenzaldehyde → Catechol

Mechanism

Step 1: Oxidation of aromatic aldehyde by hydrogen peroxide in presence of base gives aryl formate.

Phenyl formate

Step 2: Hydrolysis of aryl formate to afford phenol.

Applications

The reaction is useful for conversion of aromatic aldehydes or ketones into phenols with better yield.

a) Synthesis of Pyrogallol 1-monomethyl ether from α-vanillin.

2-(3-acetyl-4-hydroxyphenyl)acetonitrile   2-(3,4-dihydroxyphenyl)acetonitrile

2-hydroxy-3-methoxybenzaldehyde   3-methoxybenzene-1,2-diol
(α-vanillin)   (Pyrogallol 1-monomethyl ether)

b) Preparation of 3,4-dimethylbenzene-1,2-diol from 1-(2-hydroxy-3,4-dimethylphenyl)ethanone.

1-(2-hydroxy-3,4-dimethylphenyl)ethanone   3,4-dimethylbenzene-1,2-diol

59. Darzens Glycidic Ester Condensation

Principle

A reaction between carbonyls and α-hydrogen containing α-haloester in presence of a base such as sodamide ($NaNH_2$) to afford an α,β-epoxy ester (glycidic ester) is known as Darzens glycidic ester condensation. The reaction was first carried out by Erlenmeyer in 1892, but popularly known as Darzens condensation because Darzen extensively explored this reaction in 1900. Currently, this reaction has been extended to α-halogenated nitriles, sulfoxides, sulfones, sulfoximines, carboxylic amides, etc. In broad sense, all the base-catalyzed condensations between carbonyl compounds and molecules carrying activated protons at the α-halogenated position, which results in the formation of an oxirane ring and the release of a halide ion is known as Darzens condensation. The condensation reaction between carbonyl compounds and α-halogenated ketimine is referred as *aza-Darzens* reaction. Normally, this reaction would be carried out in an anhydrous aprotic nonpolar solvent (For example: benzene, ethyl ether) at low temperature, although ethanol and even aqueous dioxane have also been used as a solvent.

General Reaction

Ketone + Ethyl 2-chloroacetate (α-halo ester) $\xrightarrow{C_2H_5ONa}$ Glycidic ester (α,β-epoxy ester)

Glycidic ester (α,β-epoxy ester) $\xrightarrow{\text{Base}}$ $\xrightarrow[CO_2]{H^{\oplus}}$

Enol $\underset{\text{Tautomerism}}{\rightleftharpoons}$ Carbonyl compound

Mechanism

Step 1: Abstraction of α-hydrogen from α-haloester by sodamide

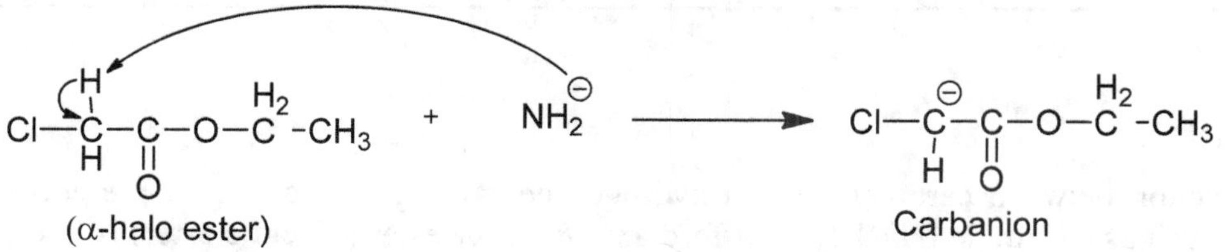

(α-halo ester) + $NH_2^{\ominus}$ → Carbanion

Step 2: Nucleophilic attack of carbanion of haloester on carbonyl carbon of aldehyde or ketone.

Step 3: Internal substitution reaction. The nucleophilic oxygen attacks the carbon bearing chlorine and thus the removal of chloride ion results in the formation of α,β-epoxy ester (glycidic ester).

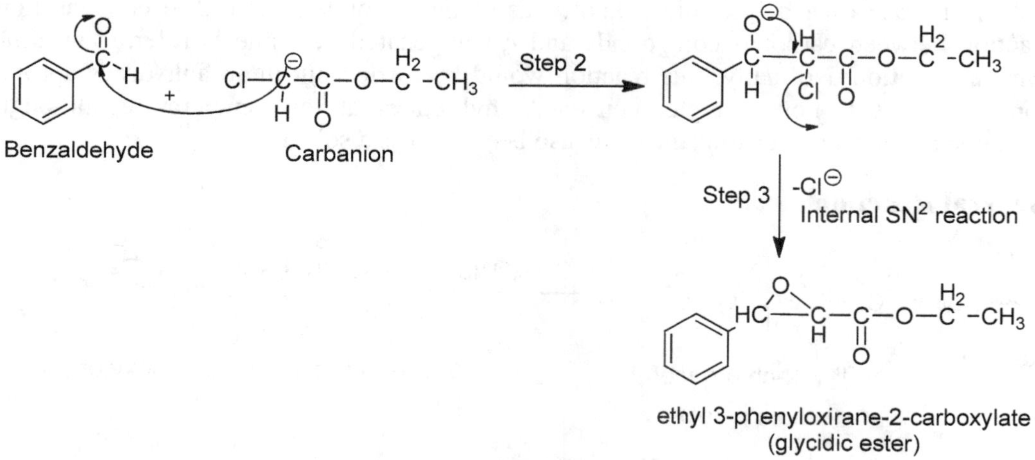

Benzaldehyde + Carbanion —Step 2→

Step 3 | $-Cl^{\ominus}$ Internal SN^2 reaction

ethyl 3-phenyloxirane-2-carboxylate
(glycidic ester)

Step 4: Conversion of glycidic ester into carbonyl compounds.

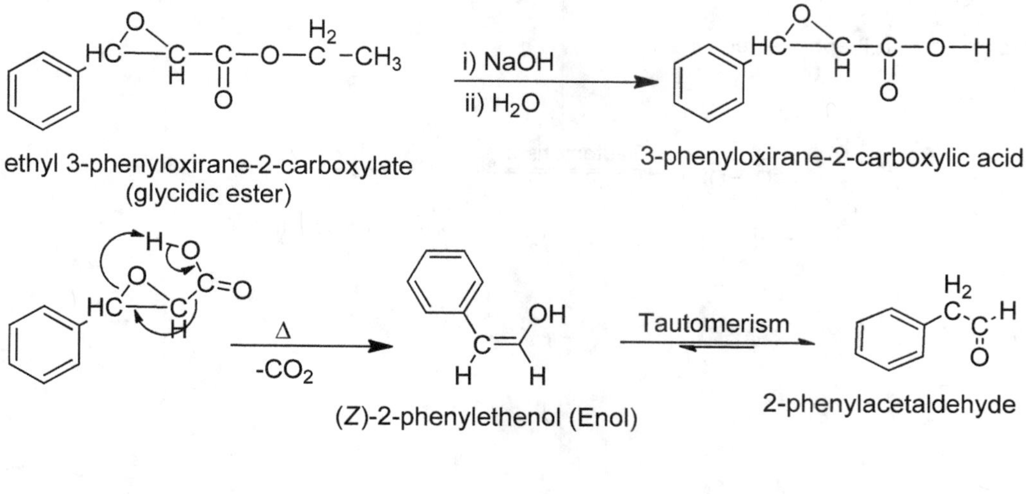

ethyl 3-phenyloxirane-2-carboxylate
(glycidic ester)

i) NaOH
ii) H_2O

3-phenyloxirane-2-carboxylic acid

Δ
$-CO_2$

(Z)-2-phenylethenol (Enol)

Tautomerism

2-phenylacetaldehyde

Applications

In the past, Darzens methodology was primarily used for the synthesis of aldehydes and ketones, as a homologation reaction without any consideration of stereocontrol in the epoxide formation. For this sequence, saponification of the α,β-epoxy ester followed by decarboxylation gives the substituted carbonyl compound. Darzens reaction has general applications in the synthesis of epoxide (or oxirane) and glycidic esters.

a) Synthesis of 2-(2,2-dimethyl-3-(1-oxopropan-2-yl)cyclobutyl)acetic acid from Ethyl pinonate.

Ethyl pinonate + Ethyl 2-chloroacetate (α-halo ester) $\xrightarrow{\text{C}_2\text{H}_5\text{ONa}}$ Glycidic ester

$\xrightarrow{\text{Hydrolysis, Acid}}$ 2-(2,2-dimethyl-3-(1-oxopropan-2-yl)cyclobutyl)acetic acid

b) Synthesis of Vitamin A1 from β-ionone.

β-ionone + Ethyl 2-chloroacetate (α-halo ester) $\xrightarrow[\text{H}^{\oplus}]{\text{C}_2\text{H}_5\text{ONa}}$ Vitamin A1

60. Diazotisation

Principle

Griess in 1858 reported the synthesis of aromatic diazonium salts from primary aromatic amines, dinitrogen trioxide, and nitric acid is known as Griess diazotization. Knoevenagel extended the work of Griess in 1890 by use of organic nitrite as diazotization agent in acidic conditions; Witt used $Na_2S_2O_5$ and nitric acid as the diazotization agents (Witt method), and Houston and Johnson invented of N_2O_4 as the diazotization agent in 1925.

Many other reagents are also known later for the diazotiazation such as NOCl, $NaNO_2/HBF4$, $NaNO_2/CuCl_2$, and $NO^+BF_4^-$ (or $NO^+PF_6^-$, $NO^+SbF_6^-$ etc.). The commonly used method for preparation of diazonium salt is to use nitrous acid generated from sodium nitrite and an acid like HCl, H_2SO_4, HCOOH, and CH_3COOH.

General Reaction

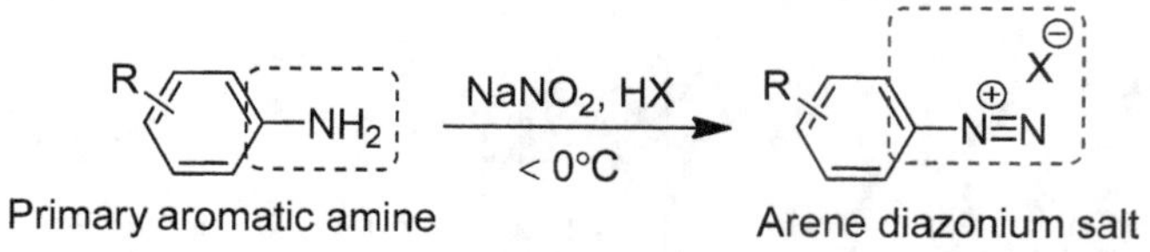

Primary aromatic amine $\xrightarrow[< 0°C]{NaNO_2, HX}$ Arene diazonium salt

Mechanism

General mechanism of diazotization from sodium nitrite and hydrochloric acid is illustrated here:

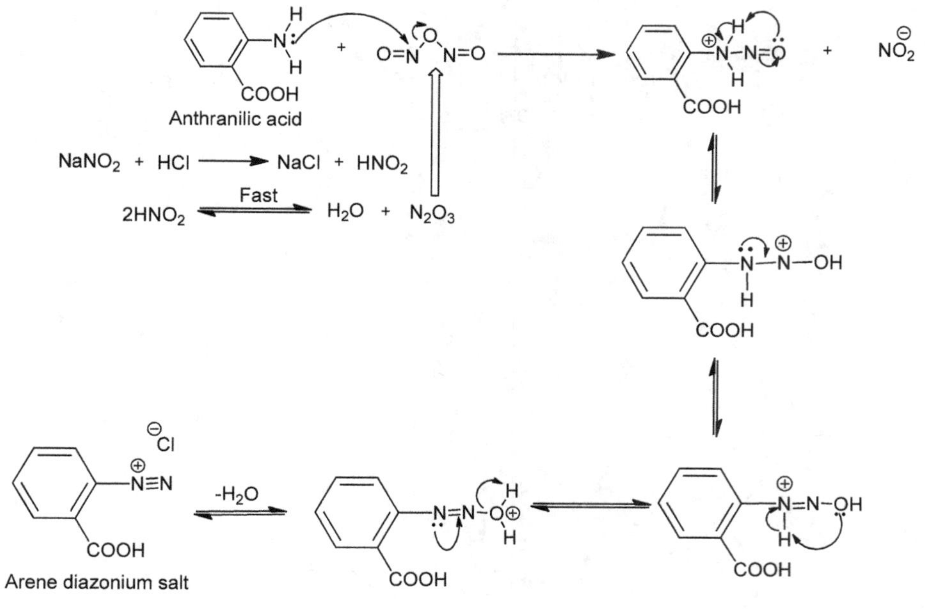

$$NaNO_2 + HCl \longrightarrow NaCl + HNO_2$$

$$2HNO_2 \underset{}{\overset{Fast}{\rightleftharpoons}} H_2O + N_2O_3$$

134|

Applications

a) Varieties of aromatic diazonium salts can be easily prepared by using this method which on Sandmeyer reaction yields halogen substituted benzoic acid. For example: Synthesis of o-chlorobenzoic acid.

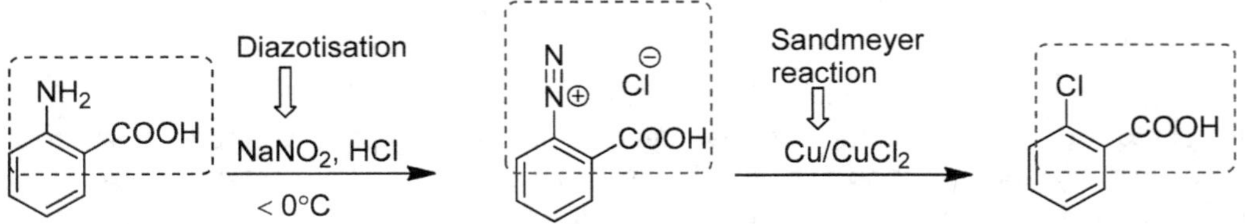

b) Synthesis of 4,4'-disulfanediylbis(3-fluorobenzoic acid).

c) Preparation of (E)-3-((2,4-dibromo-6-nitrophenyl)diazenyl)naphthalen-2-ol.

61. Dieckmann Condensation

Principle

Dieckmann in 1894 reported a base-promoted intramolecular condensation of α,ω-diesters to form cyclic β-ketoesters that can be converted into cyclic ketone on hydrolysis and decarboxylation is known as Dieckmann condensation. The reaction is also popular as Dieckmann ring closure. The cyclic β-ketoesters by a series of reactions may be converted to cycloalkanes. The reaction is useful for the synthesis of 5 or 6-membered cyclic β-keto esters and also applicable for synthesizing even larger cyclic ketones. Cyclization of various aminodiesters to 10-membered cyclic aminoketones and 16 and 20-membered cyclic diaminodiketones can be carried out Dieckmann condensation. Small cyclic β-keto esters are synthesized in alcoholic solvent by the treatment of sodium alkoxide; whereas for synthesis of large cyclic ketones an aprotic solvent (For example: Xylene) under high dilution conditions is used.

General Reaction

Diethyl adipate
(1,6-diester)

Ethyl 2-oxocyclopentanecarboxylate
(cyclic β–keto ester)

Cyclopentane

Mechanism

Step 1: Abstraction of α-hydrogen by base ($C_2H_5O^-$) to produce enolate ion.

Enolate ion

Step 2: Ring closure.

5-exo-trig ring closure

ethyl 2-oxocyclopentanecarboxylate

Step 3: Formation of cyclic β-keto ester.

+ $C_2H_5O^{\ominus}$

cyclic β–keto ester

Applications

a) *The reaction has general application in the synthesis of cyclic ketones which on series of reactions converted into cycloalkanes. For example: Preparation of cyclohexane from diethyl heptandioate.*

Diethyl heptanedioate

Ethyl 2-oxocyclohexanecarboxylate

i) $H^{\oplus}/H_2O$

ii) Zn-Hg/HCl

Cyclohexane

b) *Synthesis of heterocyclic ketones; For example: Formation of piperidone derivative (1-ethylpiperidin-4-one).*

1. H_2O, 2. $-CO_2$

1-ethylpiperidin-4-one

c) *Synthesis of bicyclic ketones.*

Tetraethyl hexane-1,3,3,6-tetracarboxylate

Bicyclic ketone

62. Diels-Alder Reaction

Principle

Diels and Alder in 1928 reported electrocyclic [4+2]-cycloaddition reaction between a conjugated diene (4Π-electron system) and a dienophile (2Π-electron system either alkene or alkyne) for the synthesis of an unsaturated six membered ring is known as Diels-Alder reaction or Diels-Alder cycloaddition. The products formed in reaction are called cycloadducts. In Diels-Alder cycloaddition, the substituents present on diene and dienophile plays vital role due to their affect on electron densities and orbital energy of reactants. A diene can react with a dienophile using diene's highest occupied molecular orbital (HOMO) and dienophile's lowest unoccupied molecular orbital (LUMO) or *vice-versa*. The reaction rate is related to magnitude of lowest HOMO-LUMO energy separation attainable by the diene-dienophile components. For the normal Diels-Alder reaction, electron-donating groups on dienes will level up HOMO energy of dienes, and electron-withdrawing group will lower LUMO energy of dienophile; as a result, the magnitude between the HOMO-LUMO gaps is reduced, hence rate of Diels-Alder reaction is enhanced. Diels-Alder reaction shows high regioselectivity, diastereoselectivity, enantioselectivity and is controlled by compatibility between HOMO-LUMO pairs of dienes and dienophiles. The introduction of an atom other than carbon to either dienes or dienophiles is considered as hetero Diels-Alder reaction.

General Reaction

a) *Preparation of cyclohexene from buta-1,3-diene and ethane.*

Buta-1,3-diene   Ethene   Cyclohexene

b) *Preparation of 4a,5,8,8a-tetrahydronaphthalene-1,4-dione from buta-1,3-diene and benzoquinone.*

Buta-1,3-diene   Benzoquinone   4a,5,8,8a-tetrahydronaphthalene-1,4-dione

c) Preparation of Cyclohex-3-enecarbaldehyde from buta-1,3-diene and acrylaldehyde.

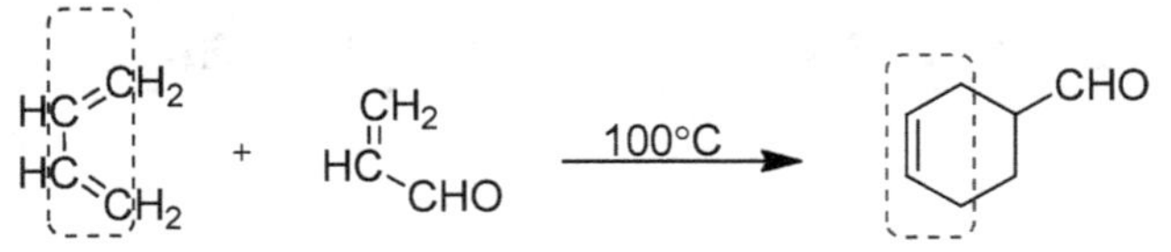

Buta-1,3-diene   Acrylaldehyde   Cyclohex-3-enecarbaldehyde

Mechanism

Diels-Alder reaction is classical example of pericyclic or concerted reaction, where all the bond formations and bond breaking occurs at the same time. During the process of cycloaddition, three Π-bonds in the reactants (diene and dienophile) are broken and two new carbon-carbon σ bonds; one new carbon-carbon Π-bond is formed.

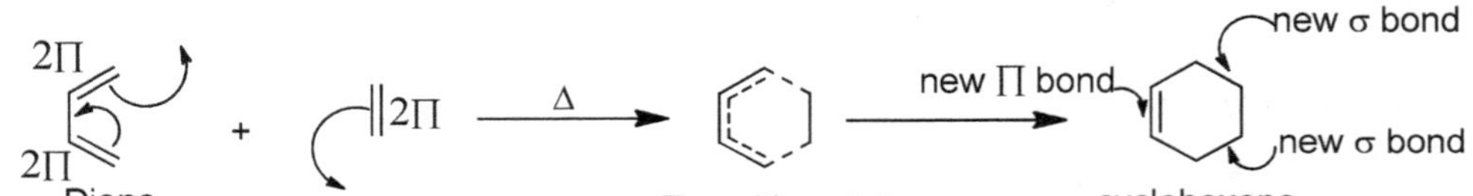

Diene
4Π electrons
s-*cis* conformation

Dienophile
2Π electrons

Transition state

cyclohexene

Applications

a) Determination of configuration: The configuration of dienophile gets retained in cyclic product of Diels-Alder reaction, hence it is easier to ascertain the configuration of cyclic adduct. As a result, Diels-Alder reaction has been used to determine the configuration of various dienophiles. *For example: Maleic and Fumaric acid gives cis and trans cinnamic acids respectively.*

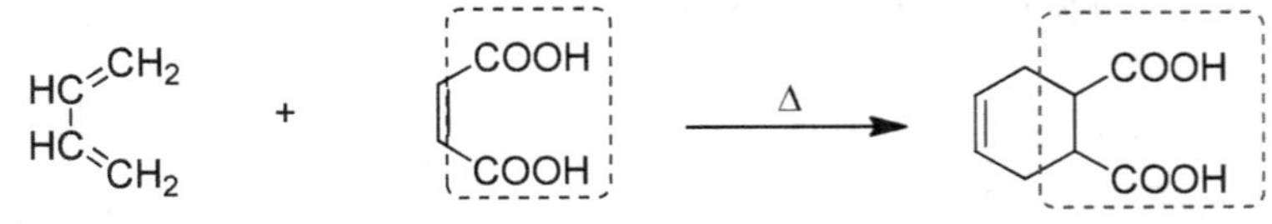

Buta-1,3-diene   Maleic acid   Cyclohex-4-ene-1,2-dicarboxylic acid
(*cis*-product)

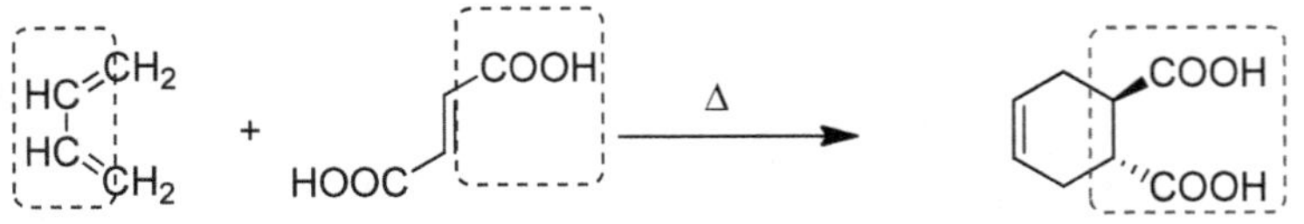

Buta-1,3-diene   Fumaric acid   (1R,2R)-cyclohex-4-ene-1,2-dicarboxylic acid
trans-product

b) Diels-Alder reaction has been used to synthesize camphene from cyclopenta-1,3-diene and acrylaldehyde.

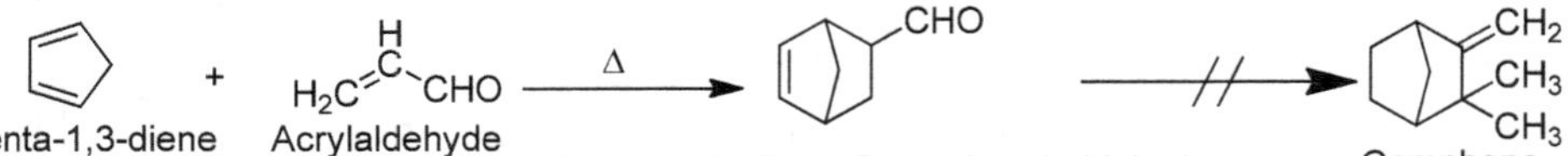

Cyclopenta-1,3-diene   Acrylaldehyde
Bicyclo[2.2.1]hept-5-ene-2-carbaldehyde   Camphene

63. Dienol-Benzene Rearrangement

Principle

An acid catalyzed transformation of 4,4-disubstituted cyclohex-2,5-dienol into substituted benzene is known as dienol-benzene rearrangement. The reaction was first reported in 1956. Substantial carbonium ion character is shown by both transitions state and intermediate in this rearrangement. Dienol-Benzene rearrangement is promoted by H^+ in acetate, formate, and phosphate buffers. In addition, the rearrangement of 4,4-dimethyl cyclohex-2,5-dienol is complicated with an isomerization to 6,6-dimethylcyclohex-2,4-dienol, leading to *o*-xylene.

General Reaction

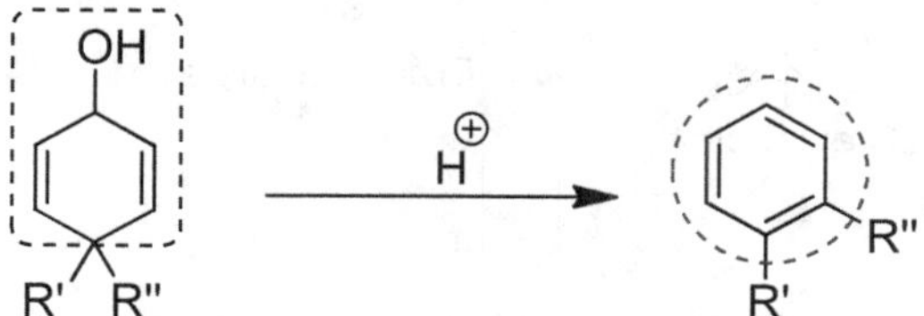

4,4-disubstituted cyclohex-2,5-dienol   Substituted benzene

Mechanism

The detailed mechanism for acid catalyzed transformation of 4,4-disubstituted cyclohex-2,5-dienol into substituted benzene is illustrated below:

Applications

a) Dienol-Benzene rearrangement is useful for transformation of 4,4-disubstituted cyclohex-2,5-dienol (or dienone) into aromatic compounds. The rearrangement is also important for introduction of aromaticity into A-ring of steroids.

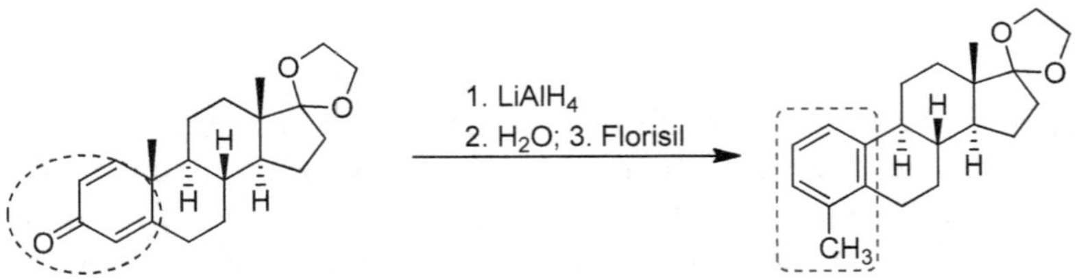

1,4-androstadiene-3,17-dione-17-ethylene ketal   4-methylestra-1,3,5(10)-triene-17-one-17-ethylene ketal

Principle

Auwers and Ziegler reported Dienone-Phenol rearrangement in 1921. Acid promoted conversion of 4,4-dialkyl cyclohexadienone in to phenol with migration of one of the alkyl group to the adjacent carbon atom is known as Dienone-Phenol rearrangement. 2,2-disubstituted cyclohexadienone and cyclohexadienone with exocyclic double bond can also be converted into corresponding disubstituted phenol under similar conditions. It is interesting that the rearrangement of endocyclic cyclohexadienone to its phenolic derivative is spontaneous unless a dichloromethyl group presents at position or the process is blocked in some manner, such as by the presence of a bulky substituent (For example: *tert*-butyl group) at both 2nd and 6th positions for cyclohexa-2,5-dienones or 6th position for cyclohexa-3,5-dienones. Aromatization is driving force for Dienone-Phenol rearrangement.

General Reaction

4,4'-dialkyl cyclohexadienone

3,4-disubstituted phenol

Mechanism

Transformation of 4,4-disubstituted cyclohexadienones to 3,4-disubstituted phenols is displayed below:

Step 1: In acidic condition, protonation of oxygen takes place to generate carbocation; a carbocation is stabilized by delocalization of the positive charge over the ring.

Resonating structures

Step 2: In one of the resonating structures, the positive charge present on carbon atom adjacent to highly substituted carbon atoms. As a result, a carbocation rearrangement with a loss of proton gives 3,4-disubstituted phenols.

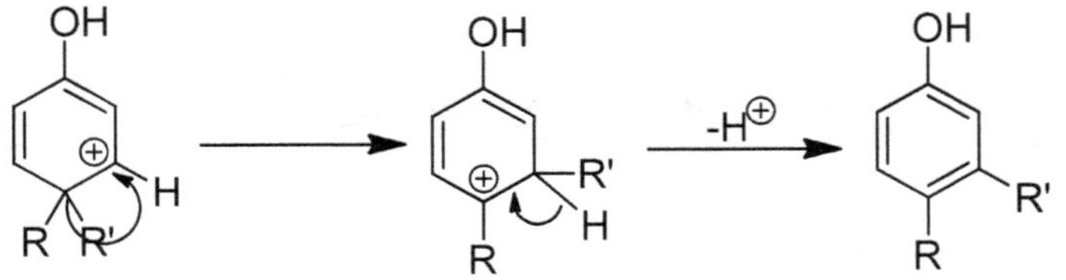

Applications

a) The reaction is very useful in synthesis of wide varieties of phenols; specifically the reaction has application in the synthesis of steroids, anthracene, and phenanthrene.

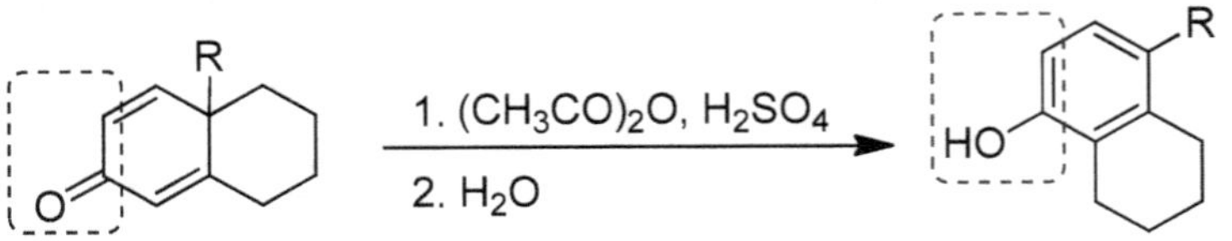

Dienone

1. $(CH_3CO)_2O$, H_2SO_4
2. H_2O

Phenol

b) Transformation of 12a-methyl-12,12a-dihydrochrysen-3(11H)-one (Dienone) into 1-methyl-11,12-dihydrochrysen-3-ol (Phenol).

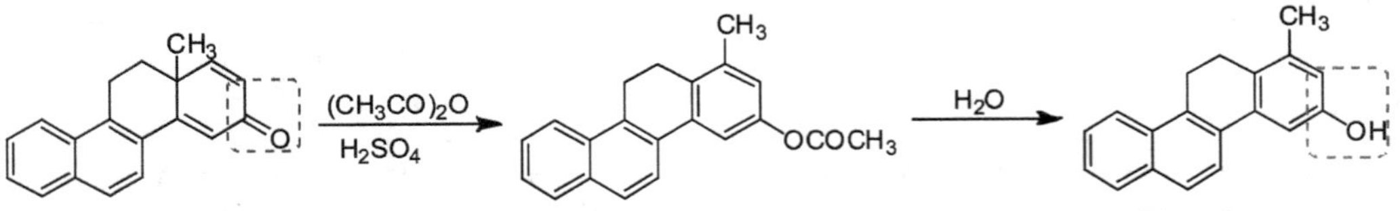

$(CH_3CO)_2O$
H_2SO_4

H_2O

(Dienone)
12a-methyl-12,12a-
dihydrochrysen-3(11*H*)-one

1-methyl-11,12-dihydrochrysen-3-yl acetate

(Phenol)
1-methyl-11,12-dihydrochrysen-3-ol

c) Conversion of 5a,9-dimethyl-3a,4,5,5a-tetrahydronaphtho[1,2-b]furan-2,8(3H,9bH)-dione into 8-hydroxy-3,6,9-trimethyl-3,3a,4,5-tetrahydronaphtho[1,2-b]furan-2(9bH)-one.

HCl

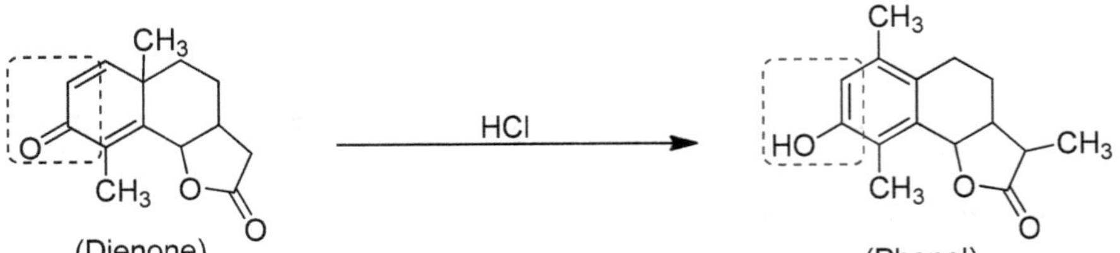

(Dienone)

5a,9-dimethyl-3a,4,5,5a-tetrahydronaphtho[1,2-b]furan-2,8(3*H*,9b*H*)-dione

(Phenol)

8-hydroxy-3,6,9-trimethyl-3,3a,4,5
-tetrahydronaphtho[1,2-*b*]furan-2(9b*H*)-one

65. Doebner Reaction

Principle

Doebner in 1887 reported the synthesis of cinchoninic acid (or quinolinic acid) analogues from reaction of aromatic amines, aldehydes, and pyruvic acid, involving the removal of hydrogen from the aromatics is known as Doebner reaction. Doebner reaction failed, when 2-chloro-5-aminopyridine, 3-aminopyridine, and 2-aminopyridine were applied as aryl amines.

General Reaction

Aniline + Aldehydes + 2-oxopropanoic acid $\xrightarrow{\Delta}$ Quinolinic acid derivatives

Mechanism

The reaction involves an Aldol condensation between an aldehyde and pyruvic acid to produce β,γ-unsaturated α-keto acid followed by Michael addition with aniline.

Step 1: Formation of β,γ-unsaturated α-keto acid.

Step 2: Michael addition with aniline.

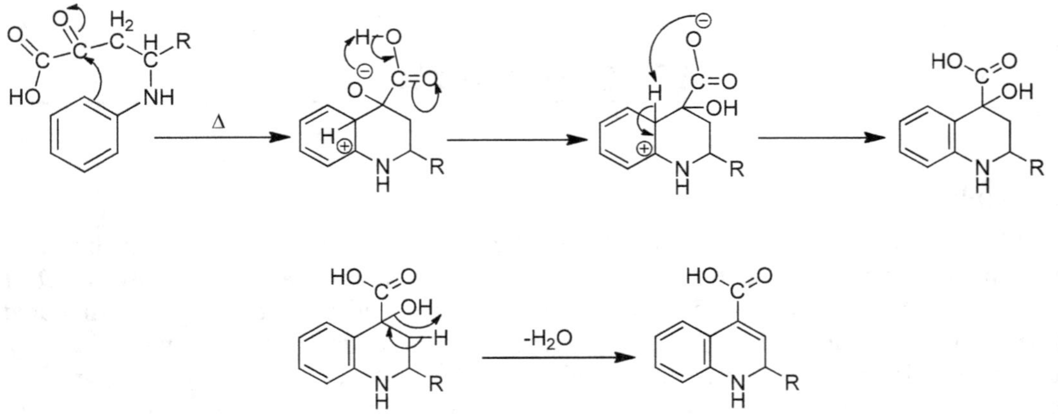

Step 3: Air oxidation of intermediate to give quinolinic acid.

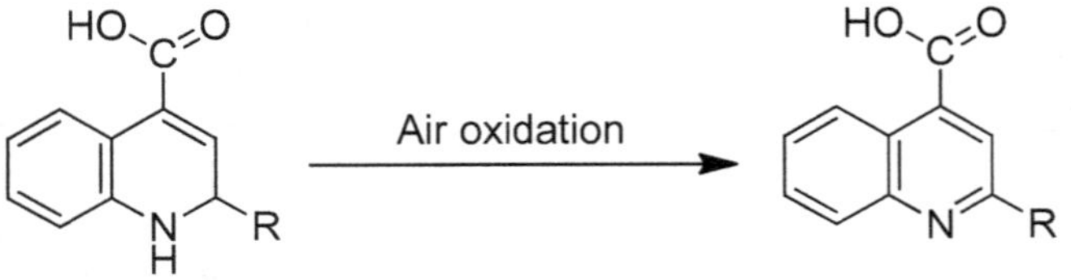

Air oxidation

Applications

a) The reaction is widely applicable for synthesis of quinolinic acids and cinchoninic acids. For example: Synthesis of 3-phenylbenzo[f]quinoline-1-carboxylic acid.

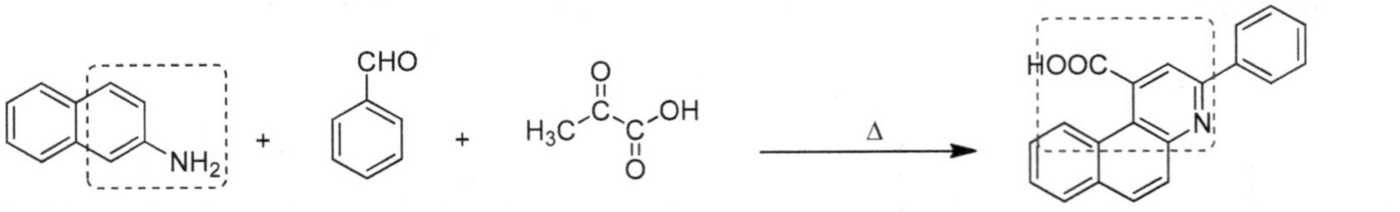

Naphthalen-2-amine   Benzaldehyde   2-oxopropanoic acid   3-phenylbenzo[*f*]quinoline-1-carboxylic acid

66. Duff Reaction

Principle

Duff and Bills in 1932 reported formylation of phenols or aromatic amines in a mixture of hexamethylenetetramine, boric acid, and glycerol. Hence, the reaction is known as Duff reaction or Duff formylation. Duff reaction is very fast and gives only *ortho*-formylated (or a small amounts of *ortho*, *para*-disubstituted) product (For example: phenols). However, this procedure gives *para*-products in the case of anilines. Hydroxypyridines and hydroxyquinolines cannot undergo formylation due to presence of electron-withdrawing groups which hinder or prevents the reaction to take place.

General Reaction

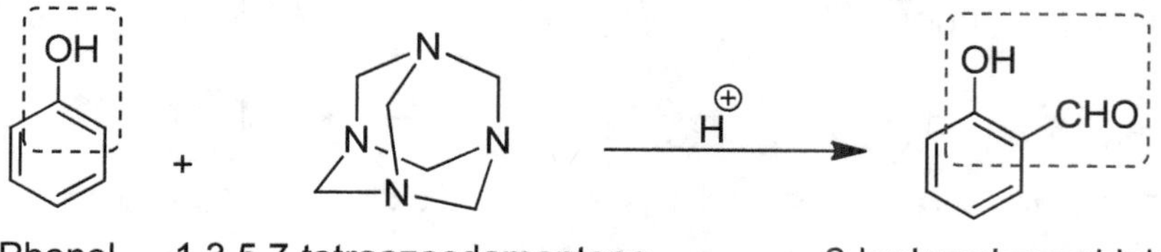

a) Formation of 2-hydroxybenzaldehyde from phenol and 1,3,5,7-tetraazaadamantane

Mechanism

The detailed mechanism for Duff reaction is illustrated below:

Applications

Duff reaction has broad applications in the formylation of aromatic compounds.

a) Conversion of benzene-1,3,5-triol into 2,4,6-trihydroxybenzene-1,3,5-tricarbaldehyde.

Benzene-1,3,5-triol → (Hexamethylenetetramine, Trifluoroacetic acid, 2 h) → 2,4,6-trihydroxybenzene-1,3,5-tricarbaldehyde

b) Conversion of 2,3,4-trimethoxyphenol into 2-hydroxy-3,4,5-trimethoxybenzaldehyde.

2,3,4-trimethoxyphenol → (Hexamethylenetetramine, Trifluoroacetic acid, Δ, 4 h) → 2-hydroxy-3,4,5-trimethoxybenzaldehyde

Principle

Elbs in 1893 reported oxidative hydroxylation of aromatic compounds *via* treatment of persulfate salt in presence of alkali to give hydroquinone is recognized as Elbs persulfate oxidation. Elbs observed that after oxidation of phenol by persulfate salt in alkali resulted into new -OH group which goes to the free para position to the hydroxyl group already present in reactant and hydroquinone is formed. However, if *para* position is occupied, the new -OH group goes to the *ortho* position to give a catechol. Electron attracting groups on the benzene ring increases the yield by activating at position *para* to hydroxyl group.

General Reaction

Potassium persulphate

For example: Transformation of phenol into hydroquinone

Phenol $\xrightarrow[\underset{OH}{\ominus}]{K_2O_8S_2}$ Pottassium salt $\xrightarrow[\text{Acidification}]{\overset{\oplus}{H}}$ Hydroquinone

Preparation of catechol derivatives

Phenol (R = Electron withdrawing/ releasing substitutents) $\xrightarrow[\underset{OH}{\ominus}]{K_2O_8S_2}$ Pottassium salt $\xrightarrow[\text{Acidification}]{\overset{\oplus}{H}}$ Catechol derivatives

Mechanism

Ionic mechanism similar to the abnormal Claisen rearrangement is proposed and displayed here for Elbs Persulphate oxidation.

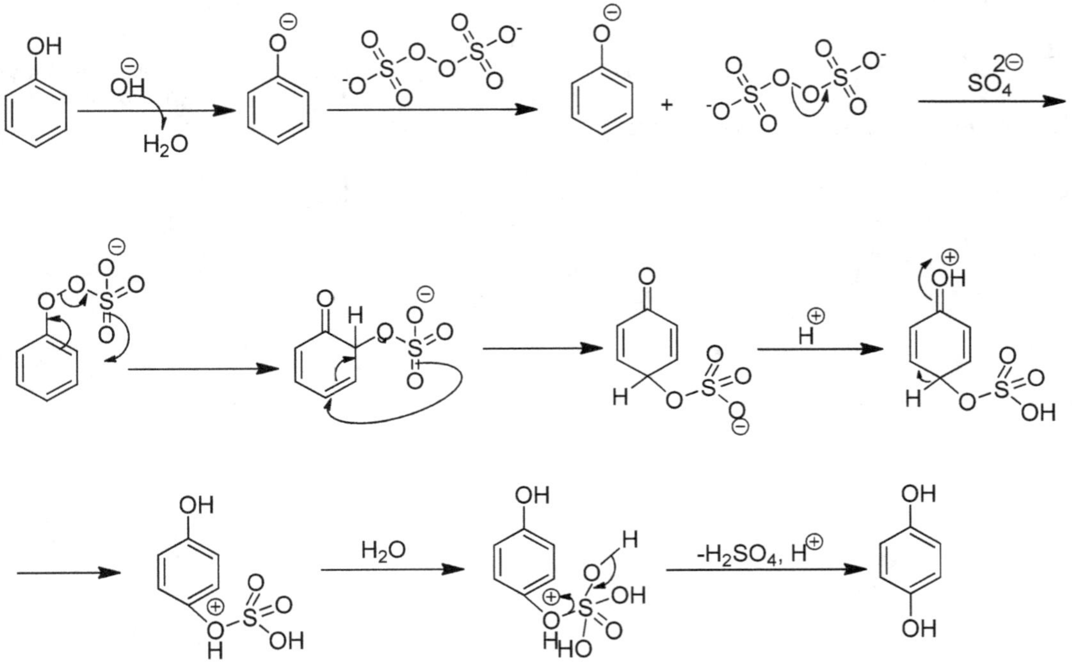

Applications

a) *The reaction is useful in synthesis of hydroxyl and alkoxylated aromatic compounds. For example: Conversion of p-cresol to 4-methylbenzene-1,2-diol.*

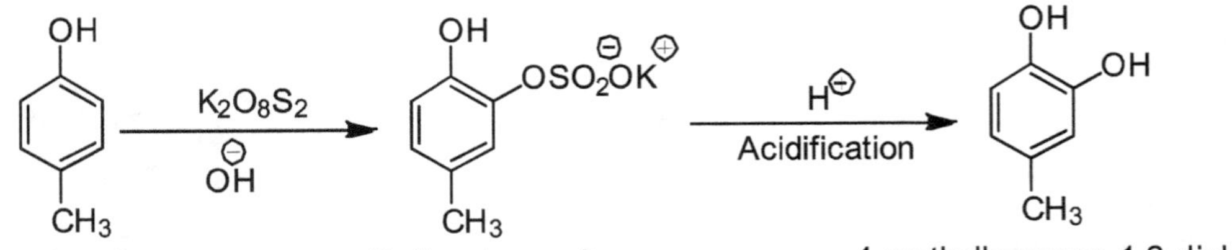

p-cresol

Pottassium salt

4-methylbenzene-1,2-diol

b) *Preparation of gentisic acid.*

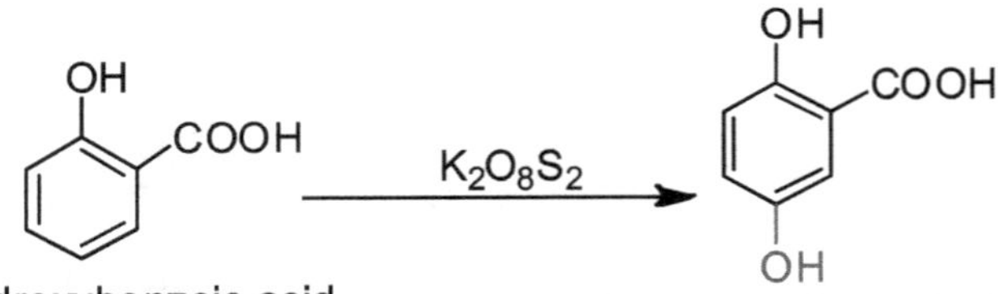

2-hydroxybenzoic acid

2,5-dihydroxybenzoic acid
(gentisic acid)

c) Synthesis of vinylhydroquinone.

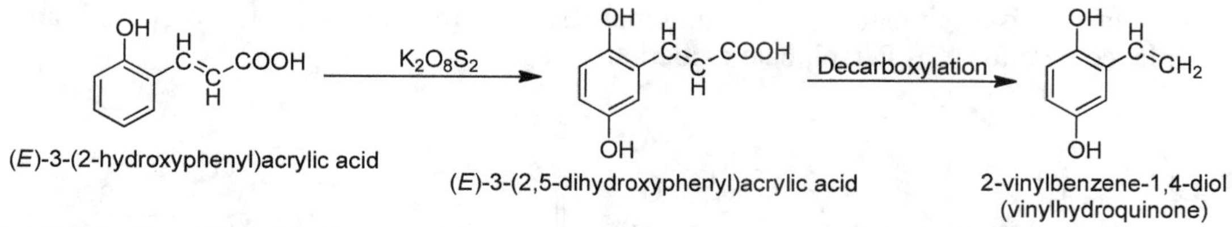

(E)-3-(2-hydroxyphenyl)acrylic acid $\xrightarrow{K_2O_8S_2}$ (E)-3-(2,5-dihydroxyphenyl)acrylic acid $\xrightarrow{\text{Decarboxylation}}$ 2-vinylbenzene-1,4-diol (vinylhydroquinone)

d) Synthesis of 2,5-dihydroxypyridine.

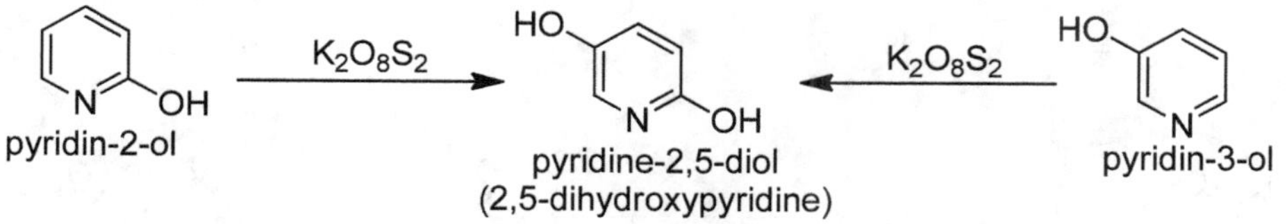

pyridin-2-ol $\xrightarrow{K_2O_8S_2}$ pyridine-2,5-diol (2,5-dihydroxypyridine) $\xleftarrow{K_2O_8S_2}$ pyridin-3-ol

68. Elimination Reaction (E-1/E-2 Reaction)

Principle

The reaction of haloalkanes with alcoholic potassium hydroxide results in dehydrohalogenation to afford an alkene is known as elimination reaction. Elimination reactions occur in presence of a strong and/or bulkier base; at high temperature; and with bulkier haloalkanes. The alcoholic potassium hydroxide results in the formation of a strong base, ethoxide ion and strong ethoxide ion is able to abstract a proton to form an alkene. The elimination reaction proceeds through a unimolecular (*E-1*) or bimolecular (*E-2*) mechanism and follows Saytzeff elimination.

General Reaction

a) Dehydrohalogenation of 2-bromobutane in presence of alcoholic potassium hydroxide to yield (*E*)-but-2-ene as major product and minor product but-1-ene.

2-bromobutane → (via Alcoholic KOH) → (*E*)-but-2-ene (Major product) + but-1-ene (Minor product)

b) Dehydrohalogenation of 2-bromo-3-methylbutane to give corresponding alkenes according to Saytzeff rule.

2-bromo-3-methylbutane → (via Alcoholic KOH) → 2-methylbut-2-ene (Major product) + 3-methylbut-1-ene (Minor product)

Mechanism

Elimination unimolecular mechanism (E1-mechanism)

Unimolecular elimination reaction occurs in two steps. The reaction proceeds by formation of carbocation intermediate.

We will consider the dehydrohalogenation of 1-bromopropane to obtain propene.

Step 1: Heterolytic fission of C-X bond from haloalkanes results in the formation of carbocation.

This step is considered as rate determining step of reaction that involves only one reactant as haloalkane. Hence reaction is known as unimolecular first order reaction.

The rate of reaction is depending upon stability of carbocation. The order of reactivity of haloalkanes towards elimination reaction is $3° > 2° > 1°$. Due to existence of carbocation intermediate; there is possibility of carbocation rearrangement.

Step 2: Abstraction of β-proton of the carbocation by strong base and formation of possible alkene (Propene).

Elimination bimolecular reaction (*E-2* mechanism)

For example: Dehydrohalogenation reaction to form prop-1-ene in presence of strong base follows *E-2* mechanism.

The reaction occurs in single step. The formation of possible alkenes takes place by simultaneous removal of β-hydrogen and halogen atom from haloalkanes.

The rate determining step involves both the reactants and follows second order kinetics; hence reaction is known as bimolecular elimination reaction.

Points should be remembered in bimolecular reaction.

i) A more substituted alkene is favored during the course of reaction.

ii) An E-2 mechanism involves anti elimination (removal of β-hydrogen and halogen atom takes place from opposite sides).

Elimination, Unimolecular, Conjugate Base Mechanism (E1cB)

The presence of electron-withdrawing substituents on β-carbon of haloalkane favors *E1cB* mechanism. In *E1cB* reaction, the first fast step involves loss of β-hydrogen of haloalkanes in presence of base to yield a carbanion. The loss of halide ion from carbanion intermediate is slow rate determining step that results in the formation of an alkene.

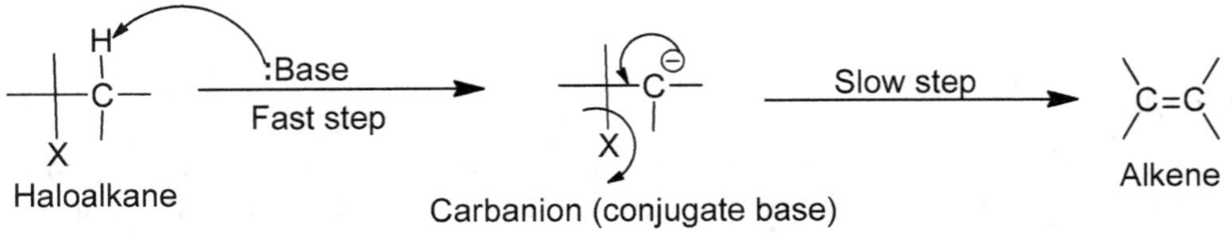

Haloalkane → (:Base, Fast step) → Carbanion (conjugate base) → (Slow step) → Alkene

69. Ester Pyrolysis

Principle

Ester pyrolysis is the unimolecular thermal decomposition of esters that hold β-hydrogen(s) on the alcoholic moiety *via cis*-elimination to form olefins without a carbon skeleton isomerization or double bond shift. Carboxylic acids are the by-products in this pyrolytic fission. Ester pyrolysis is easily carried out for esters containing volatile components of either acid or alcohol. Major controlling factor in ester pyrolysis is the difficulty to form a p-π conjugation between carbonyl oxygen and β-hydrogen in ground state of the ester, that is, the stronger the conjugation, the easier for the pyrolysis to occur. Therefore, it is found that γ-lactones are stable at 600 °C, whereas larger lactones that can form a transient six-membered ring decompose at 500 °C.

General Reaction

R' = Alkyl, Aryl
R" = H, Alkyl, Aryl

Esters   Acids   Olefins

Mechanism

Outline here is the detailed mechanism for pyrolysis of ester to generate acids and alkenes

Applications

Ester pyrolysis has wide application in thermal modification, recycling, and degradation of polyesters (poly ε-caprolactone) and the synthesis of olefins that are difficult to obtain, such as highly unsaturated compounds (For example: 2-vinylbutadiene). The reaction is used in the synthesis of a variety of olefins and other special compounds (For example: cyclic ethers).

a) Pyrolysis of (3-methylfuran-2-yl)methyl benzoate.

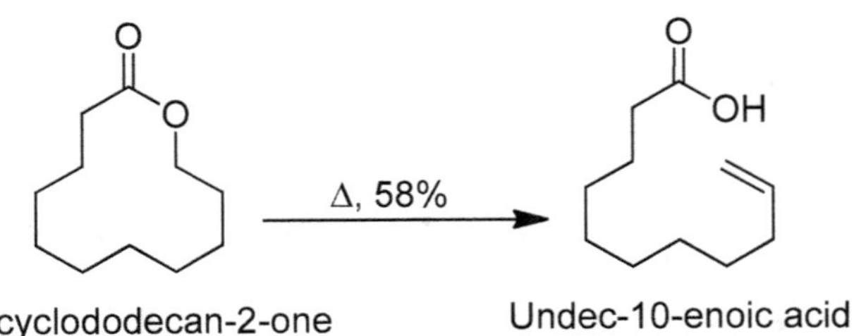

b) Pyrolysis of oxacyclododecan-2-one.

70. Etard Reaction

Principle

Etard in 1880 reported oxidation of hydrocarbons by hexavalent chromium compounds (chromatic acid chloride) to form a mixture of alcohol, carbonyl compounds, chloro-ketones, or aldehydes hence popularly known as Etard reaction. In this reaction, a highly hygroscopic brown powder is obtained which is called Etard complex and is a heterogenous, oligomeric material containing carbonyl compounds and at least 2 equivalences of chromium atom in a valence of IV and VI. It was found that the carbonyl compound functions as a ligand in Etard complex and can be displaced by other coordinating solvents, such as acetonitrile, acetone, and water. The oxidation of methylated homologs of benzene using chromyl chloride is an effective method for preparing aromatic aldehydes.

General Reaction

Oxidation of toluene in presence of dioxochromium(VI) chloride into benzaldehyde

Toluene   Dioxochromium(VI) chloride   Benzaldehyde

Mechanism

Step 1: Ene reaction between benzaldehyde and Dioxochromium(VI) chloride

Step 2: [2,3]-sigmatropic rearrangement in hydroxy(6-methylenecyclohexa-2,4-dien-1-yl)(oxo)chromium(VI) chloride intermediate with subsequent formation of benzaldehyde

Benzaldehyde   Hydroxychromium(II) chloride

Applications

a) *Etard reaction has wide application in the synthesis of aromatic aldehydes. For example: Conversion of cyclohexane into mixture of chlorocyclohexane, cyclohexanone and 2-chlorocyclohexanone.*

Cyclohexane + CrO_2Cl_2 → Chlorocyclohexane + Cyclohexanone + 2-chlorocyclohexanone

71. Favorskii Reaction

Principle

Favorskii reported transformation of α-halo ketones to esters with rearranged carbon skeleton by treating with alkoxide ions in 1894 is commonly known as Favorskii reaction. The bases such as alkalis or amines instead of alkoxides gives acids or amides respectively. Cyclic α-halo ketones produce esters with ring contraction.

Different mechanisms have been proposed for Favorskii rearrangement such as: (a) addition of alkoxide to the carbonyl group with the concomitant extrusion of a halide ion to form a final product *via* the intermediate of epoxide; (b) reaction of a nucleophile with a ketene generated by the base-induced elimination of hydrogen halide; (c) addition of alkoxide to carbonyl group followed by the *1,2*-migration of an alkyl group (quasi-Favorskii rearrangement); (d) Unimolecular dissociation of an α-halo ketone to a carbonium ion, which tautomerizes to isomeric carbonium then rearranges to product; (e) Generation of a carbanion by deprotonation of α-hydrogen followed by a nucleophilic substitution to form cyclopropanone, which undergoes rearrangement to a final product during alkoxide attacks (Loftfield mechanism or cyclopropanone mechanism); (f) Formation of an oxyallylic anion that undergoes [4+3] cycloaddition. Semi-benzilic acid and cyclopropanone mechanisms are widely known mechanism for Favorskii rearrangement. In presence of α-hydrogens, the cyclopropanone mechanism is more favorable and semi-benzilic acid is acceptable in absence of α-hydrogens or when the formation of cyclopropanone is prevented.

General Reaction

1-chloro-3-phenylpropan-2-one
(α-haloketone)

Rearranged ester

For example: Conversion of 2-chlorocyclohexanone into Cyclopentane ester

2-chlorocyclohexanone

Cyclopentane ester

Transformation of 2-chlorocycloheptanone into Cyclohexane ester

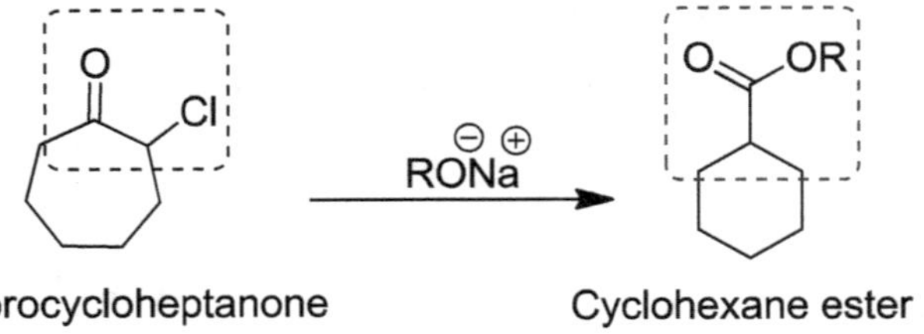

2-chlorocycloheptanone   Cyclohexane ester

Mechanism

Only cyclopropanone mechanism (**Scheme 1**) and benzilic acid mechanism (**Scheme 2**) are illustrated.

Scheme 1: *Loftfield mechanism or cyclopropanone mechanism*

Step 1: The generation of a carbanion by deprotonation of α-hydrogen followed by a nucleophilic substitution to form cyclopropanone intermediate.

Cyclopropanone intermediate

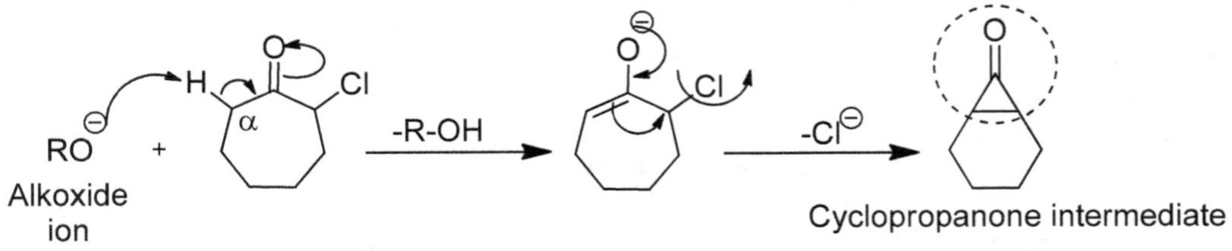

Step 2: Subsequent attack of the alkoxide ion on the carbonyl carbon opens the ring with equal ease on either side of the carbonyl carbon so that product has 50% of ^{14}C at the α-position and 50% at the β-position.

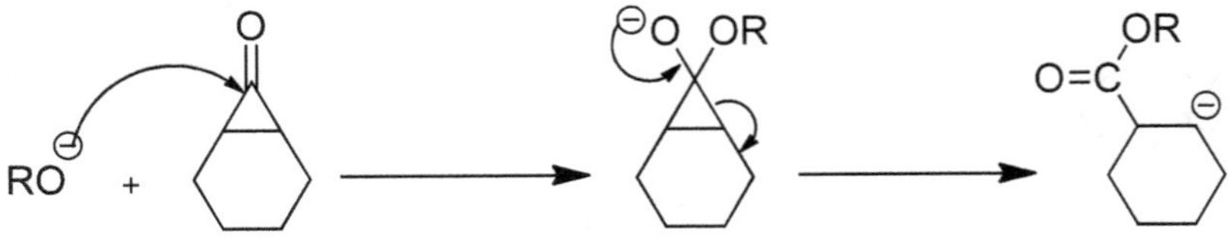

For example: In unsymmetrical ketones, the unsymmetrical cyclopropanone ring is formed which opens to acquire the most stable carbanion. Thus two isomeric ketones give the same cyclic cyclopropanone intermediate which open on either side of the carbonyl group to afford two carbanions.

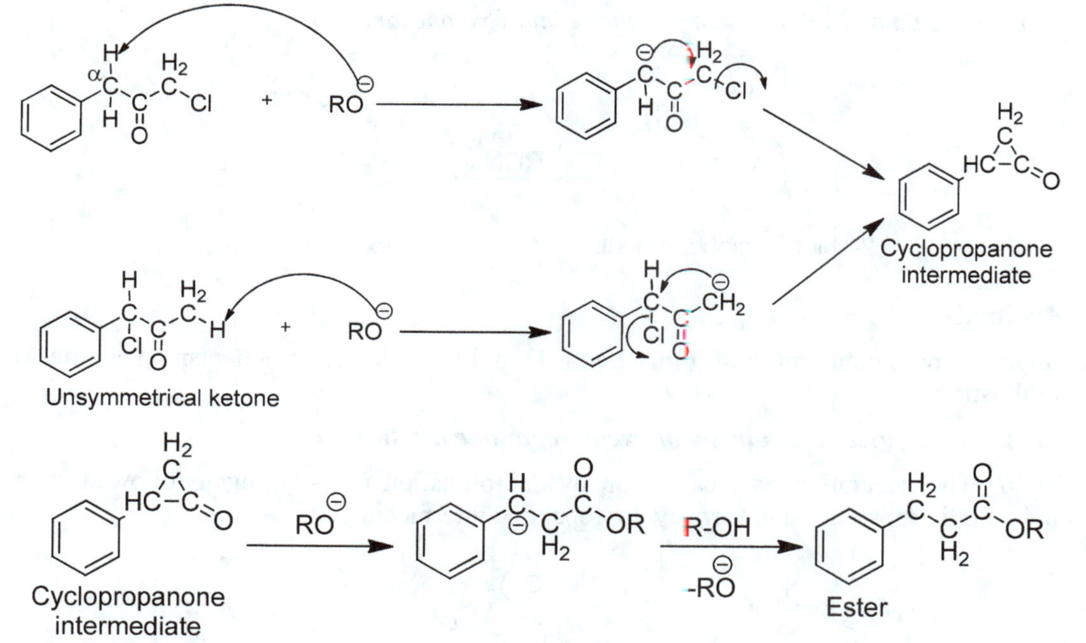
Cyclopropanone intermediate
Unsymmetrical ketone
Cyclopropanone intermediate
R-OH
-RO
Ester

Step 3: Generation of final product as cyclohexane ester.

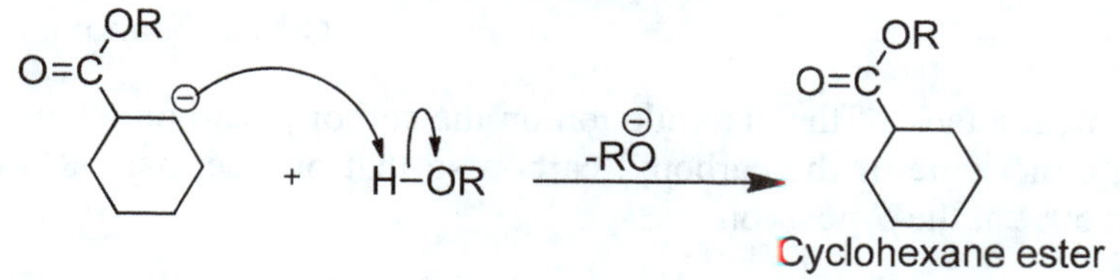
-RO
Cyclohexane ester

Scheme 2: ***Favorskii rearrangement proceeds through benzilic acid mechanism (Quasi-Favorskii rearrangement)***

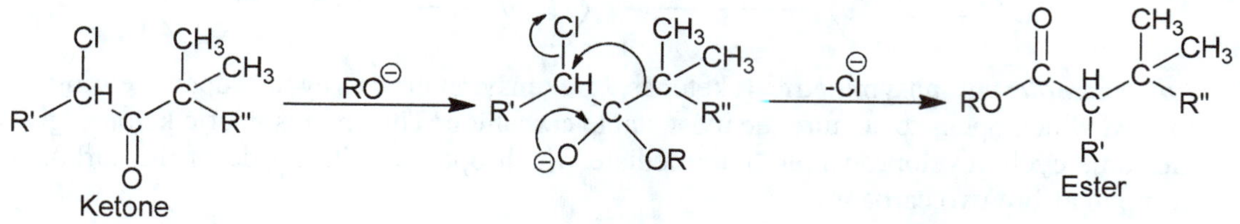
Ketone
-Cl
Ester

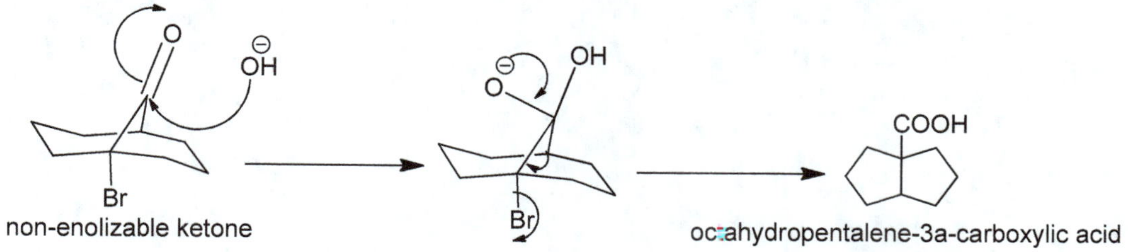
non-enolizable ketone
octahydropentalene-3a-carboxylic acid

Applications

a) The reaction is useful for synthesis of bicyclic esters, steroids, norsteroids, and pyrollidine.

(*E*)-2,2,2-trifluoroethyl 7-(furan-2-yl)hept-2-enoate

1,1-dichloro-7-furan-3-yl-heptan-2-one

$(C_2H_5)_3N$ / CF_3CH_2OH

+

2,2,2-trifluoroethyl 7-(furan-2-yl)-3-oxoheptanoate

b) Conversion of bromo ketone into bis(4-methoxyphenyl)(tricyclo[4.2.1.02,5]non-7-en-2-yl)methanol.

Bromo ketone

+

(4-methoxyphenyl)lithium

THF (91%)

bis(4-methoxyphenyl)(tricyclo[4.2.1.0^{2,5}]non-7-en-2-yl)methanol

72. Fischer Indole Synthesis

Fischer and Jourdan in 1883 reported synthesis of substituted indoles. The reaction between aryl hydrazones and carbonyl compounds (ketones or aldehydes) followed by treatment with mineral or Lewis acid results into formation of substituted indoles. The reaction is popularly known as Fischer indole synthesis. The reaction is catalyzed by acids, including mineral acids (For example: Polyphosphoric acid) and Lewis acids (For example: BF_3, $ZnCl_2$). Synthesis of indole by using 1% $ZnCl_2$ as catalyst is known as the Fischer-Arbuzov reaction.

General Reaction

Phenylhydrazine

Phenylhydrazone having α methylene group

Indole analogues

Phenylhydrazine   2-oxopropanoic acid

Phenylhydrazone

Indole-2-carboxylic acid

Indole

Mechanism

Step 1: Acid-catalyzed tautomerization of an aromatic hydrazone to an *ene*-hydrazine and a *[3,3]*-sigmatropic rearrangement of *ene*-hydrazine to a *bis*-imine intermediate.

Phenylhydrazone having
α methylene group

tautomerization

[3,3] sigmatropic shift
of N-N α bond

Bis-imine intermediate

Step 2: Re-aromatization to aniline and intramolecular nucleophilic attack to form aminal.

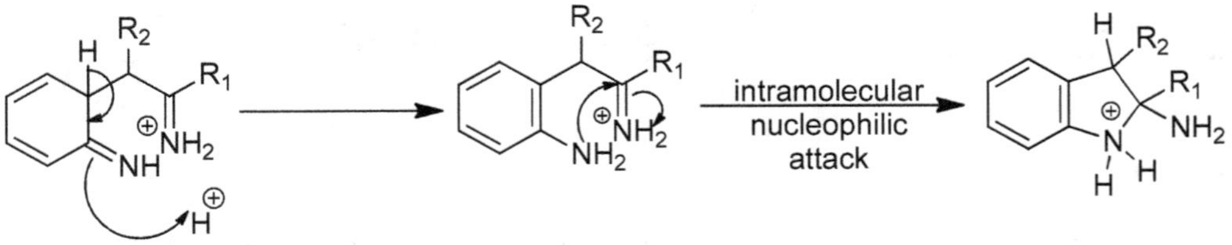

Step 3: Extrusion of an ammonia to yield the indole analogues.

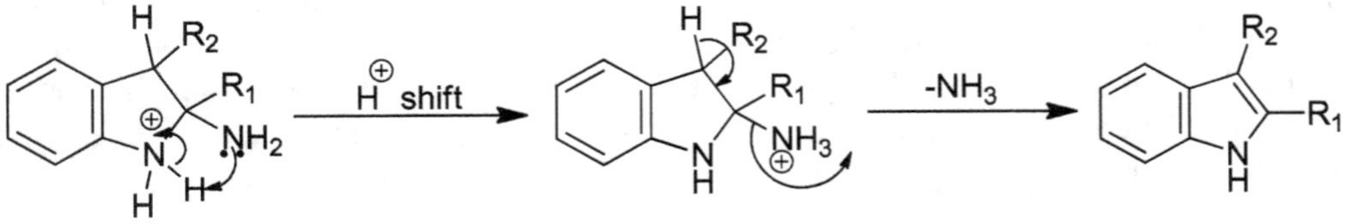

Applications

a) *Synthesis of indole-3-acetic acid from (Z)-3-(2-phenylhydrazono) propanoic acid.*

(Z)-3-(2-phenylhydrazono)propanoic acid

2-(1*H*-indol-3-yl)acetic acid

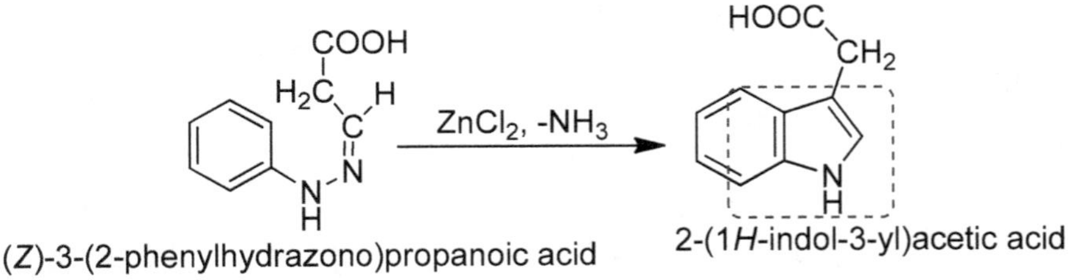

b) *Synthesis of 3-methyloxindole from N'-phenylpropionohydrazide.*

N'-phenylpropionohydrazide

3-methylindolin-2-one

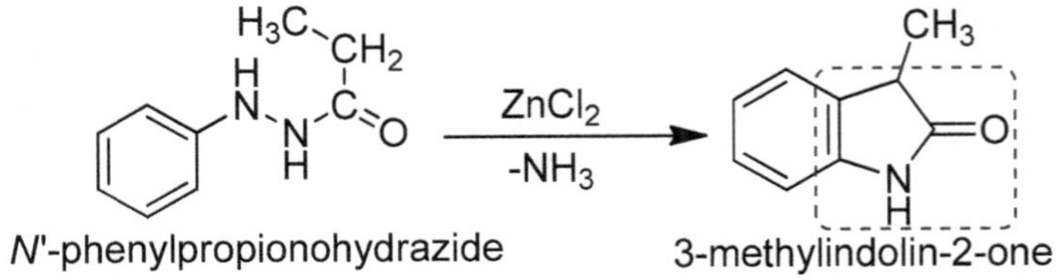

Principle

A synthesis of oxazoles by treatment of equimolar amounts of aldehyde cyanohydrin and aromatic aldehyde in anhydrous ether with dry hydrochloride is called as Fischer oxazole synthesis or Fischer synthesis. A synthesis of oxazoles by this method was given by Fischer in 1896. By using this method 2,5-diaryloxazoles has been synthesized in dry ether by dissolving aromatic aldehydes and aldehyde cyanohydrins followed by passing dry HCl gas. The synthesized oxazole precipitates as its hydrochloride salt, which can be converted into free oxazoles by the addition of water or by boiling with alcohol.

General Reaction

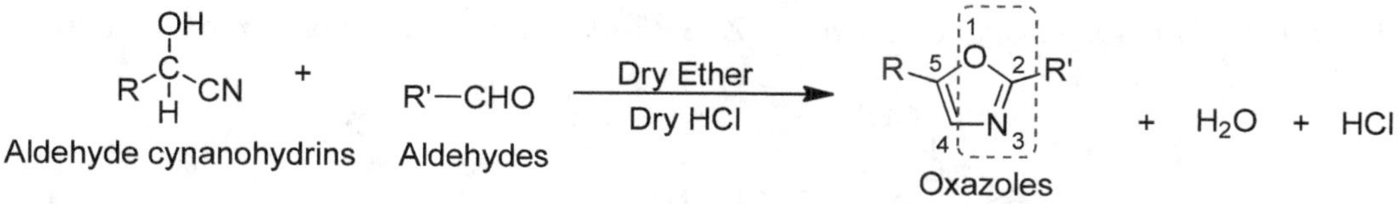

Aldehyde cynanohydrins + Aldehydes $\xrightarrow[\text{Dry HCl}]{\text{Dry Ether}}$ Oxazoles + H_2O + HCl

Mechanism

Step 1:

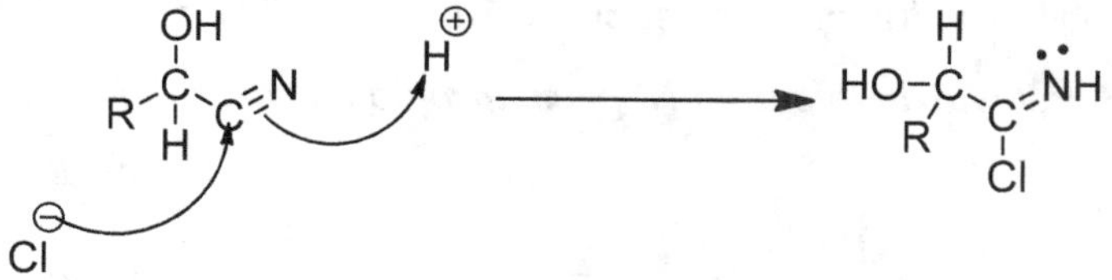

Step 2:

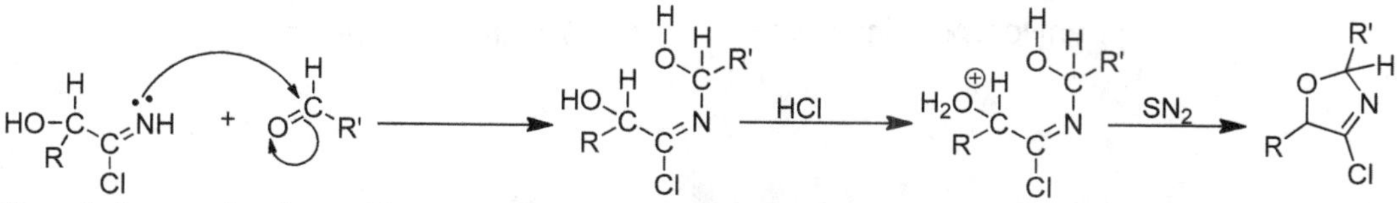

Step 3: Isomerization of intermediate followed by elimination of hydrochloric acid to give substituted oxazoles.

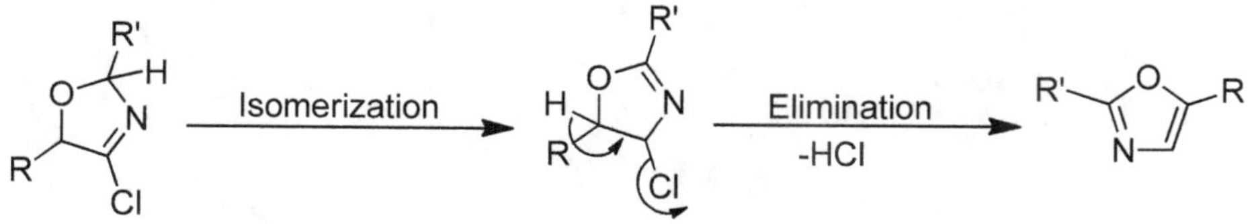

Applications

a) Synthesis of 2,5-diaryl-oxazoles. For example: Synthesis of 4-(2-(pyridin-3-yl)oxazol-5-yl)phenol.

2-hydroxy-2-(4-hydroxyphenyl)acetonitrile + nicotinaldehyde $\xrightarrow[\text{Dry HCl}]{\text{Dry Ether}}$ 4-(2-(pyridin-3-yl)oxazol-5-yl)phenol

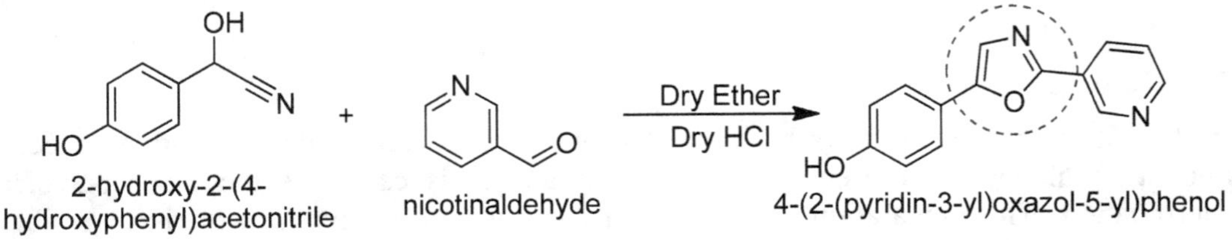

b) Synthesis of 4-chloro-2,5-diphenyloxazole.

benzoyl cyanide + benzaldehyde $\xrightarrow[\text{Dry HCl}]{\text{Dry Ether}}$ 4-chloro-2,5-diphenyloxazole

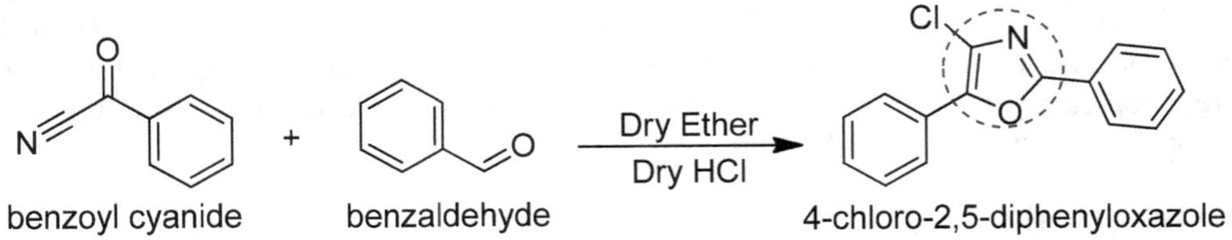

74. Fischer-Speier Esterification

Principle

Fischer and Speier in 1895 synthesized ester by refluxing carboxylic acid with an excess amount of alcohol in presence of an acidic catalyst is called as the Fischer-Speier esterification. The preparation of ester from alcohol and carboxylic acid is reversible reaction, and equilibrium always exists between ester and carboxylic acid. To promote the equilibrium towards ester side, azeotropic refluxing is helpful to remove water molecule. An acidic catalyst is added to force the esterification, often by passing dry hydrochloric acid gas into solution. Compared to a sulfuric acid promoted esterification, the reaction catalyzed by hydrochloric acid does not produce olefin, and does not undergo electrophilic substitution with benzene.

General Reaction

Carboxylic acid + Alcohol $\rightleftharpoons$ Ester + H_2O

a) Preparation of ethyl ethanoate

Ethanoic acid + Ethanol $\rightleftharpoons$ Ethyl ethanoate + H_2O

b) Preparation of ethyl propanoate

Propanoic acid + Ethanol $\rightleftharpoons$ Ethyl propanoate + H_2O

Mechanism

Step 1: Protonation of carboxylic acid

In presence of acidic medium, protonation of carbonyl oxygen occurs to enhance the electrophilicity of carbon atom.

Step 2: Attack of alcohol on carbonyl carbon atom (nucleophilic addition)

Even though alcohol is a weak nucleophile, it easily attacks at carbonyl carbon due to enhanced electrophilicity.

Step 3: Deprotonation and elimination of leaving group (formation of esters)

The loss of proton occurs from oxonium ion. This proton is captured by the oxygen of hydroxyl group and it is converted into a better leaving group.

Applications

a) *Conversion of 3-(4-hydroxyphenyl)propanoic acid to methyl 3-(4-hydroxyphenyl)propanoate.*

3-(4-hydroxyphenyl)propanoic acid   methanol   methyl 3-(4-hydroxyphenyl)propanoate

b) *Conversion of adipic acid to dimethyl adipate.*

Adipic acid   Methanol   2,2-dimethoxypropane   Dimethyl adipate

Principle

The 4-phenylbut-3-enoic acid, on heating with concentrated sulfuric acid undergoes ring closure to form α-napthol which on distillation over zinc dust gives naphthalene.

General Reaction: Synthesis of Napthalene from (E)-4-phenylbut-3-enoic acid.

(E)-4-phenylbut-3-enoic acid   i) Conc. H_2SO_4,   ii) Zn, dust   →   Naphthalene

Mechanism

Step 1: Ring closure of (E)-4-phenylbut-3-enoic acid in presence of dehydrating agent concentrated sulfuric acid to yield Naphthalen-1(2H)-one which on oxidation gives Naphthalen-1-ol.

$-H_2O$ → Naphthalen-1(2H)-one → Naphthalen-1-ol

Step 2: Reduction of Naphthalen-1-ol gives Naphthalene.

Naphthalen-1-ol   Zn, Dust Reduction →   Naphthalene

Applications

a) *The reaction is applicable for the synthesis of different substituted naphthalenes. For example: Conversion   of   (E)-4-(4-chlorophenyl)but-3-enoic   acid   into 2-chloronaphthalene.*

(E)-4-(4-chlorophenyl)but-3-enoic acid   i) Conc. H_2SO_4,   ii) Zn, dust ⟶ 2-chloronaphthalene

Principle

Fukuyama and his colleagues developed two-step transformation of primary amines into secondary amines through *o*-nitrobenzenesulfonation in combination with Mitsunobu reaction and subsequent removal of *o*-nitrobenzenesulfonyl group by thiophenol is commonly known as Fukuyama amine synthesis. *O*-nitrobenzenesulfonyl protected amine is alkylated with alcohol in presence of triphenylphosphine and diethyl azodicarboxylate (DEAD) or diisopropyl azodicarboxylate (DIAD), and the deprotection occurs in neutral condition. The *o*-nitrobenzenesulfonyl group is referred as Fukuyama sulfonamide protecting group.

General Reaction

2-nitrobenzene-1-sulfonyl chloride + Primary amines → *o*-nitrobenzene sulfonated product

o-nitrobenzene sulfonated product + R"—OH (Alcohol) + Triphenylphosphine → [DEAD]

+ Thiophenol → [K_2CO_3] → Secondary amines

Mechanism

The detailed mechanism for synthesis of secondary amines by using Fukuyama's protocol is illustrated below:

Step 1:

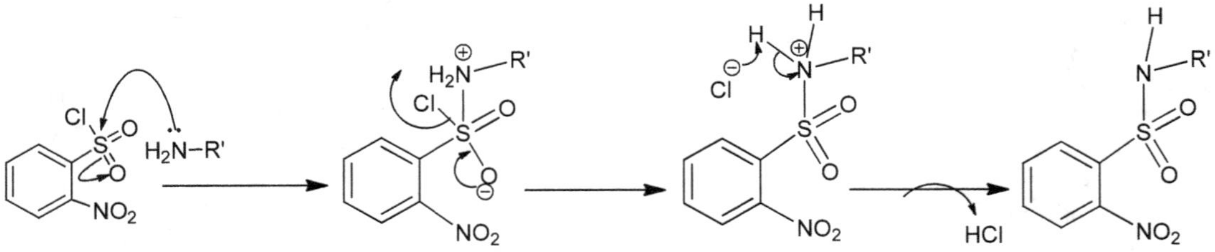

Step 2:

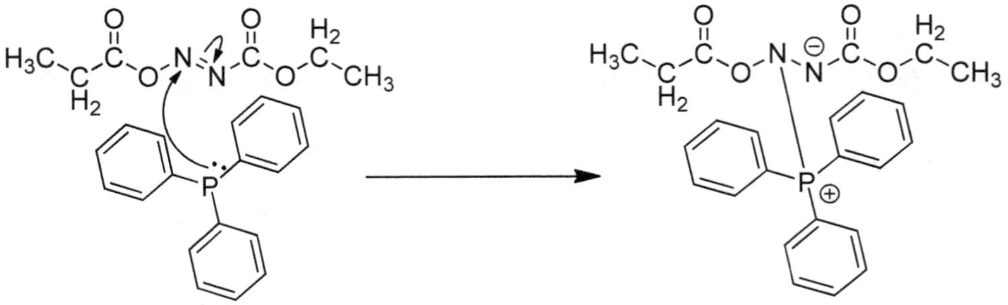

Step 3:

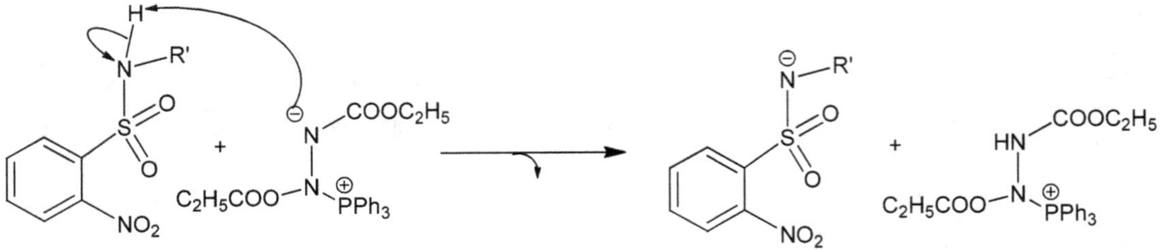

Step 4:

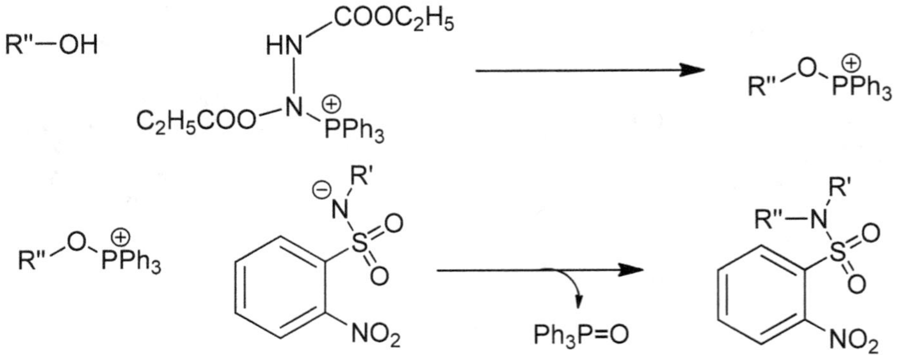

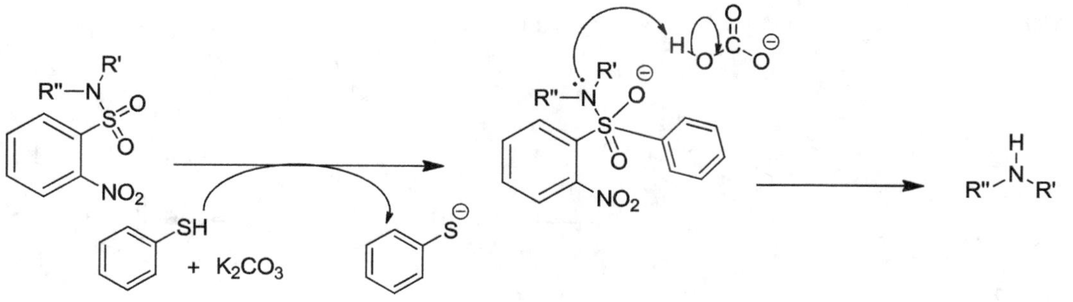
R'
R''—N
S=O
O
NO2
SH
+ K2CO3
S⊖
R'
R''—N
O⊖
S
O
NO2
H
O
C
O⊖
O
H
N
R'' R'

Applications

The wide varieties of secondary amines have been synthesized by using this method.

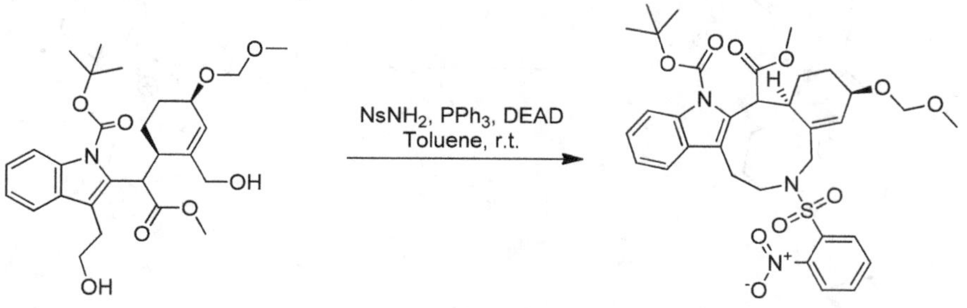
NsNH2, PPh3, DEAD
Toluene, r.t.

tert-butyl 3-(2-hydroxyethyl)-2-(1-((1*R*,4*R*)-2-(hydroxymethyl)-4-(methoxymethoxy)cyclohex-2-en-1-yl)-2-methoxy-2-oxoethyl)-1*H*-indole-1-carboxylate

(3*R*,14a*R*)-13-*tert*-butyl 14-methyl 3-(methoxymethoxy)-6-((2-nitrophenyl)sulfonyl)-2,3,5,6,7,8,14,14a-octahydrobenzo [7,8]azonino[5,4-*b*]indole-13,14(1*H*)-dicarboxylate

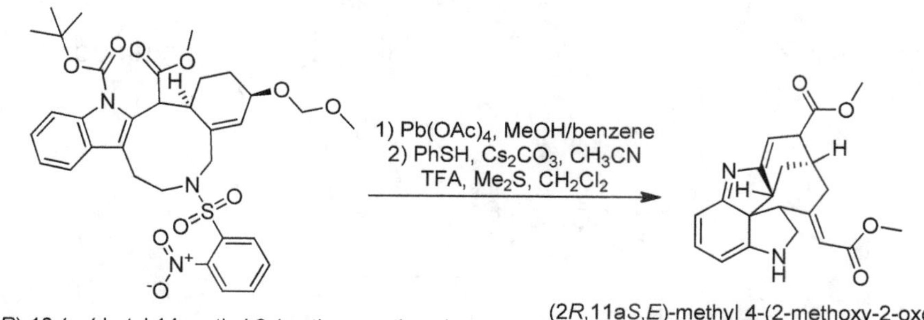
1) Pb(OAc)4, MeOH/benzene
2) PhSH, Cs2CO3, CH3CN
TFA, Me2S, CH2Cl2

(3*R*,14a*R*)-13-*tert*-butyl 14-methyl 3-(methoxymethoxy)-6-((2-nitrophenyl)sulfonyl)-2,3,5,6,7,8,14,14a octahydrobenzo[7,8]azonino[5,4-*b*]indole-13,14(1*H*)-dicarboxylate

(2*R*,11a*S*,*E*)-methyl 4-(2-methoxy-2-oxoethylidene)-1,2,3,4,4a,5,6,11a-octahydro-2,11-(epiethan[2]yl[1]ylidene)pyrrolo[3',2':2,3]benzo[1,2-*b*]pyrrolo[3',4':1,7]cyclohepta[1,2-*c*]pyrrole-13-carboxylate

77. Fukuyama Indole Synthesis

Principle

A synthesis of 3-substituted or 2,3-disubstituted indoles from a tributyltin-based intramolecular radical cyclization of *o*-alkenylphenylisocyanides (2-isocyanostyrenes) is reported by Fukuyama in 1994. Hence the reaction is known as Fukuyama indole synthesis. Fukuyama's protocol involves the treatment of 2-isocyanostyrenes with tributyltin hydride in presence of a radical initiator [For example: Azobisisobutyronitrile (AIBN)], giving a (2-alkenyl) phenylstannoimidoyl radical, which undergoes intramolecular cyclization and tautomerizes subsequently to afford 3-substituted-2-tributylstannyl indoles. Acidic treatment of the indole tin derivatives will afford simple 3-alkyl-substituted indole analogues. The synthesis of indoles takes place by using both acid and base sensitive functionalities at positions 2 and 3 from readily accessible phenylisonitriles; however, it would be very difficult to form 2-alkyl-substituted indoles directly.

General Reaction

Mechanism

Fukuyama indole synthesis involves the intramolecular radical cyclization.

Step 1: Formation of tributyltin hydride radical in presence of a radical initiator AIBN.

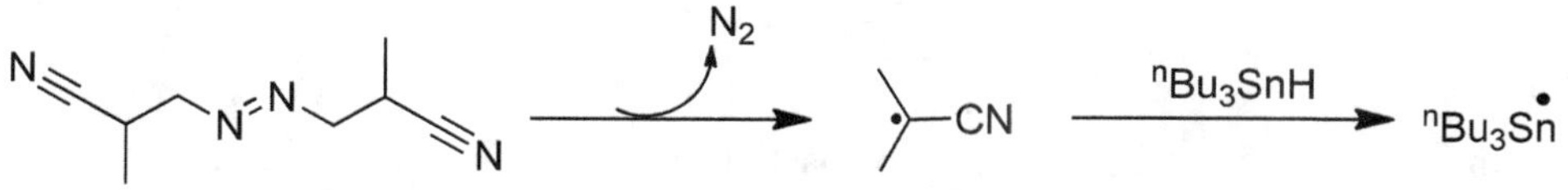
N$_2$
CN
^{n}Bu$_3$SnH
^{n}Bu$_3$Sn

(*E*)-3,3'-(diazene-1,2-diyl)bis(2-methylpropanenitrile)
(Azobisisobutyronitrile)

Step 2: Reaction of 2-isocyanostyrenes with tributyltin hydride radical to give a (2-alkenyl) phenylstannoimidoyl radical which undergoes intramolecular cyclization to afford intermediate.

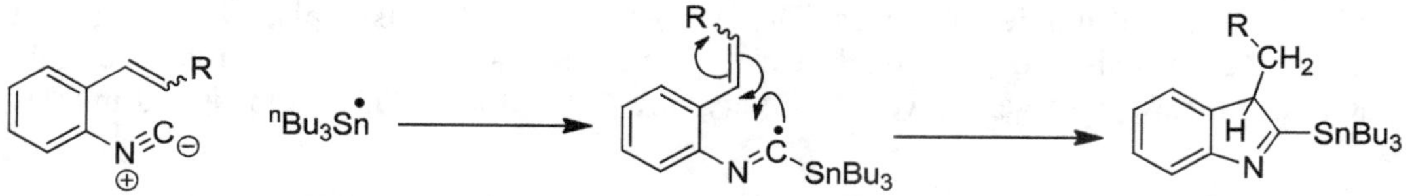
^{n}Bu$_3$Sn
SnBu$_3$
CH$_2$
SnBu$_3$

Step 3: Tautomerization of intermediate followed by acid treatment to yield 3-alkyl-substituted indoles.

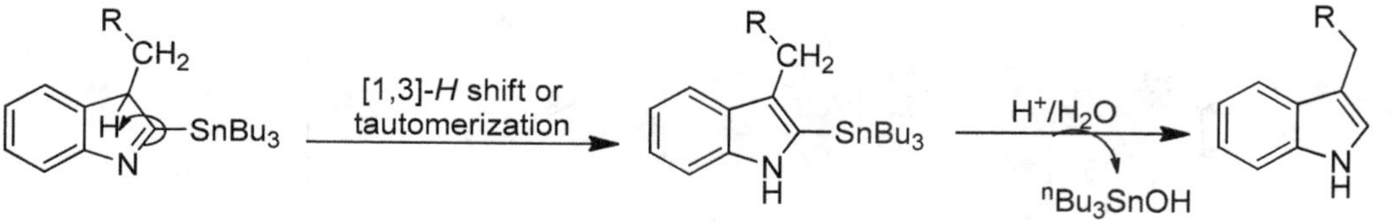
[1,3]-*H* shift or
tautomerization
H$^+$/H$_2$O
^{n}Bu$_3$SnOH

Applications

a) Indole-containing alkaloids can be prepared by using Fukuyama indole synthesis.

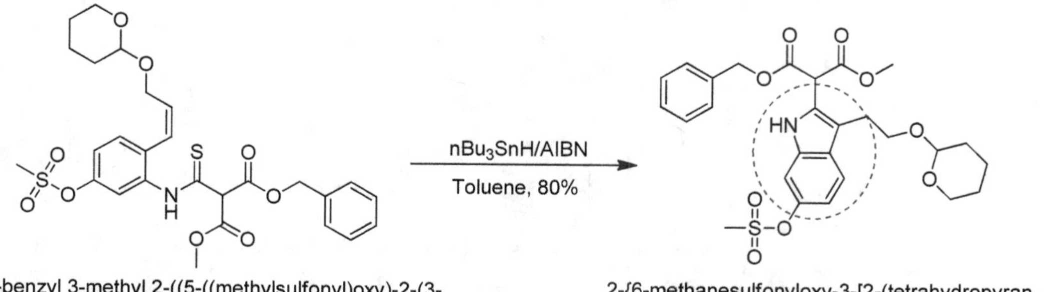
nBu$_3$SnH/AIBN
Toluene, 80%

(*Z*)-1-benzyl 3-methyl 2-((5-((methylsulfonyl)oxy)-2-(3-((tetrahydro-2*H*-pyran-2-yl)oxy)prop-1-en-1-yl)phenyl)carbamothioyl)malonate

2-{6-methanesulfonyloxy-3-[2-(tetrahydropyran-2-yloxy)ethyl]-1H-indol-2-yl} malonic acid benzyl ester methyl ester

78. Friedel-Crafts Acylation

Principle

The reaction of benzene with acid halides or acid anhydride in presence of Lewis acid catalyst such as $AlCl_3$ results in the introduction of an acyl group on benzene ring to give arylketone. Acylation is carried in presence of solvents such as carbondisulfide or nitrobenzene. Friedel and Crafts in 1877 given this electrophilic aromatic substitution reaction. Hence, the reaction is known as Friedel-Crafts acylation. Aromatic ring with electron-donating groups promotes the reaction, whereas electron withdrawing groups hinders the reaction. Nitrobenzene, with its strong electron withdrawing nitro group impede acylation, is often used as solvent for Friedel-Crafts acylation. Acylating agents are also helpful to enhance the rate of acylation; a more electrophilic acylating agent, faster the acylation proceeds. Commonly used acylating agents are acyl halides and acyl anhydrides along with other derivatives of carboxylic acids, esters, and lactams useful for direct acylation. To increase an electrophilicity of acylating agents, at least 1 equivalent of Lewis acid is used in this reaction. The Lewis acids used in acylation with acyl halides include $AlCl_3$, $AlBr_3$, $BeCl_2$, $CuCl_2$, $FeBr_3$, $HgCl_2$, $MbCl_5$, $SbBr_3$, $SbCl_5$, and $ZrCl_4$. When acyl anhydrides are used as acylating agents, effective catalysts are $AgClO_4$, BF_3, CF_3CO_2H, $(CF_3CO)_2O$, $HClO_4$, HF, H_3PO_4, $SnCl_4$, $SOCl_2$, $ZnCl_2$, etc.

General Reaction

a) Benzene reacts with ethanoyl chloride to give acetophenone

b) Benzene reacts with acetic anhydride to yield acetophenone

Benzene + Acetic anhydride $\longrightarrow$ Acetophenone + CH_3COOH

Mechanism

Step 1: Generation of electrophile (acylium ion)

The carbonyl oxygen of acid chloride forms complex with $AlCl_3$. This complex forms corresponding acylium ion that behaves as strong electrophile.

$R\text{-}CO\text{-}Cl$ + $AlCl_3$ $\xrightleftharpoons{\text{complexation}}$ $R\text{-}\overset{\oplus}{C}(H)\text{=}O\text{-}\overset{\ominus}{AlCl_3}$ $\rightleftharpoons$ $\left[\ \overset{\oplus}{O}\text{≡}C\text{-}R \longleftrightarrow R\text{-}\overset{\oplus}{C}\text{=}O\ \right]\ \overset{\ominus}{AlCl_4}$

Acylium

Step 2: Attack of acylium ion on benzene Π-electron system (formation of arenium ion)

The attack of acylium ion on benzene ring results in the formation of arenium ion which is stabilized through delocalization.

Loss of aromaticity as one of the carbons is sp^3 hybridized.

Step 3: Loss of proton (regeneration of catalyst and formation of ketone)

In order to regain aromaticity, the arenium ion undergoes loss of proton that is captured by $AlCl_4$ (*Step 1*) to form $AlCl_3$ and simultaneous removal of HCl. The formation of aryl ketone is an exothermic process.

$\xrightarrow[\text{Aromatization}]{\overset{\ominus}{AlCl_4}}$ + HCl + $AlCl_3$

Applications

a) The reaction is generally applied for the preparation of aromatic ketones.

3-phenylpropanoyl chloride $\xrightarrow{\text{AlCl}_3}$ 2,3-dihydro-1*H*-inden-1-one

b) The products of acylation can be sometimes cyclized under mild conditions. This can be used in the preparation of higher carbocyclic rings (Howarth synthesis).

Isobenzofuran-1,3-dione + Benzene $\xrightarrow[\text{2. H}_2\text{SO}_4]{\text{1. AlCl}_3}$ Anthracene-9,10-dione

c) 2-napthyl ketone can be synthesized from naphthalene.

Naphthalene $\xrightarrow[\text{C}_6\text{H}_5\text{NO}_2,\ 95°\text{C}]{\text{CH}_3\text{COCl, AlCl}_3,}$ 1-(naphthalen-2-yl)ethanone
(2-napthyl ketone)

79. Friedel-Crafts Alkylation

Principle

Friedel-Craft alkylation is the work reported by Friedel and Crafts reported in 1877. It is an example of electrophilic aromatic substitution reaction in which an electrophile is a carbocation. In Friedel-Crafts alkylation, the electron-donating groups facilitate the reaction, whereas the electron-withdrawing groups hinder an alkylation. An acidic catalyst (Lewis acid) is needed to enhance an electrophilicity of an alkylating agent. More often alkylation is carried out between an aromatic compound and haloalkanes in presence of a Lewis acid. The Lewis acid polarizes the haloalkane, results into more positive charge on hydrocarbon part and become more electrophilic, which then saturates its electron deficiency from electron-rich benzene ring by forming a π-complex that subsequently convert into a σ-complex resulted into loss of aromaticity.

The generally used Lewis acid catalysts in alkylation are arranged in the following order of decreasing Lewis acidity: $Al_2Br_6 > Al_2Cl_6 > Al_2I_6 > Fe_2Cl_6 > SbCl_5 > ZrCl_4$, $SnCl_4 > BCl_3, BF_3, SbCl_3$.

General Reaction

For example: Benzene reacts with chloromethane to give toluene (methylbenzene)

Mechanism

Step 1: Generation of electrophile (alkyl carbocation)

The Lewis acid enhances the electrophilic character on carbon of alkyl group in the complex.

Step 2: Attack of alkyl carbocation on benzene Π-electron system (formation arenium ion)

Attack of electrophile is the rate determining step; formation of arenium ion is an endothermic process. Continuous delocalization ceases, as one of the carbon becomes sp^3 hybridized, this results in the loss of aromaticity in arenium ion. The four Π-electrons are delocalized among five sp^2 hybrid carbons of the ring.

Arenium ion
(Resonance stablization of arenium ion)

Step 3: Loss of proton (formation of alkylbenzene and regeneration of catalyst)

In order to regain aromaticity, the arenium ion undergoes loss of proton. The proton is taken by $AlCl_4$ (*Step 1*) to form $AlCl_3$ and simultaneous removal of HCl.

Applications

a) *The reaction has been widely used for the synthesis of substituted aromatic compounds.*

Benzene   Chloroethane   Ethylbenzene

b) *In the synthesis of polynuclear hydrocarbons. For example: Preparation of anthracene.*

(chloromethyl)benzene   9,10-dihydroanthracene   Anthracene

80. Friedlander Quinoline Synthesis

Principle

The Friedlander quinoline synthesis clubs α-amino aldehyde or ketone with another aldehyde or ketone containing at least one α-methylene group (active) adjacent to the carbonyl group to give substituted quinoline derivatives. The reaction was reported by Friedlander in 1882. The reaction is carried out in presence of acid, base or heat. The acidic catalysts used in reaction are HCl, H_2SO_4, p-TsOH, and polyphosphoric acid (PPA); where as NaOH and $NaOOC_2H_5$ are useful as basic catalysts.

General Reaction

o-aminobenzaldehyde + Aldehyde/Ketone $\xrightarrow{\text{Alkali}}$ Quinoline analogues

a) For example: o-aminobenzaldehyde reacts with acetaldehyde to give quinoline

o-aminobenzaldehyde + Acetaldehyde $\xrightarrow{\text{NaOH}}$ Quinoline

Mechanism

The stepwise mechanism of Friedlander quinoline synthesis involves following steps: It is assumed that aldol type condensation is involved at a certain stage of the overall condensation.

An irreversible intramolecular condensation between an active methylene or α-methyl group and the carbonyl group attached to the aromatic ring. A condensation reaction between an amino group and carbonyl group of aldehyde or ketone to form intermediate compound. Intramolecular ring closure followed by removal of water molecule gives quinoline derivatives.

Step 1:

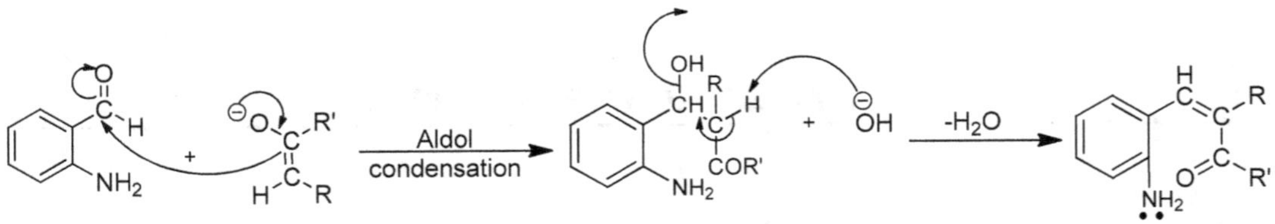

Step 2: Intramolecular nucleophilic attack at carbonyl carbon with subsequent ring closure to yield quinoline analogues.

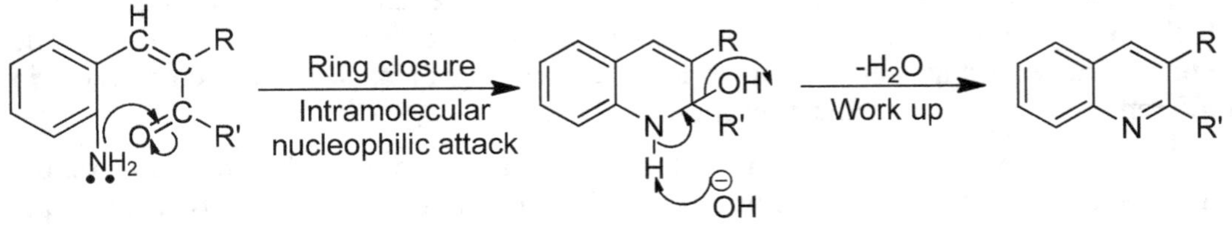

Applications

a) *The reaction has been used in the synthesis of quinolines, naphthyridines, and other polycyclic heterocycles.*

b) *Friedlander quinoline synthesis is one of the most successful methods for synthesizing quinoline analogues with high yield. For example: Synthesis of Ethyl 2,4-dimethylquinoline-3-carboxylate.*

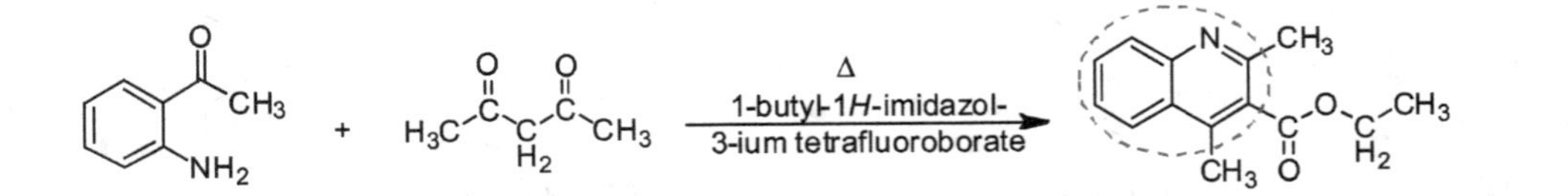

1-(2-aminophenyl)ethanone   Pentane-2,4-dione   Ethyl 2,4-dimethylquinoline-3-carboxylate

81. Fries Rearrangement

Principle

Phenols on treatment with acid chloride or acid anhydride form phenolic esters. Phenolic esters, when heated with anhydrous aluminium chloride undergo a rearrangement in which the acyl group migrates towards aromatic ring to give principally ortho-phenolic ketones. This rearrangement is known as Fries rearrangement. Fries and co-workers in 1908 reported this rearrangement reaction.

If the ortho position is not vacant or the reaction is carried out at low temperature; the para product predominates. Electron withdrawing groups in the substrate lowers the rate of reaction as in the case of Friedel-Crafts reaction. Generally, low temperature favors the para product and high temperature favors the ortho product. Fries rearrangement can also be carried out photochemically by exposure to UV light.

General Reaction

Mechanism

It is an electrophilic substitution reaction where acylium ion (**RCO$^+$**) is an electrophile. The reaction may proceed either by intermolecular or by intramolecular mechanism.

Intermolecular mechanism

Step 1: Complex formation with Lewis acid AlCl$_3$; Formation of acyl cation as an electrophile

Step 2: Attack of electrophile at ortho and para positions of the aromatic ring; resonance stabilization of arenium ion.

a) Attack at ortho position of aromatic ring.

Acyl cation   Electrophile

Resonance stabilization of arenium ion

b) Attack at para position of aromatic ring.

Acyl cation   Electrophile

Resonance stabilization of arenium ion

Step 3: Removal of proton from resonance stabilized arenium ion; regain of aromaticity.

Step 4: Hydrolysis to give ortho and para acyl phenols.

σ complex   →(Step 3, -HCl)→   →(Step 4, H_2O)→   o-hyroxyacetophenone

σ complex   →(Step 3, -HCl)→   →(Step 4, H_2O)→   p-hyroxyacetophenone

Intramolecular mechanism

The intramolecular rearrangement explains the formation of ortho isomer.

Step 1: Complex formation of phenolic ester with Lewis acid

Step 2: Intramolecular shift of acyl group at ortho position

Step 3: Hydrolysis of intermediate results in the formation of ortho-acylphenols

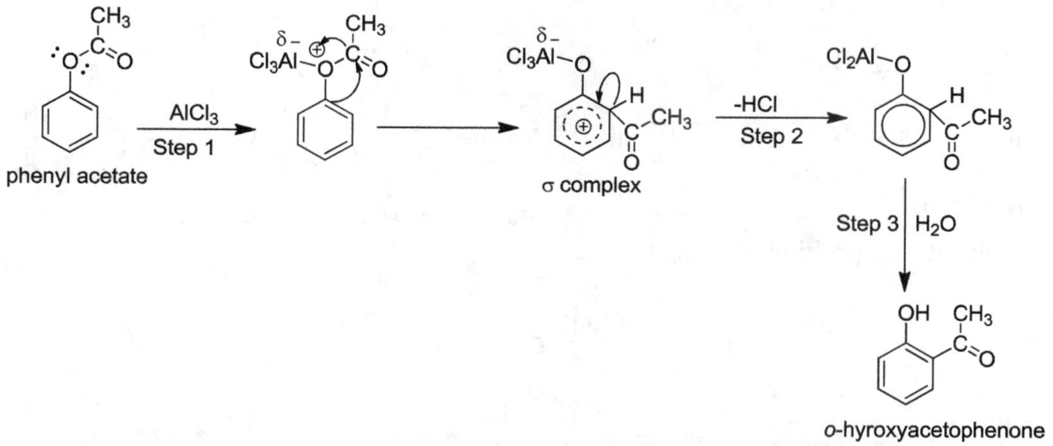

The ortho isomer formed by hydrolysis is stabilized by intramolecular hydrogen bonding.

Applications

a) Synthesis of adrenaline.

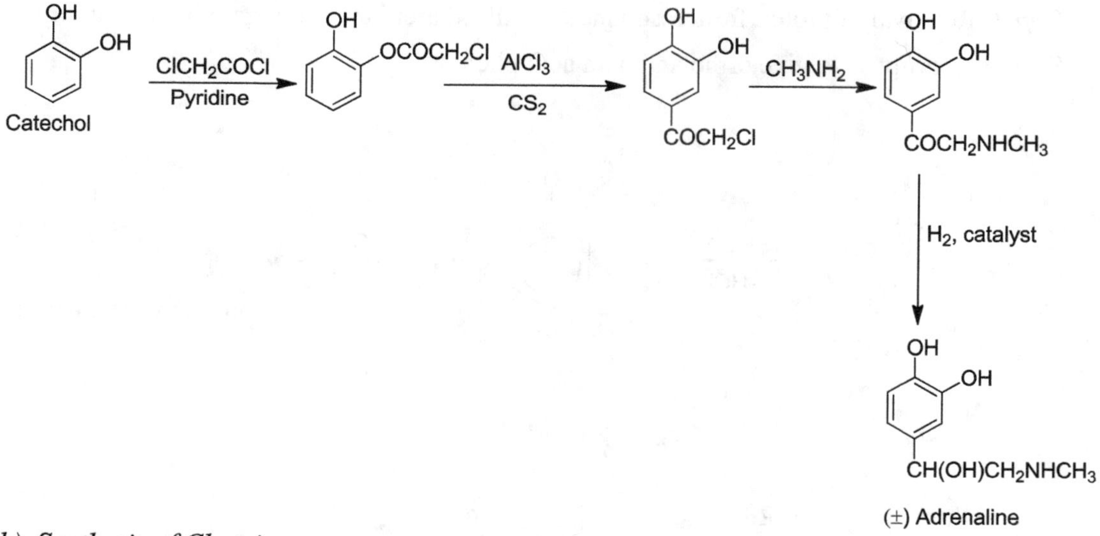

b) Synthesis of Chrysin.

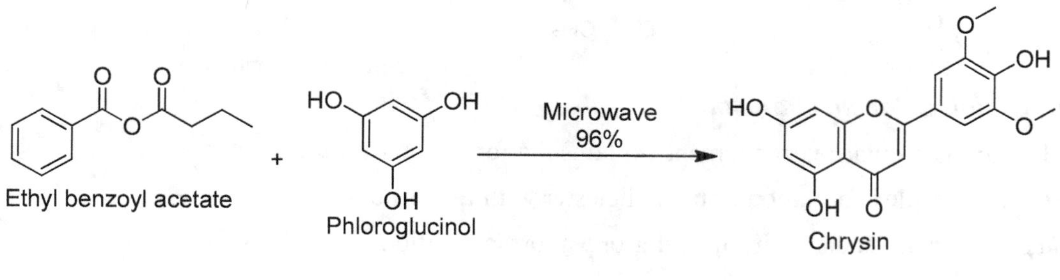

82. Gabriel Phthalimide Synthesis (Gabriel Primary Amine Synthesis)

Principle

The synthesis of primary aliphatic amines involving the *N*-alkylation of phthalimide followed by acidic or basic hydrolysis of *N*-alkyl phthalimides is known as Gabriel phthalimide synthesis. Gabriel in 1887 reported this two step reaction to yield primary aliphatic amines. The major advantage of reaction is to impede over alkylation on the nitrogen atom; because primary or secondary amine is more nucleophilic than ammonia. Gabriel phthalimide synthesis involving ammonia and an alkyl halide may produce mixtures of primary amine, secondary amine, tertiary amine, and small concentration of quaternary ammonium salt.

General Reaction

Phthalimide   *N*-alkylphthalimide   1°-amine   Phthalic acid

Mechanism

The hydrolysis of *N*-alkyl phthalimide is proposed here to illustrate the reaction mechanism. In Gabriel primary amine synthesis, phthalimide is easily deprotonated by potassium hydroxide/sodium hydroxide due to presence of two electron-withdrawing carbonyl groups, results nucleophile; after next step of *N*-alkylation, *N*-alkyl phthalimide can be decomposed to provide pure primary amines.

Applications

a) *The reaction is very useful for synthesis of primary aliphatic amines. For example:*
Synthesis of 2-(3,4-dihydroquinolin-1(2H)-yl)ethanamine.

2-(2-bromoethyl)isoindoline-1,3-dione

1,2,3,4-tetrahydroquinoline

2-(3,4-dihydroquinolin-1(2*H*)-yl)ethanamine

b) *Dihaloalkane produces monophthalamide like haloalkanes. Finally, phthaloyl group*
is removed to obtain primary aliphatic amine.

Potassium 1,3-
dioxoisoindolin-2-ide

1° amine

c) *α-amino acids are synthesized through reaction of malonic esters with potassium*
1,3-dioxoisoindolin-2-ide.

Potassium 1,3-dioxoisoindolin-2-ide

HCl, 200°C

R'CH(NH$_2$)COOH

α-amino acid

Principle

Gassman and his colleagues (1973) reported synthesis of substituted indoles through reaction between anilines, *tert*-butyl chlorite, and β-keto sulfide in presence of triethylamine; this reaction is commonly known as Gassman indole synthesis. Raney nickel is used; if desulfurization is needed. It is one pot synthesis of 1- or 2- substituted and 2,7-disubstituted indoles in high yields. This reaction is advantageous over Fischer indole synthesis due to cheap and readily available starting material; accessibility for 1-, 2-, 4-, 5-, 6-, or 7-substituted indoles; mild reaction conditions without involving a strong acid or base; and higher yield.

General Reaction

Mechanism

Step 1: Nucleophilic substitution reaction between *N*-chloroaniline and β-keto sulfide to generate sulfonium ion.

Step 2: [2,3]-sigmatropic rearrangement to produce intermediate compound.

Step 3: Intramolecular cyclization to form 3-thioalkoxyindoles.

Applications

Many types of substituted indoles have been synthesized in higher yield by using this method.

a) Synthesis of 2-methyl-3-(methylthio)-1H-indole.

Aniline

2-methyl-3-(methylthio)-1*H*-indole

b) Synthesis of 2,3,4,9-tetrahydro-1H-carbazole.

Aniline

4a-(methylthio)-2,3,4,
4a-tetrahydro-1*H*-carbazole

2,3,4,9-tetrahydro-1*H*-carbazole

84. Gassman Oxindole Synthesis

Principle

Gassman and his colleagues synthesized oxindoles in 1973 by using one pot synthesis of aniline, tert-butyl hypochlorite, ethyl (methylthio) acetate, triethylamine, hydrochloric acid, and final treatment with zinc or raney nickel required for the desulfurization. It is a multistep process (one pot reaction) for synthesis of oxindole with an electron withdrawing group involving sequential treatment of aniline with different reactants. Therefore, this reaction is often known as the Gassman oxindole synthesis. For the anilines with an electron-donating group, the nucleophilic attack by a sulfur atom on the nitrogen atom is disfavored, thus an alternative approach is used to prepare the corresponding oxindoles by treatment of the anilines with chlorosulfonium salt. Further oxidation of the resulting oxindole by N-chlorosuccinamide/HgO or by direct air oxidation produces possible isatin.

General Reaction

Subsituted aniline + β-carbonyl sulfide derivatives → 3-substituted oxindole

Mechanism

Step 1: Nucleophilic substitution reaction between N-chloroaniline and β-keto sulfide to generate sulfonium ion.

N-chloroaniline + → Sulfonium ion

Step 2: [2,3]-sigmatropic rearrangement to produce intermediate compound.

Step 3: Intramolecular cyclization.

Step 4: Desulfurization by raney nickel to yield oxindole.

Applications

a) *The reaction is useful for synthesis of many substituted oxindoles and isatins. For example: Synthesis of Indoline-2,3-dione.*

3-(methylthio)indolin-2-one

Indoline-2,3-dione

85. Gatterman Aldehyde Synthesis

Principle

Gattermann in 1898 synthesized aromatic aldehydes by sequential treatment of aromatic hydrocarbons with hydrogen cyanide and hydrogen chloride in anhydrous solvent with or without presence of Lewis acid as a catalyst, the reaction is known as Gatterman aldehyde synthesis. The reaction is applicable for preparation of aromatic aldehydes containing hydroxyl or alkyloxyl groups on the aromatic nucleus. In this reaction, the aldimine hydrochloride functions as an intermediate. Aldimine intermediate after hydrolysis yields corresponding amine. Phenolic aldehydes and aromatic aldehydes with multi-alkyl groups, such as mesitaldehyde can be synthesized by using this method.

General Reaction

$$\text{Benzene} + \text{HCN} + \text{HCl} \xrightarrow[\text{2. } H_2O]{\text{1. } AlCl_3} \text{Benzaldehyde}$$

Benzene   Benzaldehyde

Mechanism

The reaction proceeds through electrophilic substitution on the aromatic ring as follows:

Step 1: The imidoformyl chloride is primarily formed from HCN and HCl.

$$H-C\equiv N \ + \ HCl \longrightarrow HN=\underset{H}{C}-Cl$$

imidoformyl chloride

Step 2: Imidoformyl chloride forms complex with lewis acid ($AlCl_3$)

$$HN=\underset{H}{C}-Cl \ + \ AlCl_3 \longrightarrow Cl_3\overset{\ominus}{Al}-Cl-\underset{}{\overset{\oplus}{C}}\overset{H}{=}NH$$

imidoformyl chloride complex

Step 3: Imidoformyl chloride complex attacks on the benzene ring (electrophilic attack) results in the formation of benzenonium ion (resonance stabilized) and loss of proton to afford an imine of aromatic aldehyde. This after hydrolysis gives the aldehyde.

benzenonium ion

imine of
aromatic aldehyde

Applications

a) The reaction is widely applicable for the preparation of aromatic aldehydes with hydroxyl, alkoxyl, and bulkier alkyl groups. The use of hydrogen cyanide is avoided by using HCl and zinc cyanide which produces hydrogen cyanide and zinc chloride.

Resorcinol + $Zn(CN)_2$ (Dicyanozinc) + ClH $\xrightarrow[\text{2. } H_2O/H^{\oplus}]{\text{1. } ZnCl_2}$ 2,4-dihydroxybenzaldehyde + NH_4Cl

b) Alkoxy substituted aldehydes are synthesized by using phenolic ethers.

Anisole + HCN + HCl $\xrightarrow[\text{2. } H_2O/H^{\oplus}]{\text{1. } AlCl_3}$ 4-methoxybenzaldehyde + NH_4Cl

c) Preparation of alkyl groups containing aromatic aldehydes.

1,2,3-triisopropylbenzene + $Zn(CN)_2$ + HCl $\xrightarrow[\text{2. } H_2O/H^{\oplus}]{\text{1. } ZnCl_2}$ 2,4,6-triisopropylbenzaldehyde

86. Gattermann-Koch Reaction

Principle

The synthesis of aromatic aldehydes by reaction of aromatic compounds with CO and HCl in presence of $AlCl_3$ and cuprous chloride is known as Gattermann-Koch formylation reaction. The reaction is given by Gattermann and Koch in 1897. The reaction involves direct introduction of the formyl group on to aromatics.

Note: Important points to be remembered during the course of reaction; a) at atmospheric pressure of CO with Cu_2Cl_2 present and, b) at high-pressure in absence of Cu_2Cl_2, in which the optimal temperature for high-pressure synthesis of benzaldehyde is 35°C.

General Reaction

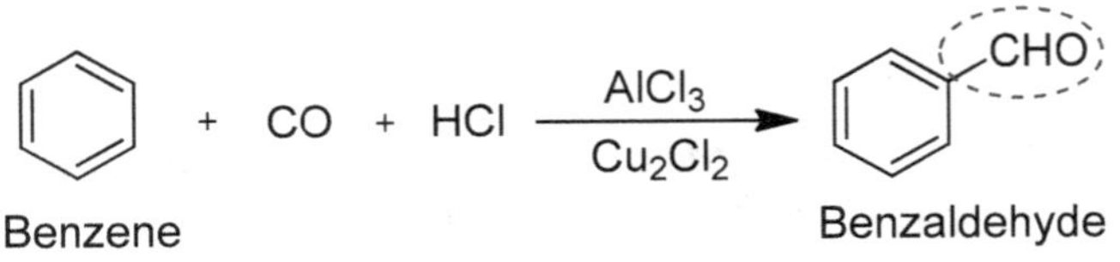

Benzene + CO + HCl $\xrightarrow[Cu_2Cl_2]{AlCl_3}$ Benzaldehyde (CHO)

Mechanism

The reaction is an example of electrophilic substitution reaction closely related to the Friedel Craft's reaction.

Step 1: The reaction proceeds *via* formation of formyl chloride from CO and HCl followed by a Friedel Craft acylation in presence of Lewis acid. Formyl chloride could not be isolated during the reaction. Hence, the reaction is considered to be take place through the formation of an electrophile (Acylium ion; generated by protonation of CO).

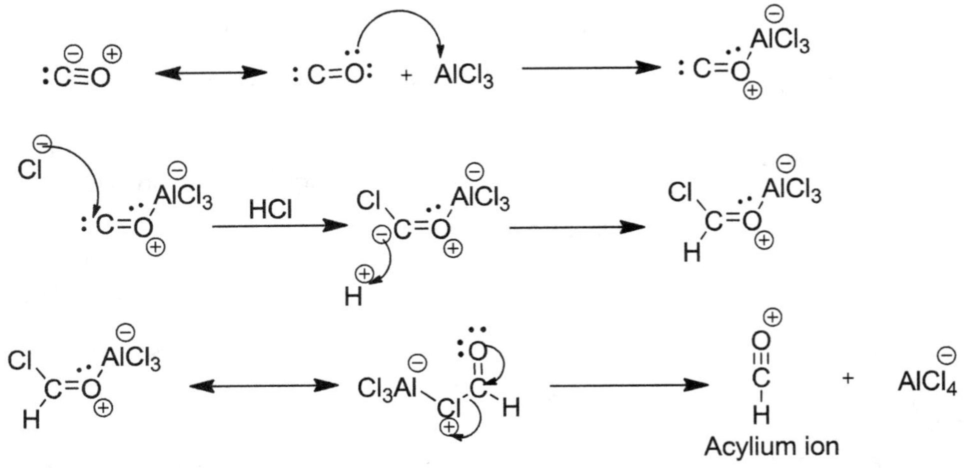

Acylium ion

Step 2: Formylation of aromatics (electrophilic substitution) by acylium ion produces aromatic aldehyde.

Acylium ion

Benzaldehyde

Applications

a) The reaction has been widely used to synthesize homologues of aromatic aldehydes. For example: Synthesis of 4-methylbenzaldehyde from toluene.

Toluene

4-methylbenzaldehyde

b) Synthesis of 2,4-dimethylbenzaldehyde from p-xylene.

p-xylene

2,5-dimethylbenzaldehyde

c) Preparation of dialdehydes from 1,2-diphenylethane.

1,2-diphenylethane

4,4'-(ethane-1,2-diyl)dibenzaldehyde

87. Gomberg-Bachmann Reaction

Principle

The preparation of diaryls through condensation of aryl diazonium salts with aromatic or heteroaromatics in presence of aqueous sodium hydroxide is known as Gomberg-Bachmann alkaline reaction. Extensive studies on mechanism proposed that the reaction involves generation of free radicals; hence sometimes this reaction is also referred as Gomberg-Bachmann-Hey aryl coupling. The aromatic diazonium salts possessing an electron-withdrawing group undergoes Gomberg-Bachmann reaction very rapidly, but accompanying by certain drawbacks such as an instability of diazonium salt and side reactions to form tar residue. The reaction has been modified to use stable aromatic diazonium tetrafluoroborate under phase transfer conditions.

General Reaction

Benzenediazonium + Arene $\xrightarrow{\text{NaOH}}$ 1,1'-biphenyl + N_2 + HCl

Benzenediazonium + Anisole $\xrightarrow{\text{NaOH}}$ 2-methoxy-1,1'-biphenyl

Mechanism

Step 1: Reaction between Benzenediazonium and (*E*)-1-hydroxy-2-phenyldiazene to form (1*E*, 1'*E*)-2,2'-oxybis(1-phenyldiazene).

Benzenediazonium + (*E*)-1-hydroxy-2-phenyldiazene $\longrightarrow$ (1*E*,1'*E*)-2,2'-oxybis(1-phenyldiazene)

Step 2: Generation of free radicals.

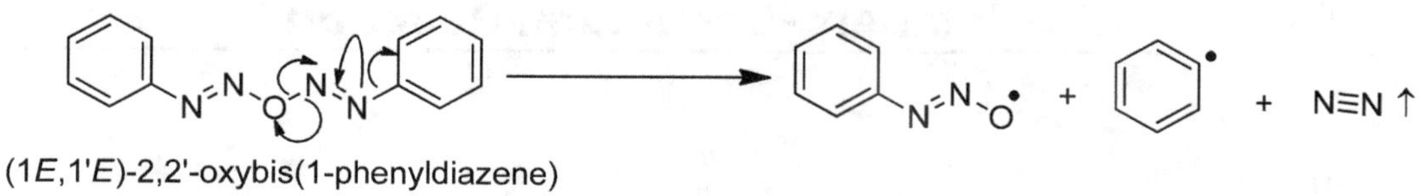

(1*E*,1'*E*)-2,2'-oxybis(1-phenyldiazene)

Step 3: Formation of 1,1'-biphenyl.

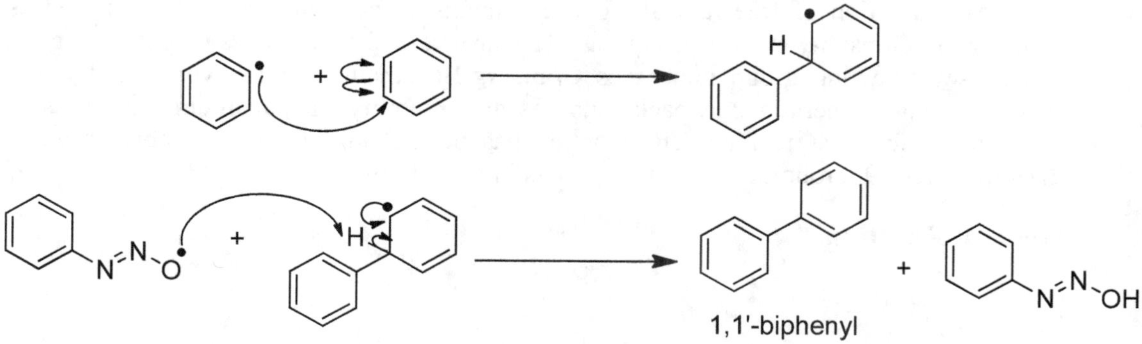

1,1'-biphenyl

Applications

The reaction is widely applicable for synthesis of polycyclic aromatics and heteroaromatics.

a) Synthesis of 4-bromo-1,1'-biphenyl by reaction of 4-bromobenzenediazonium and bezene in alkaline medium.

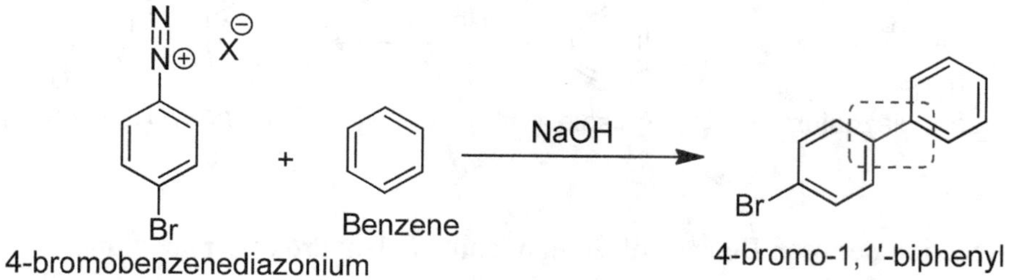

4-bromobenzenediazonium   +   Benzene $\xrightarrow{\text{NaOH}}$ 4-bromo-1,1'-biphenyl

b) Pschorr's synthesis of phenanthrene.

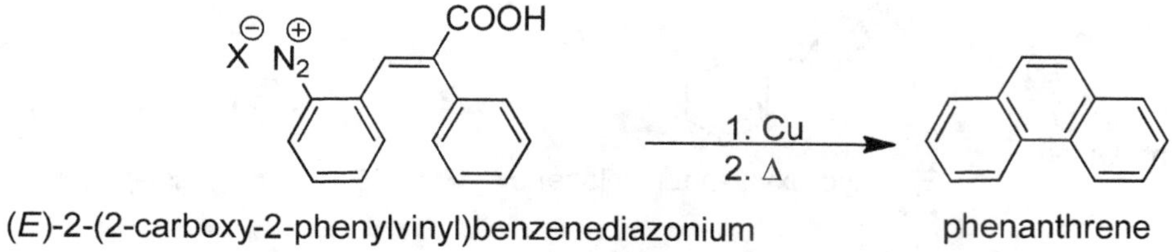

(*E*)-2-(2-carboxy-2-phenylvinyl)benzenediazonium $\xrightarrow[\text{2. }\Delta]{\text{1. Cu}}$ phenanthrene

88. Grignard Degradation

Principle

The dehalogenation of polyhalogen containing heterocyclic compounds was first reported by Steinkopf in 1934. The dehalogenation of polyhalo-compounds with the help of corresponding Grignard reagents in presence of water to produce the compounds having one halogen atom less than starting reactant is referred as Grignard degradation. This reaction was primarily works for aromatic compounds. In addition, the dehalogenation of aromatic halides also occurs when the haloalkanes are treated with other Grignard reagents in the presence of catalytic amounts of cobaltous chloride.

General Reaction

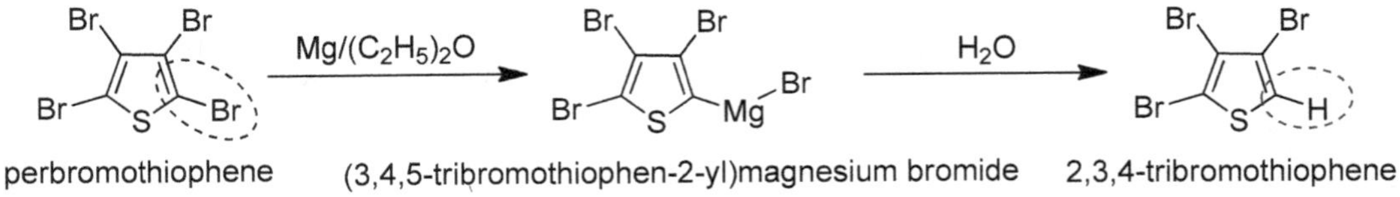

perbromothiophene   (3,4,5-tribromothiophen-2-yl)magnesium bromide   2,3,4-tribromothiophene

Mechanism

Step 1:

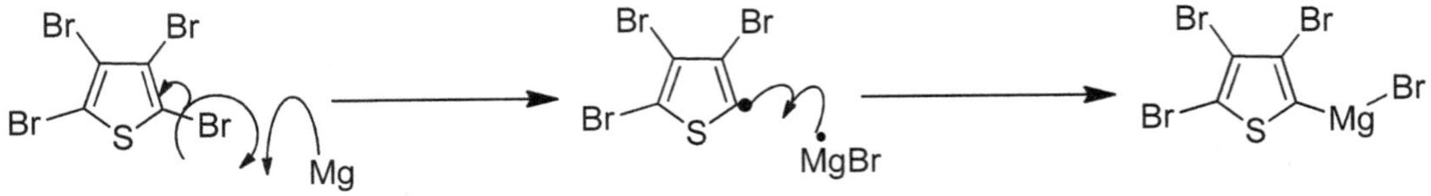

Step 2:

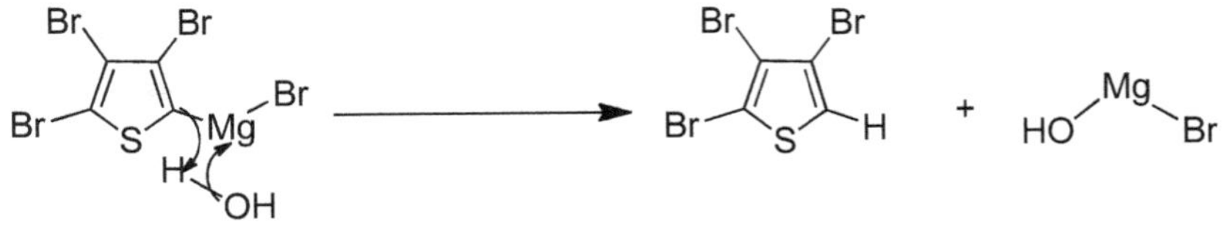

Applications

a) *The reaction is applicable to dehalogenation of aryl or alkyl halides. For example: The reaction between 1-bromonaphthalene and butylmagensium bromide in presence of ether gives naphthalene.*

1-bromonaphthalene   Butylmagnesium bromide   Naphthalene

89. Grignard Reaction

Principle

The preparation of alkyl or aryl magnesium halides (R-Mg-X or Ar-Mg-X) from corresponding organanic halides and magnesium in ethereal solvents and their nucleophilic addition to carbonyl compounds to produce alcohols is known as Grignard reaction. The reaction was initially given by Grignard in 1900. The alkyl magnesium halides are commonly referred as Grignard reagents; Grignard reagents interacts with different electrophiles; such as CO_2 to produce carboxylic acid; nitriles or acyl halides to obtain ketones; carbonyl compounds to give alcohols (primary alcohol from formaldehyde, secondary alcohols from higher aldehydes, and tertiary alcohols from ketones). Because the Grignard reagent is equivalent to a carbanion, the preparation of Grignard reagent must be carried out in an anhydrous solvent that can dissolve but not react with the generated carbanion species. Only ethereal solvents that can combine with the Grignard reagent are useful for this reaction and diethylether is the best solvent for Grignard reaction.

General Reaction

$$R-X \xrightarrow{\text{Mg (O)}} R-Mg-X \; + \; R'\overset{\overset{\displaystyle O}{\|}}{\underset{}{C}}R'' \longrightarrow R-\overset{\overset{\displaystyle OH}{|}}{\underset{\underset{\displaystyle R'}{|}}{C}}-R''$$

Haloalkanes   Grignard reagents   Carbonyl compounds   Alcohols

Mechanism

Formation of Grignard reagent

The formation of Grignard reagents is heterogeneous. The reaction rate for formation of Grignard reagent is proportional to the concentration of haloalkane and the surface area of magnesium, and all organic iodides and many secondary bromoalkane react at a mass transport or diffusion-controlled rate in ether; other less reactive organic bromides (neopentyl, phenyl, etc.) and most organic chlorides react at lower rates. The formation of Grignard reagents involves a free radical or single-electron transfer process. The concept that the radicals diffuse freely in solution is consistent with the formation of Grignard reagents from primary haloalkane.

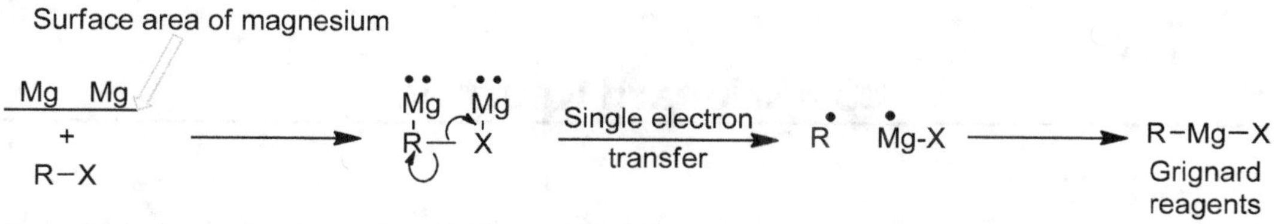

Surface area of magnesium

$$Mg + R-X \longrightarrow \underset{R-X}{Mg\;Mg} \xrightarrow[\text{transfer}]{\text{Single electron}} R^\bullet \quad Mg\text{-}X \longrightarrow \underset{\text{Grignard reagents}}{R-Mg-X}$$

Reaction proceeds via nucleophilic addition mechanism

In Grignard reagent, the alkyl group behaves as a nucleophile (electron rich) and attack at the electron deficient carbon of carbonyl compounds (aldehyde or ketone) to give the addition product. Hydrolysis of adduct gives alcohol.

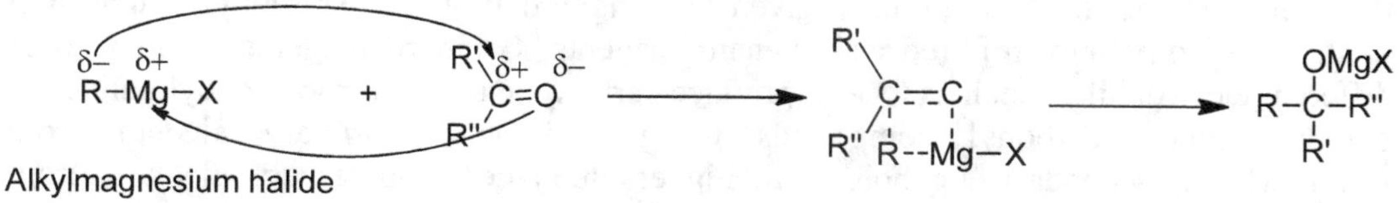

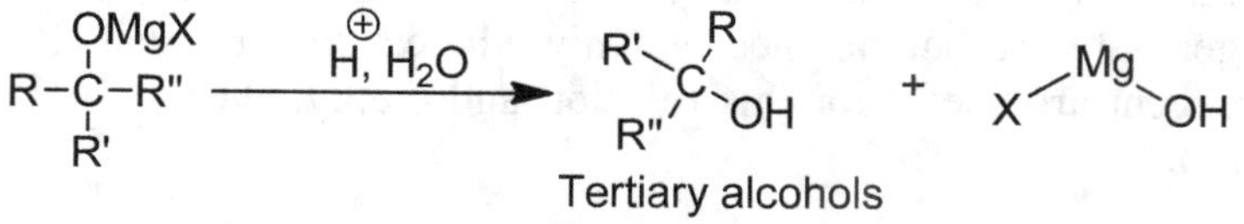

Alkylmagnesium halide

Tertiary alcohols

Reaction proceeds through radical mechanism

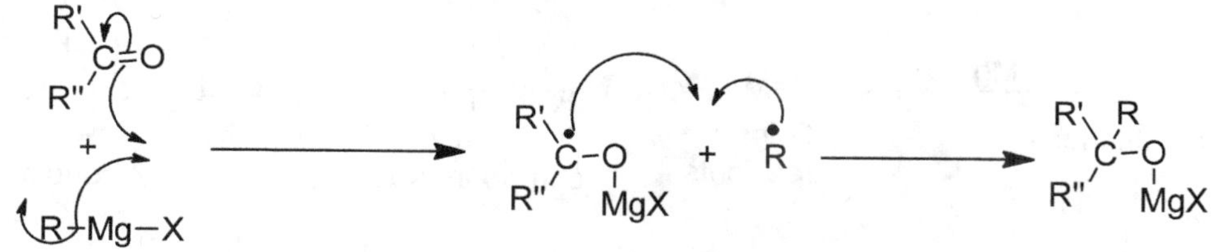

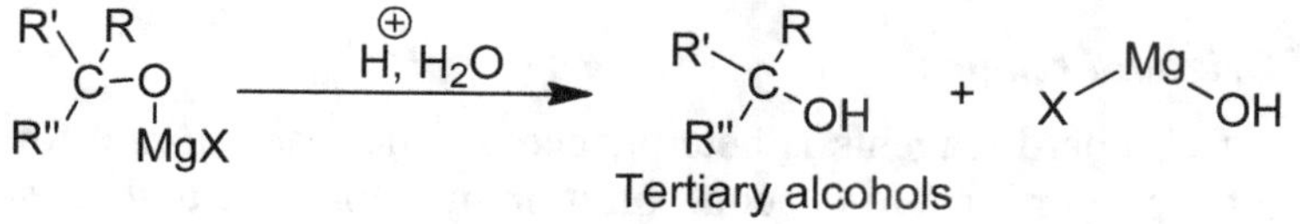

Tertiary alcohols

Applications

a) The reaction is applicable for the preparation of many substituted alcohols. For example: Preparation of propan-1-ol.

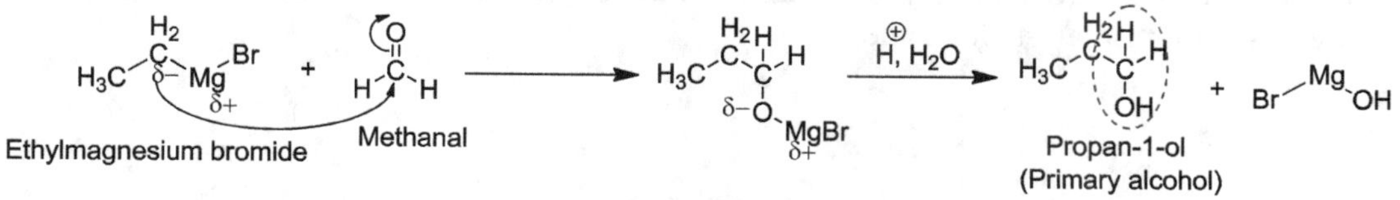

Ethylmagnesium bromide   Methanal   Propan-1-ol
(Primary alcohol)

Preparation of butan-2-ol

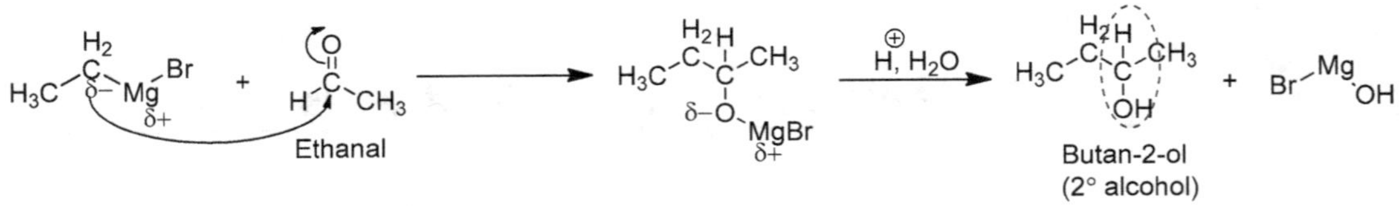

Preparation of 2-methylbutan-2-ol

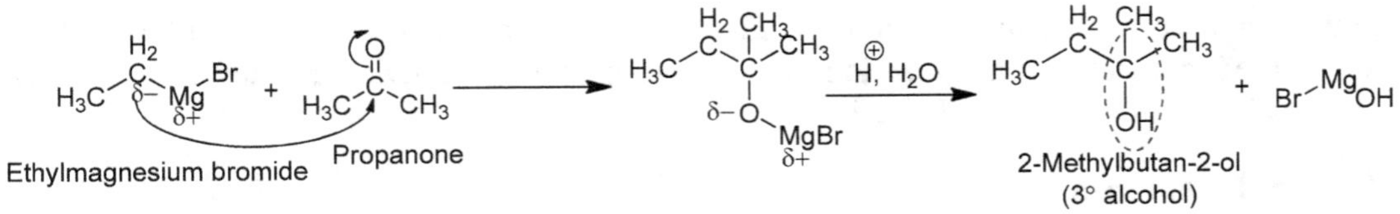

b) *One of the important reactions to form carbon-carbon bonds. For example: Synthesis of 4-ethyl-5-azatetracyclo[5.3.1.13,9.04,6]dodecane.*

(Z)-tricyclo[4.3.1.1^{3,8}]undecan-4-one oxime

4-ethyl-5-azatetracyclo[5.3.1.1^{3,9}.0^{4,6}]dodecane

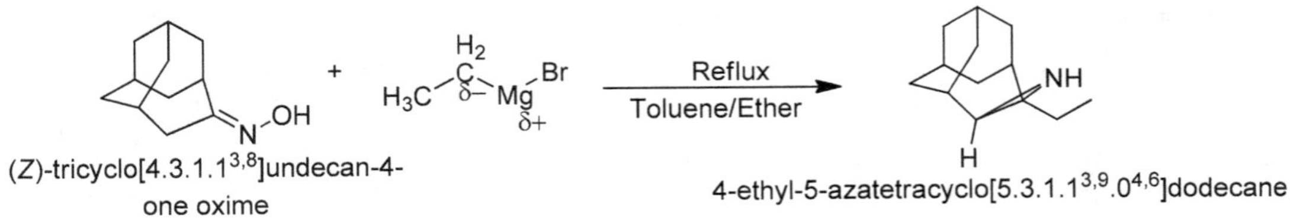

90. Hantzsch Pyridine Synthesis

Principle

Hantzsch reported (1882) condensation of β-dicarbonyl derivative with an aldehyde and ammonia to obtain dihydropyridine followed by oxidation with nitric acid to afford substituted pyridines. This reaction for synthesis of substituted pyridines is known as Hantzsch dihydropyridine synthesis.

General Reaction

β–diketo compound   acetaldehyde   β–diketo compound

+

ammonia

Δ

Dihydropyridine derivative

Conc. HNO$_3$

Substiuted pyridine

Mechanism

The reaction includes condensation of two important intermediates: the α-alkylidene or α-arylidene β-keto ester condensed from the β-keto ester and aldehyde, and the ester enamine originated from the β-keto ester and amine, as illustrated here.

Step 1: Formation of α-alkylidene β-keto ester.

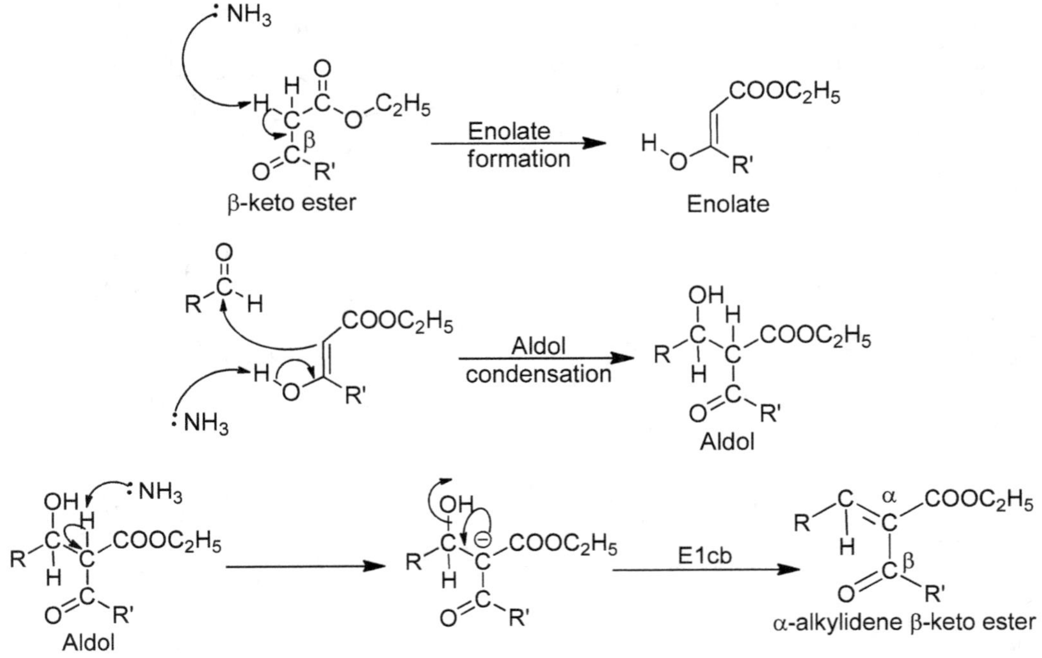

β-keto ester → Enolate formation → Enolate

Aldol condensation → Aldol

Aldol → E1cb → α-alkylidene β-keto ester

Step 2: Formation of ester enamine.

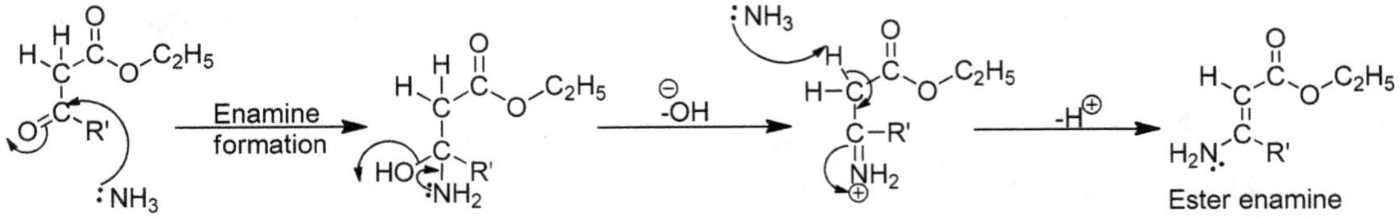

Enamine formation → $-OH$ → $-H^{\oplus}$ → Ester enamine

Step 3: Conjugate addition to form pyridine derivative.

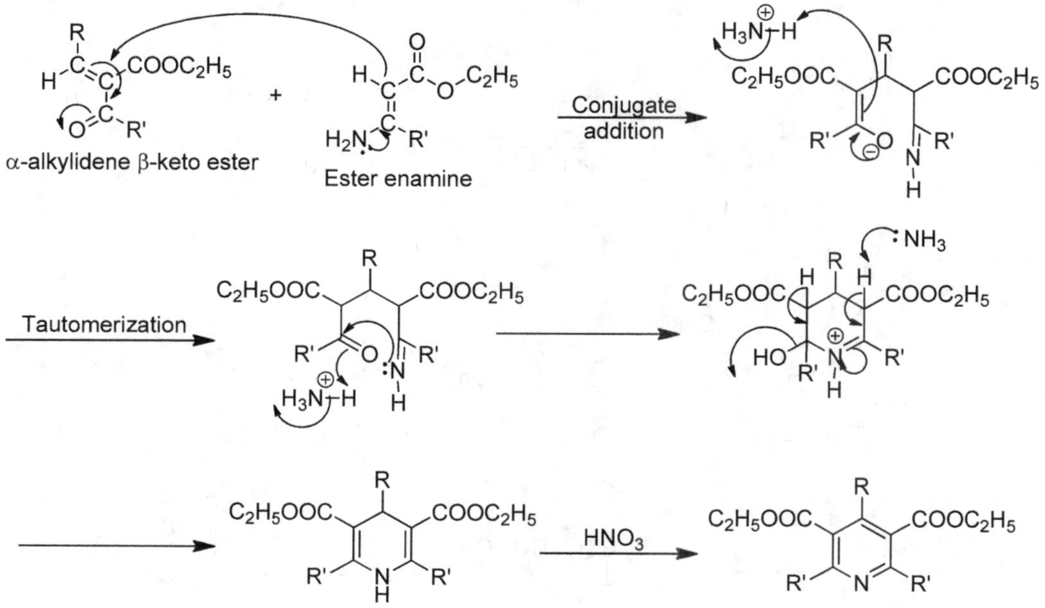

Applications

a) The reaction is useful in synthesis of substituted 1,4-dihydropyridines and corresponding pyridines. For example: Synthesis of diethyl 2,4,6-trimethylpyridine-3,5-dicarboxylate.

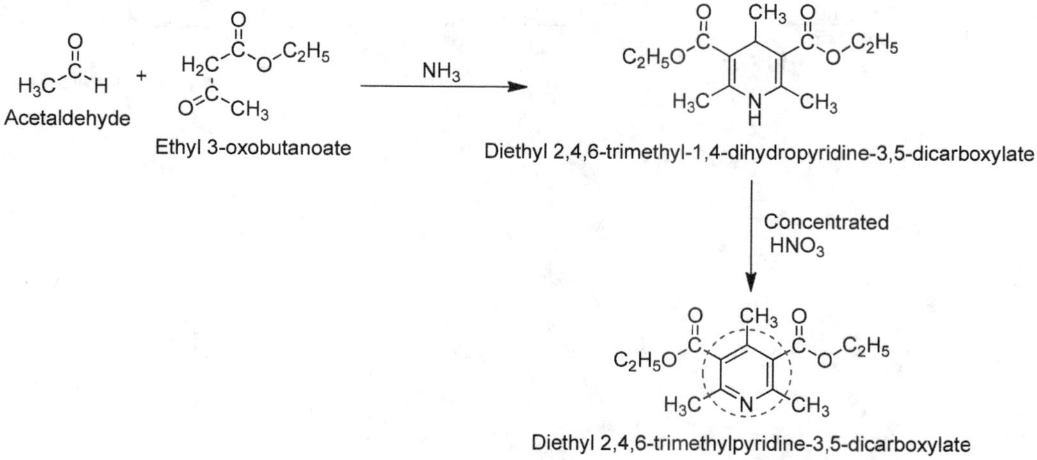

b) *The Hantzsch ester, as a calcium channel blocker.*

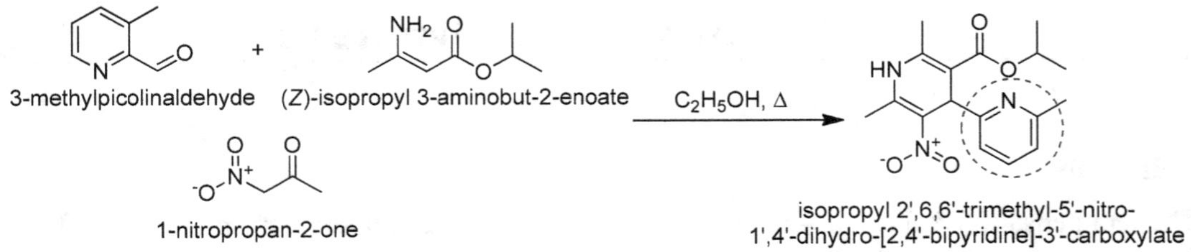

3-methylpicolinaldehyde   (*Z*)-isopropyl 3-aminobut-2-enoate C_2H_5OH, Δ

1-nitropropan-2-one

isopropyl 2',6,6'-trimethyl-5'-nitro-
1',4'-dihydro-[2,4'-bipyridine]-3'-carboxylate

91. Hantzsch Pyrrole Synthesis

Principle

The condensation of α-halo-ketones, β-ketoesters and ammonia or amines to afford 2,5-dialkyl or 2,4,5-trialkylpyrrole derivatives is commonly known as Hantzsch pyrrole synthesis. Synthesis of pyrrole derivatives was first reported by Hantzsch in 1890. In this reaction, ammonia or amine reacts with β-keto esters to produce enamine esters or 3-amino crotonates that cyclize with α-haloketones to obtain pyrrole derivatives upon heating. If the reaction is carried out using aromatic amines results in the formation of indoles, or carbazoles, if cyclized with α-halocyclohexanones.

General Reaction

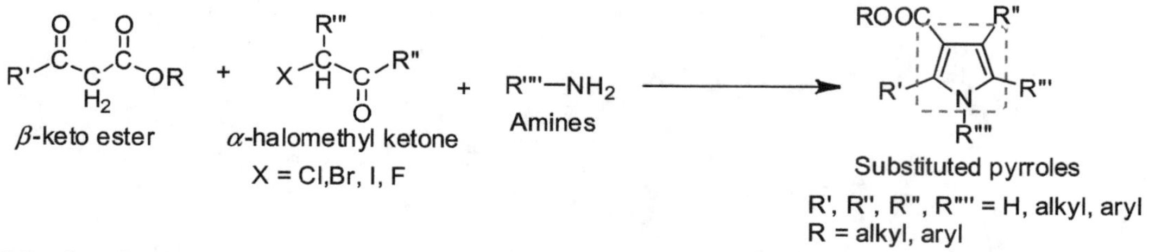
β-keto ester
α-halomethyl ketone
X = Cl,Br, I, F
Amines
Substituted pyrroles
R', R'', R''', R'''' = H, alkyl, aryl
R = alkyl, aryl

Mechanism

Step 1: Reaction between β-keto ester and primary amines; followed by reaction of intermediate with α-halomethyl ketone.

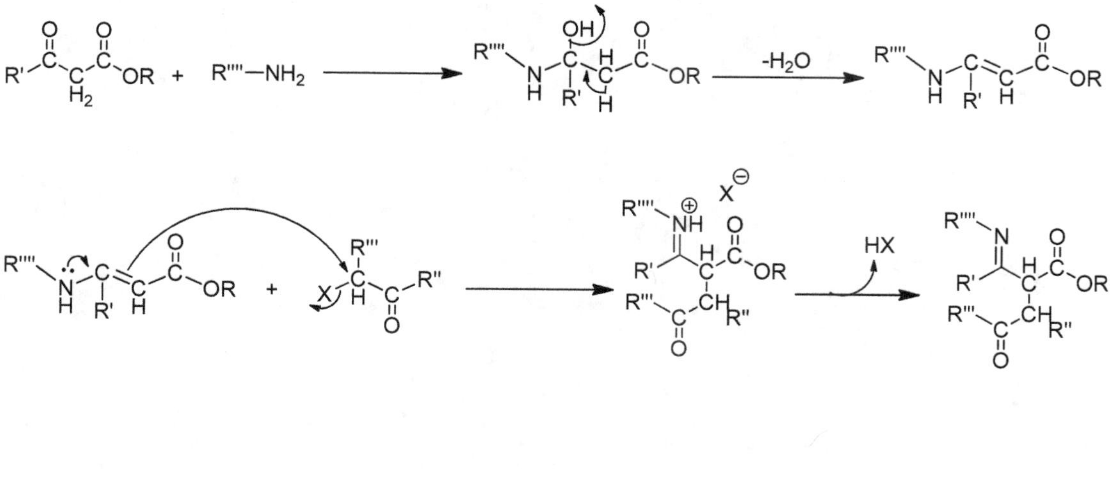
-H₂O
HX

Step 2: The formed intermediate undergoes isomerization and subsequent intramolecular cyclization to yield substituted pyrroles.

Applications

a) *The reaction has general applications in synthesis of 2,5-dialkyl- or 2,4,5-trialkylsubstituted pyrrole derivatives. The synthesized pyrroles have wide application in medicinal chemistry, conducting polymers, molecular optics, sensors, etc. For example: Synthesis of (3-methyl-1H-indol-2-yl)(phenyl)methanone.*

1-(2-aminophenyl)ethanone   2-chloro-1-phenylethanone   (3-methyl-1*H*-indol-2-yl)(phenyl)methanone

92. Hantzsch Thiazole Synthesis

Principle

Hantzsch and Weber (1887) reported synthesis of thiazoles *via* condensation of α-haloketones (or aldehydes) and thioamides, the reaction is known as Hantzsch thiazole synthesis. Other thio-ketone derivatives such as thiourea, dithiocarbamates, and ketone thiosemicarbazone also give different thiazoles by condensation with α-halo ketones (or aldehydes). The nucleophilicity of a sulfur atom in thioamides or thioureas is mainly responsible for high yield in this reaction for simple thiazoles but low yields are obtained for some substituted thiazoles, as of dehalogenation. At different reaction conditions formation of racemized thiazoles takes place that contain an enolizable proton at their chiral center, and it is the intermediate not the final product that is involved in the racemization. Therefore, some modifications have been made to reduce or even eliminate the epimerization upon thiazole formation.

General Reaction

R_1 = H, alkyl, aryl
R_2 = H, alkyl, aryl
R_3 = alkyl, aryl, NH_2, etc.

Thiazole analogues

Mechanism

The reaction has been proven to be a multistep process, and the intermediates have been isolated at low temperatures, in which the dehydration of cyclic intermediates seems to be the slow step.

Step 1: Isomerization of thioamide followed by reacting with base to give intermediate ion.

Step 2: Electron rich nitrogen of intermediate ion attacks at carbonyl carbon of α-haloketone followed by dehydration of cyclic intermediate to give thiazoles.

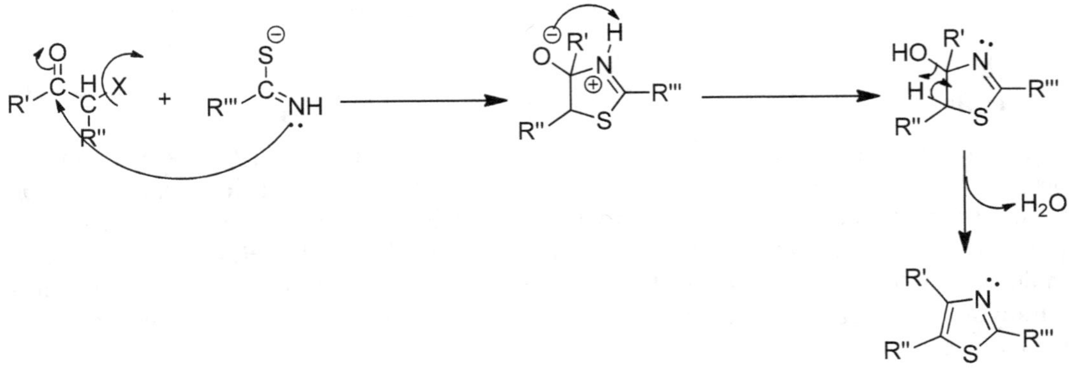

Applications

a) The synthesis of many substituted thiazoles is conveniently carried out by using Hantzsch protocol.

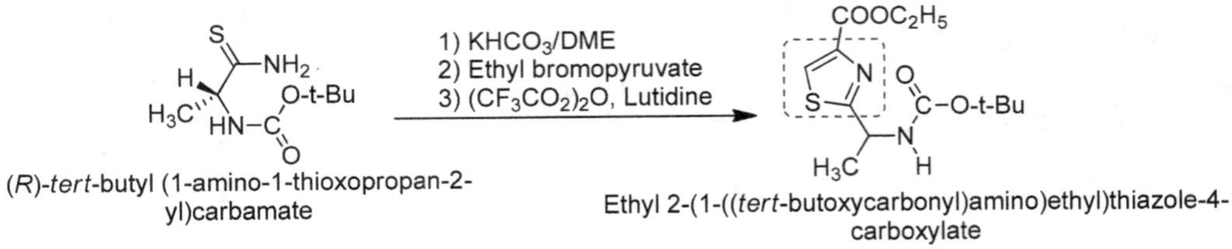

1) $KHCO_3$/DME
2) Ethyl bromopyruvate
3) $(CF_3CO_2)_2O$, Lutidine

(R)-*tert*-butyl (1-amino-1-thioxopropan-2-yl)carbamate

Ethyl 2-(1-(((*tert*-butoxycarbonyl)amino)ethyl)thiazole-4-carboxylate

Principle

The reaction was first reported by Haworth in 1932. The Friedel Crafts acylation of benzene with succinic anhydride gives 3-benzoylpropanoic acid (**I**), which on Clemmensen reduction forms 4-phenylbutanoic acid (**II**). Heating compound (**II**) in presence of concentrated H_2SO_4 results in the ring closure by elimination of water molecule to form α-tetralone (**III**). The Clemmensen reduction of α-tetralone yields tetrahydronapthalene (**IV**; tetralene), which after dehydrogenation with selenium gives napthalene.

General Reaction

Benzene + Succinic anhydride $\xrightarrow{AlCl_3}$ 4-oxo-4-phenylbutanoic acid (I) $\xrightarrow{Zn\text{-}Hg,\ HCl}$ 4-phenylbutanoic acid (II)

$\xrightarrow{H_2SO_4}$ 3,4-dihydronaphthalen-1(2H)-one (tetralone; III) $\xrightarrow{Zn\text{-}Hg,\ HCl}$ 1,2,3,4-tetrahydronaphthalene (tetralene; IV) $\xrightarrow{Se}$ Naphthalene

Mechanism

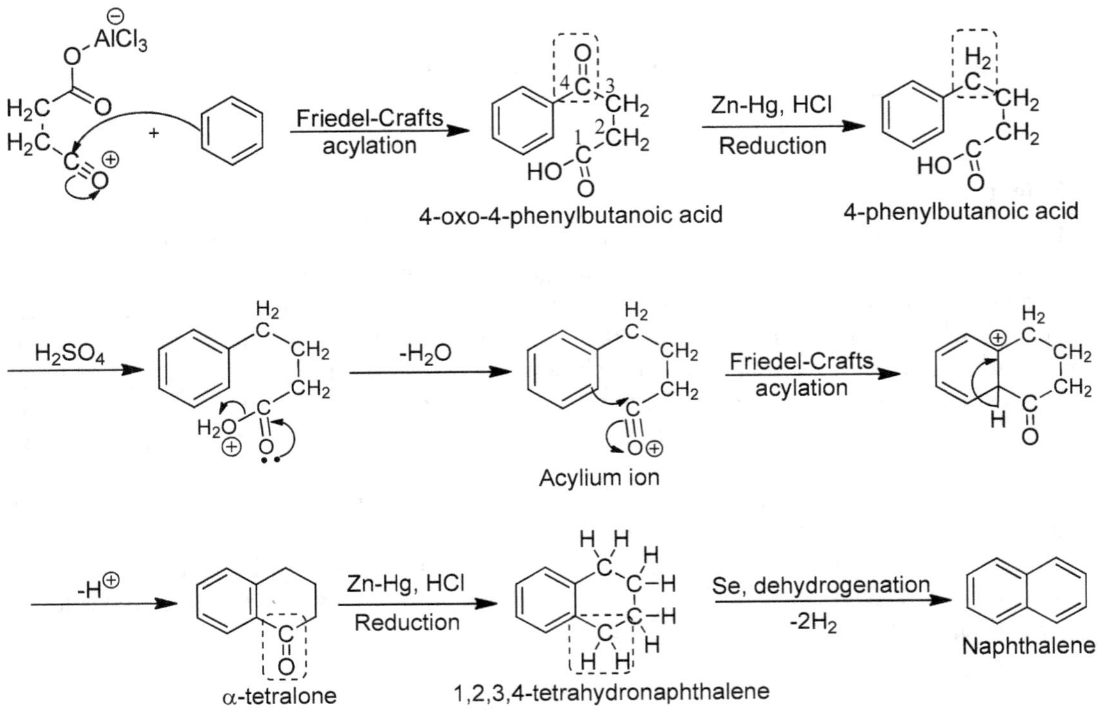

4-oxo-4-phenylbutanoic acid

4-phenylbutanoic acid

Acylium ion

α-tetralone

1,2,3,4-tetrahydronaphthalene

Naphthalene

Applications

a) This reaction is one of the classic methods for preparation of fused aromatics, including naphthalenes and phenanthrenes. For example: Synthesis of phenanthrene from naphthalene and succinic anhydride.

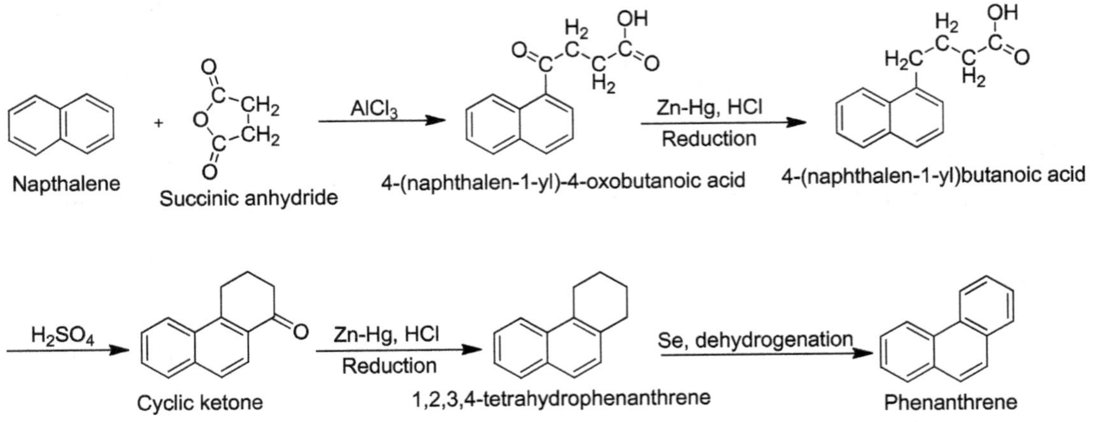

Napthalene

Succinic anhydride

4-(naphthalen-1-yl)-4-oxobutanoic acid

4-(naphthalen-1-yl)butanoic acid

Cyclic ketone

1,2,3,4-tetrahydrophenanthrene

Phenanthrene

94. Heck Reaction

Principle

Heck in 1968 carried out the palladium-catalyzed coupling between an aryl (or vinyl) halide and an alkene without allylic hydrogen to produce olefins is generally known as Heck reaction or Heck olefination. In addition, the intramolecular reaction to generate cyclic olefins is called as Heck cyclization. The Heck reaction is an influential method for formation of carbon-carbon bond in a single transformation using different aryl or vinyl halides and alkenes, which tolerate many functionalities, including amino, hydroxy, aldehyde, ketone, carboxy, ester, cyano, and nitro groups. The traditional Heck reaction uses 1-5mol% of a palladium catalyst along with a triphenylphosphine in the presence of a base (For example: soluble as triethylamine or insoluble as potassium carbonate or silver carbonate), which search for generated hydrogen halide in β-elimination step of a catalytic cycle. The reaction proceeds *via* Pd(0)/Pd(II) catalytic cycle, in which Pd(II) is reduced by a phophsine to Pd(0), a few cases that involve the catalytic cycle of Pd(II)/Pd(IV) also exist with some ligands.

General Reaction

Aryl halide + Alkene $\xrightarrow[\text{Base}]{\text{Pd-catalyst, Ligand}}$ Aromatic olefines

$(X = I, Br, Cl, SiR_3, SnR_3, CO_2H,$
$P(O)(OH)_2, B(OH)_2, OTf, N_2BF_4,$ etc.)

Mechanism

The mechanism of Heck reaction catalyzed by a neutral palladium complex is displayed below.

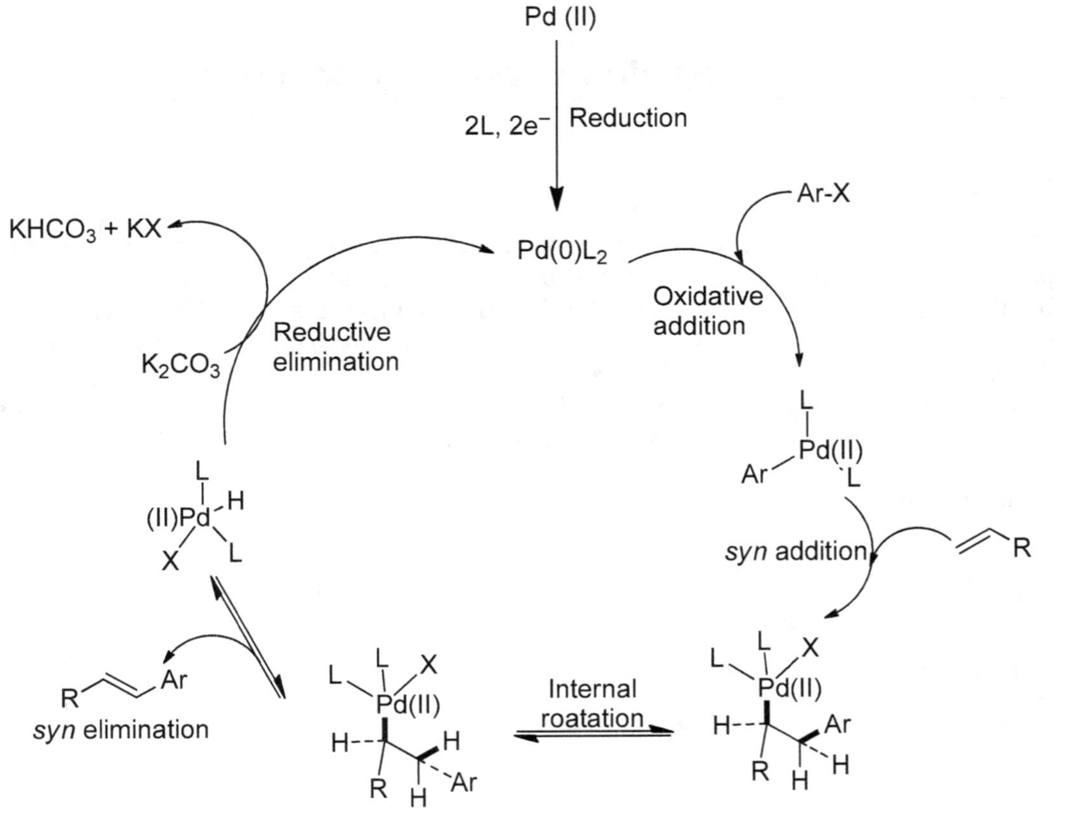

Pd (II)

2L, 2e⁻ | Reduction

Pd(0)L₂

Ar–X

Oxidative addition

Reductive elimination

KHCO₃ + KX

K₂CO₃

syn addition

Internal roatation

syn elimination

Applications

a) The reaction is useful for construction of carbon-carbon bonds to yield olefins. For example: Synthesis of methyl cinnamate.

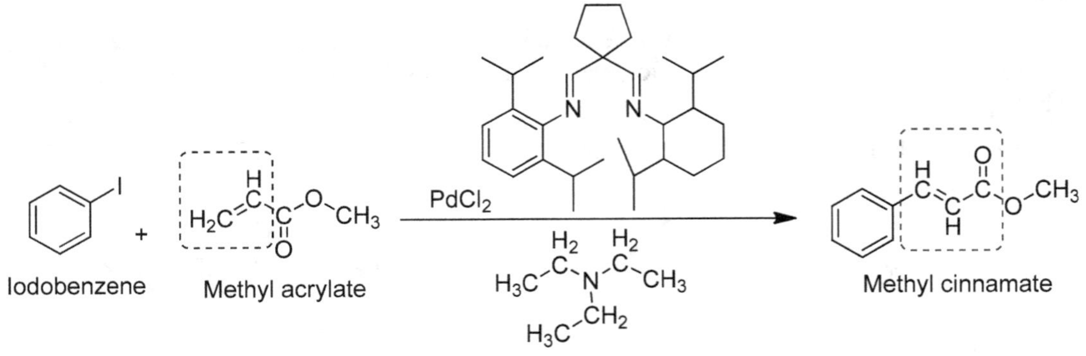

Iodobenzene   +   Methyl acrylate $\xrightarrow{\text{PdCl}_2}$ Methyl cinnamate

95. Hell-Volhard-Zelinsky Reaction

Principle

The reaction of aliphatic acid containing at least one α-hydrogen with bromine in presence of a catalytic amount of phosphorus or phosphorus tribromide to yield α-bromocarboxylic acid is known as Hell-Volhard-Zelinsky reaction. The reaction was reported by Hell in 1881 and subsequently explored by Volhard and Zelinsky in 1887. Reaction is synthetically useful because α-halogen can easily be replaced by other functional groups, such as cyanide, and the resulting α-cyano carboxylic acid yields malonic ester in presence of aqueous acid and ethanol.

General Reaction

Carboxylic acid   Tribromophosphine

α–bromocarboxylic acid

For example: Synthesis of 2-bromopropanoic acid

Propionic acid   Tribromophosphine   2-bromopropanoyl bromide   2-bromopropanoic acid

Mechanism

In this reaction, the formation of an enol takes place instead of ketene as intermediate analogous to the acid-catalyzed halogenation of ketones. The reaction is not a free radical chain process.

Step 1: Phosphorus tribromide or trichloride helps to convert the acid into an acyl halide.

Step 2: The acyl bromide undergoes enolization more readily than carboxylic acid. The reaction of bromine with enol form of acyl bromide to form α-bromo acid bromide is an important step of reaction.

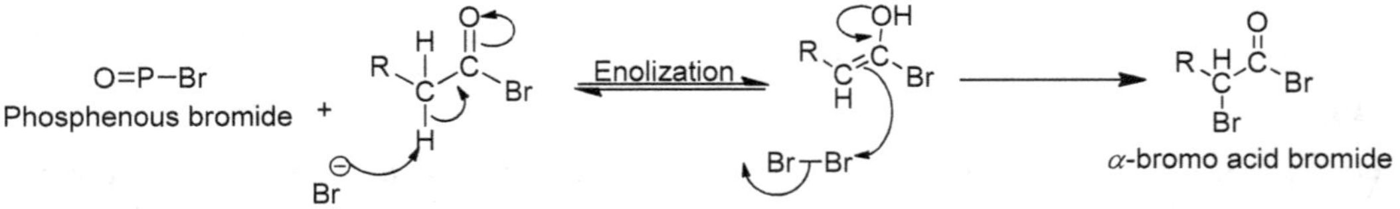

Step 3: Finally, the hydrolysis of α-bromo acid bromide gives α-bromo carboxylic acid.

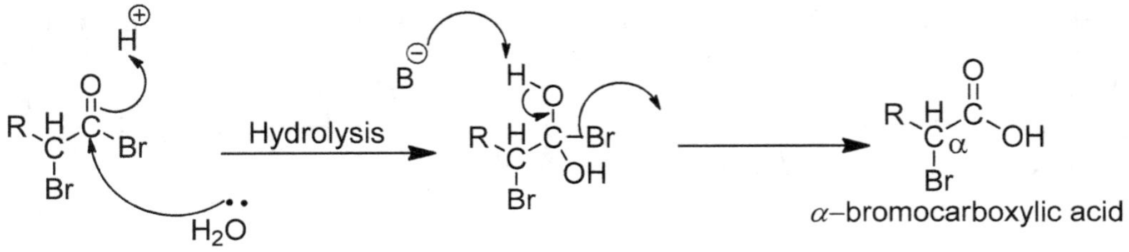

Applications

a) The synthesis α-halo aliphatic acids have been carried out by using this reaction. For example: Preparation of (1R,2R,4S)-2-bromobicyclo[2.2.1] heptane-2-carboxylic acid.

(1R,2S,4S)-bicyclo[2.2.1]heptane-2-carboxylic acid

tribromophosphine

(1R,2R,4S)-2-bromobicyclo[2.2.1] heptane-2-carboxylic acid

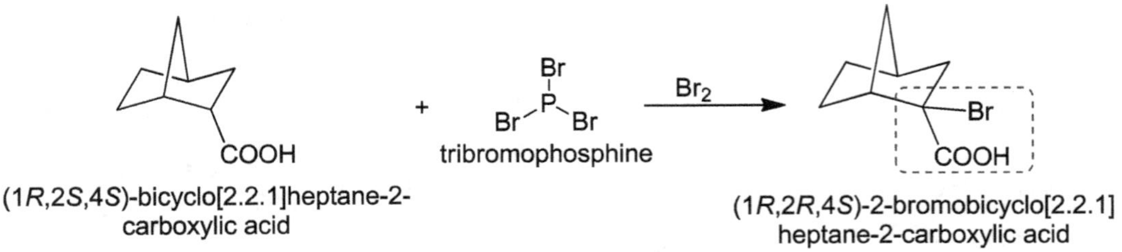

b) Synthesis of 2-bromohexadecanoic acid.

palmitic acid

tribromophosphine

2-bromohexadecanoic acid

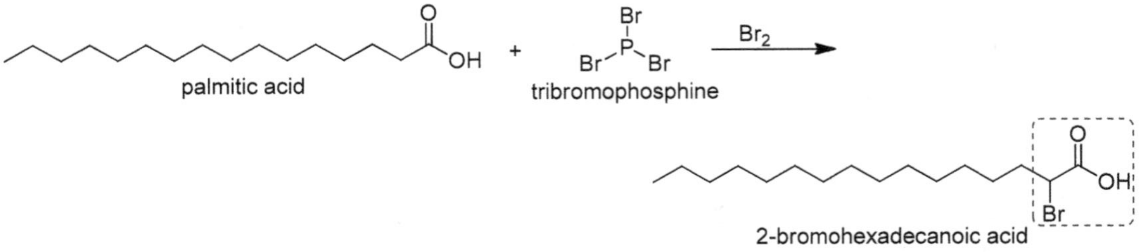

96. Henry Reaction (Nitro Aldol Reaction)

Principle

Henry (1895) reported the base-catalyzed condensation between a carbonyl compounds and a nitroalkane having at least one α-hydrogen atom to obtain 1,2-nitroalcohols, the reaction is known as Henry reaction and commonly referred to as nitroaldol reaction. The condensation between nitroalkane and imine is known as *aza*-Henry reaction or nitro-Mannich reaction. The resulting 1,2-nitroalcohol may further undergo reactions such as oxidation, reduction, and elimination. Due to the strong electron-withdrawing ability of nitro group, α-hydrogen can easily be deprotonated, even in presence of weak base, so that triethylamine or pyrrolidine is often used for this purpose.

General Reaction

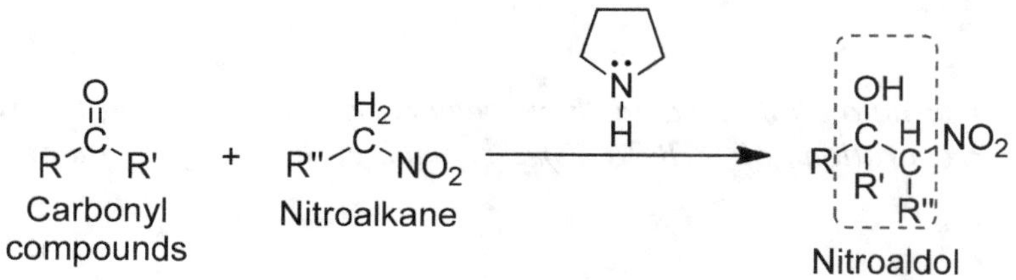

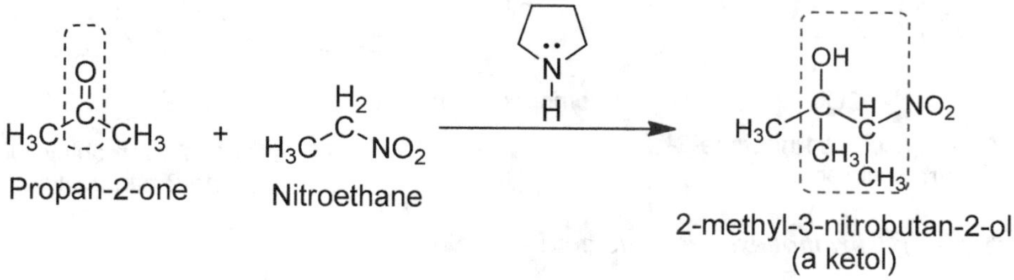

Mechanism

Step 1: Generation of carbanion.

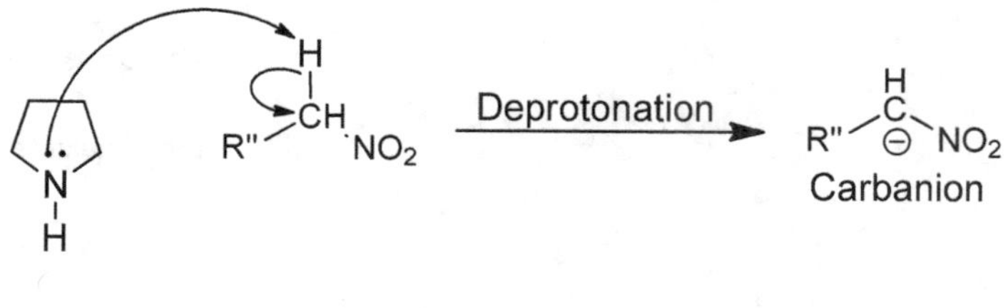

Step 2: Nucleophilic addition reaction of carbanion at carbonyl carbon atom to yield nitroaldol.

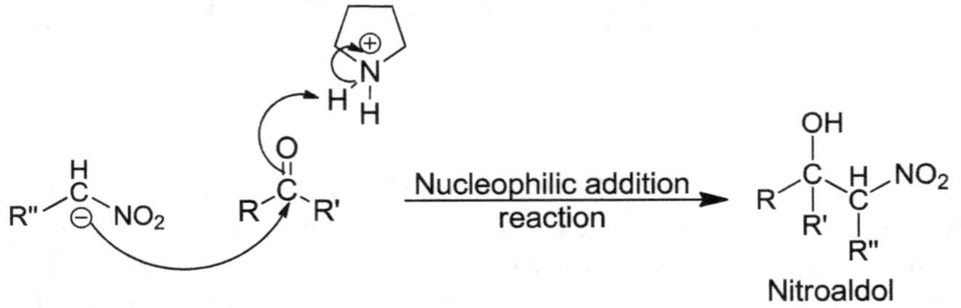

Nitroaldol

Applications

a)   The reaction is used for synthesis of nitroaldols, For example: Preparation of 1,2-nitro alcohol and 1,2-amino alcohol. Ethanal reacts with (nitromethyl)benzene in presence of pyrrolidine to give 1-nitro-1-phenylpropan-2-ol.

Ethanal

(nitromethyl)benzene

1-nitro-1-phenylpropan-2-ol
(an aldol)

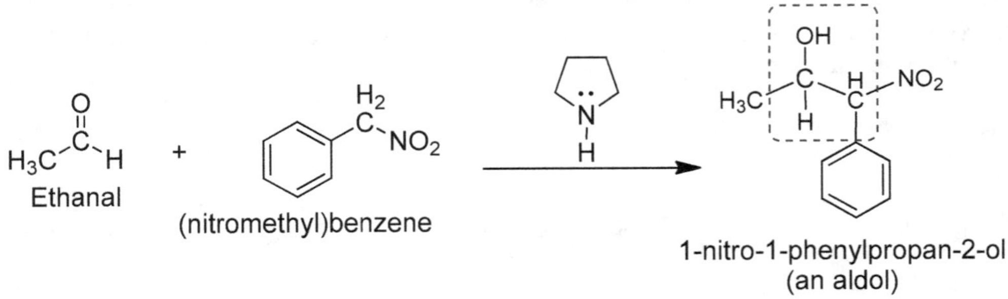

Principle

The reaction was reported by Glover at the Heron Island conference on reactive intermediates and unusual molecules, Heron Island, Australia in 1994, hence referred as Heron rearrangement. Heron is an acronym for heteroatom rearrangements on nitrogen. It is rearrangement of amides, with two heteroatoms substituted at nitrogen (RC(O)NXY), to esters and 1,1-diazenes through migration of oxygen from the nitrogen to carbonyl carbon. Amides undergoing rearrangement are referred as Heron amides. In this rearrangement, heteroatom (X) with pair of electrons initiates rearrangement from nitrogen to carbon through strong anomeric overlap with σ^* orbital of an adjacent N-heteroatom bond (N-Y), and anomeric interactions at nitrogen are enhanced when second heteroatom (Y) is highly electronegative. Derivatives of N,'N-diacyl-N,'N-dialkoxyhydrazines decompose on heating to esters and diazines through two consecutive rearrangements.

General Reaction

Amide   Secondary amine   →   Ester   1,1-diazines

X = Cl, OCOCH$_3$

Mechanism

Step 1:

Step 2:

Ester   1,1-diazenes

Applications

a) Heron rearrangement is useful for the synthesis of esters. For example: Synthesis of tert-butyl benzoate from N-(tert-butoxy)-N-chlorobenzamide.

Reagents: NaN_3, CH_3CN/H_2O; 87%

N-(*tert*-butoxy)-*N*-chlorobenzamide → *Tert*-butyl benzoate

Principle

Herz reported synthesis of 1,3,2-benzothiazathiolium chlorides in 1914. The reaction between primary aromatic amines and their salts, or *N*-acetyl analogues with sulfur monochloride either in presence or absence of an inert solvent (benzene, nitrobenzene, or acetic acid, etc.) to produce 1,3,2-benzothiazathiolium chlorides is commonly called as Herz reaction. The compounds formed in this reaction are known as Herz compounds or Herz salts. As Herz salts thermally decompose and are difficult to purify, hence converted into other derivatives. For example, acidic hydrolysis of Herz compounds results in 1,2,3-benzodithiazole 2-oxide, whereas basic hydrolysis by NaOH results in *o*-amino thiophenol, and treatment with acetic-formic anhydride results synthesis of substituted benzothiazoles.

General Reaction

Mechanism

Step 1:

Step 2:

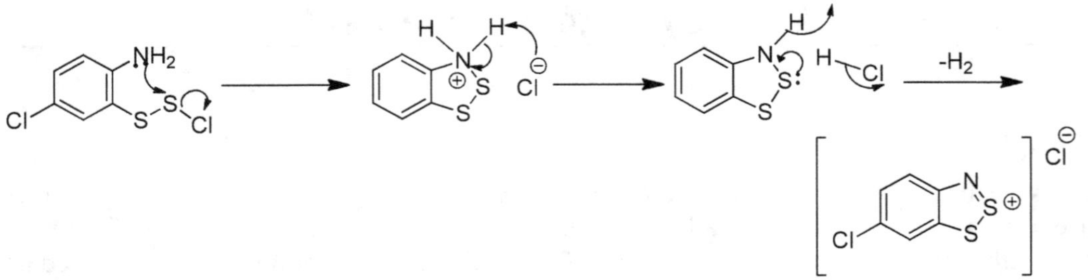

Applications

a) *The reaction is useful for synthesis of o-amino thiophenols and benzothiozoles. For example: Synthesis of 4,6-dichlorobenzo[d]thiazole from 2,4-dichloroaniline and Sulfurothious dichloride.*

2,4-dichloroaniline   Sulfurothious dichloride

4,6-dichlorobenzo[d]thiazole

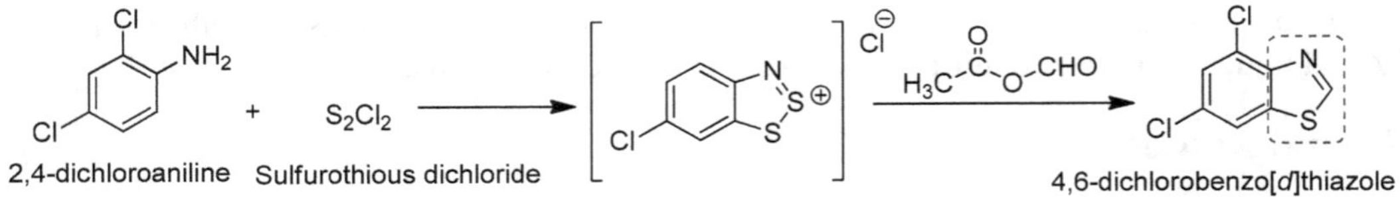

99. Heyns Rearrangement

Principle

Heyns reported the rearrangement of α-hydroxy imines (including α-hydroxy Schiff bases) into stable α-amino carbonyl compounds under non-reducing conditions is commonly called as Heyns rearrangement. The obtained α-ketoamines are also referred as Heyns products. The formation of imines or Schiff bases in this rearrangement is thermodynamically reversible process; however α-hydroxyl group facilitates the Heyns rearrangement, leading to the formation of stable α-ketoamines. However, under reducing conditions, such as in the presence of sodium cyanoborohydride, the imine or Schiff base is reduced to secondary α-hydroxyamine and no Heyns rearrangement occurs. In addition, the α-ketoamines can be destroyed by acidic hydrolysis, whereas α-hydroxyamines formed under reducing conditions are stable in acidic hydrolysis.

General Reaction

α-hydroxy carbonyl compounds + Primary amine $\rightleftharpoons$ α-hydroxy imines $\rightarrow$ α-amino carbonyl compounds

R, R', R" = Alkyl or aryl

Mechanism

Formation of Schiff base is rate-limiting step in this rearrangement followed by fast Heyns rearrangement.

Hydroxy enamine

Applications

a) Heyns rearrangement is widely applicable in food chemistry. For example: Synthesis of 3-hydroxy-17a-(2-methoxyethylamino)estra-1,3,5(10)trine-16-one.

3,16α-dihydroxy-estra-1,3,5(10)-trien-17-one

2-methoxyethanamine

33% CH$_3$OH/H$_2$O

51.9%

3-hydroxy-17α-(2-methoxyethylamino)
estra-1,3,5(10)trine-16-one

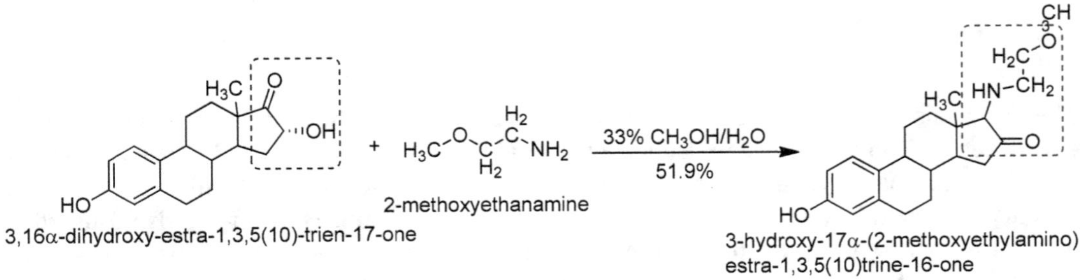

Principle

The synthesis of sulfonamides involving the treatment of primary or secondary amines with arenesulfonyl chloride (Benzenesulfonyl chloride) followed by alkylation of resulting sulfonamides from primary amines, and the hydrolysis of sulfonamides into primary or secondary amines is known as Hinsberg reaction. Hinsberg in 1890 use this method to synthesize secondary amines and explored the application of this reaction for separation of primary, secondary and tertiary amines.

Following are the results obtained after treatment of mixture of amines with benzenesulfonyl chloride (popularly known as Hinsberg reagent):

1) The primary amine reacts with benzenesulfonyl chloride to give sulfonamide, which on reaction with aqueous potassium hydroxide solution forms a water soluble salt.

2) The secondary amine also forms a sulfonamide but on treatment with aqueous sodium hydroxide it remains as an insoluble derivative.

3) The tertiary amine does not react with benzenesulfonyl chloride.

These arenesulfonyl chlorides have been also used to protect the amino groups. Because the formed sulfonamides are very stable and different methods have been tried to hydrolyze the sulfonamides; For examples: Reactions of sulfonamides with chlorosulfonic acid, 80% sulfuric acid, sulfuric acid, acetic acid, and hydrochloric acid in presence of phenol.

General Reaction

a) Reaction of primary amine with Benzene sulfonyl chloride

b) Reaction of secondary amine with Benzene sulfonyl chloride

c) Reaction of tertiary amine with Benzene sulfonyl chloride

3° Amine

Benzene sulfonyl chloride
(Hinsberg reagent)

Mechanism

The mechanism for preparation of secondary amines from primary amines is given below:

Step 1: Attack of electron rich primary amine over electron deficient sulfur on benzenesulfonyl chloride to give *N*-alkylbenzenesulfonamide.

Step 2: Treatment of *N*-alkylbenzenesulfonamide with base followed by S_N2 reaction with haloalkane gives *N,N'*-dialkylbenzenesulfonamide. Acidic workup of reaction separates primary amine and benzenesulfonic acid.

Applications

a) The reaction is widely helpful to protect primary and secondary amino groups. The reaction is also applicable for selective conversion of primary amines into secondary amines.

2-nitrobenzene-1-
sulfonyl chloride

1,2,3,4-tetrahydroquinoline

1-((2-nitrophenyl)sulfonyl)-
1,2,3,4-tetrahydroquinoline

8-((2-nitrophenyl)sulfonyl)-1,2,3,4-
tetrahydroquinoline

101. Hinsberg Sulfone Synthesis

Principle

The reaction between sulfinic acids and quinines to produce sulfonylquinol analogues was first reported by *Hinsberg* in 1894; hence this reaction is known as Hinsberg sulfone synthesis. Aqueous solutions and nonaqueous aprotic solvents can be used to carry out synthesis of sulfone. However, the reaction of unsymmetric quinone will form two isomeric products in *various* ratios, depending on the solvent and acidity of the reaction system. For example, the treatment of phenylsulfinic acid with 2-methyl quinone results in 2-methyl-5-phenylsulfonylhydroquinone (2,5-adduct) and 2-methyl-6-phenyl sulfonyl hydroquinone (2,6-adduct), in which 2,5-*adduct* predominates in acidic buffer at pH 4.5, and 2,6-adduct ruled in either more acidic (pH = 1) or more basic (pH=5.5) conditions. In addition, 2,6-adduct is also the major product in strong acidic two-phase systems and aprotic solvent systems (For example: acetone, or acetonitrile). In hydroalcoholic solution, the percentage of 2,6-adduct increases along with the decreasing of alcohol percentage in solution. It is assumed that protonated quinone is the actual species added to sulfinic acid, and the product ratio reflects the selectivity of protonation on carbonyl group.

General Reaction

Mechanism

Step 1: Formation of 2,5-adduct

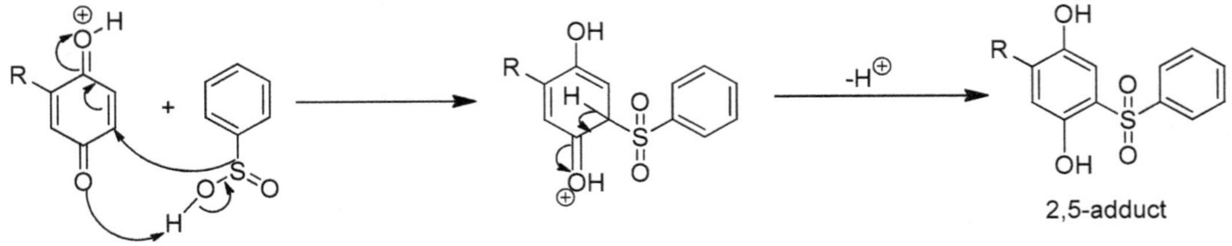

2,5-adduct

Step 2: Formation of 2,6-adduct

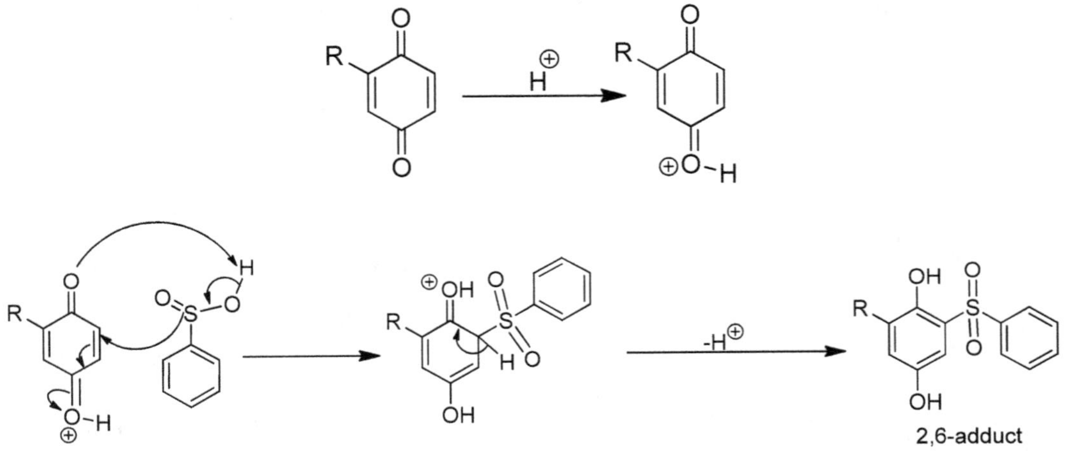

2,6-adduct

Applications

a) *The reaction is widely applicable for the synthesis of arylsulfonyl hydroquinones.*

For example: Synthesis of mixture of 2-((4-(2-bromoethyl)phenyl)sulfonyl)-6-methylbenzene-1,4-diol and 2-((4-(2-bromoethyl)phenyl)sulfonyl)-5-methylbenzene-1,4-diol

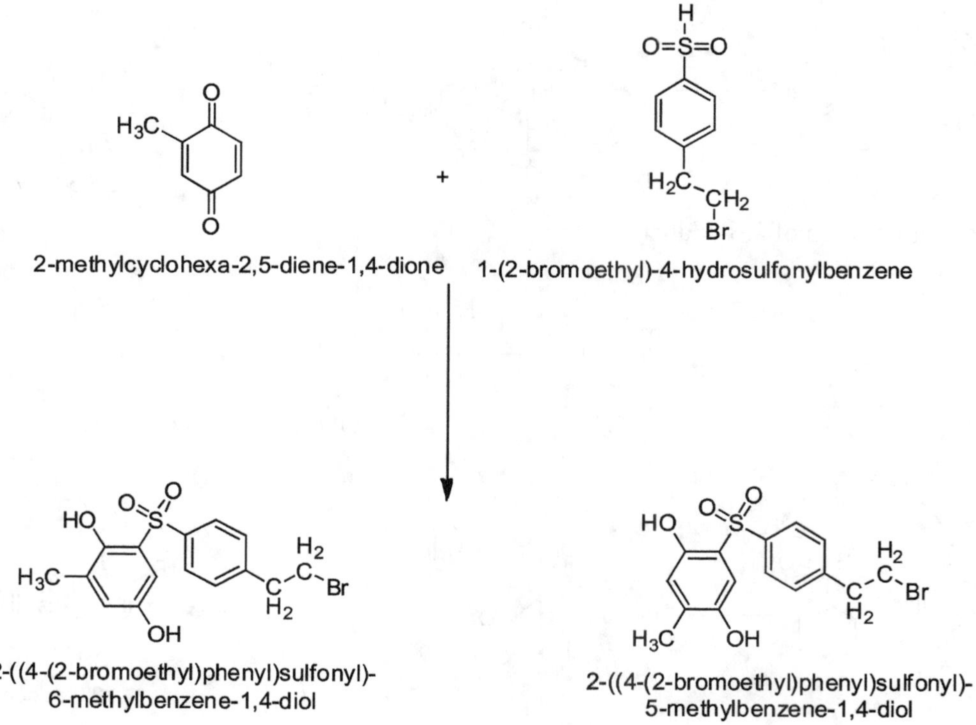

2-methylcyclohexa-2,5-diene-1,4-dione

+

1-(2-bromoethyl)-4-hydrosulfonylbenzene

2-((4-(2-bromoethyl)phenyl)sulfonyl)-
6-methylbenzene-1,4-diol

2-((4-(2-bromoethyl)phenyl)sulfonyl)-
5-methylbenzene-1,4-diol

102. Hinsberg Thiophene Synthesis

Principle

The reaction of 1,2-diketones and dialkyl thiodiacetate to afford thiophene carboxylic acids in presence of sodium or potassium alkoxide was reported by Hinsberg in 1910 and commonly known as Hinsberg thiophene synthesis. The reaction proceeds through formation of δ-lactone. The product was extracted by using hot hydroalcoholic alkaline solution. In addition, the reaction has been explored using 1,4-diketones; For example: the reaction between *o*-phthalaldehyde and diethyl thiodiacetate to yield 3-benzothiepin-2,4-dicarboxylic acid followed by desulfurization in ethanol to give naphthalene-2,3-dicarboxylic acid.

General Reaction

1,2-diketone   Diethyl 2,2'-thiodiacetate   3,4-disubstituted thiophene-2,5-dicarboxylic acid

Preparation of 3,4-diphenylthiophene-2,5-dicarboxylic acid

Benzil   Diethyl 2,2'-thiodiacetate   3,4-diphenylthiophene-2,5-dicarboxylic acid

Mechanism

Step 1: Diethyl 2,2'-thiodiacetate in presence of base lose proton to give carbanion which further attack at one of the carbonyl carbon of 1,2-diketone.

Step 2: Intramolecular cyclization to give δ-lactone.

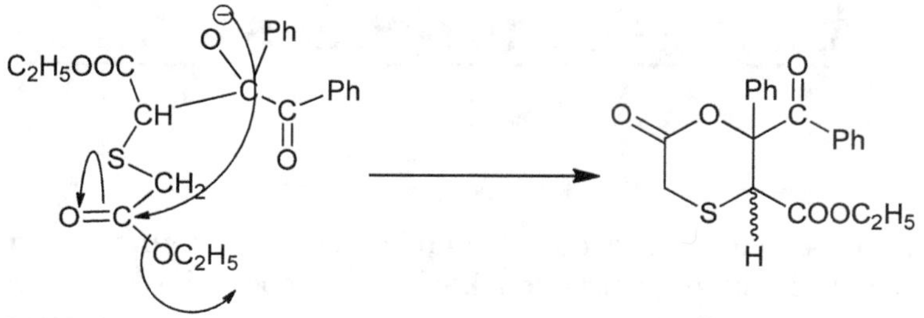

Step 3: Ethoxide ion abstracts proton results into ring opening of δ-lactone followed by ring closure to give 3,4-disubstituted thiophene-2,5-dicarboxylic acid.

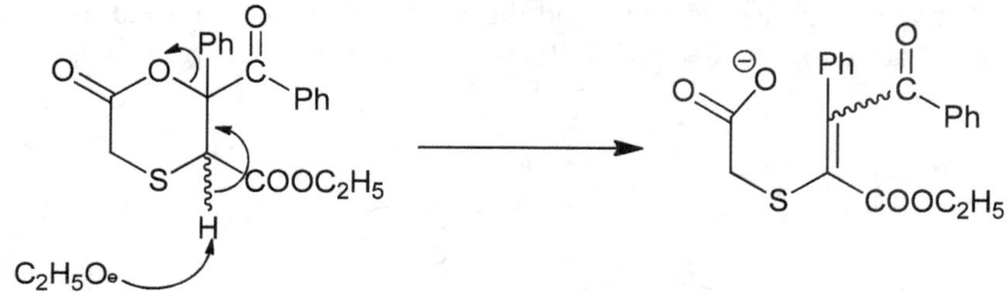

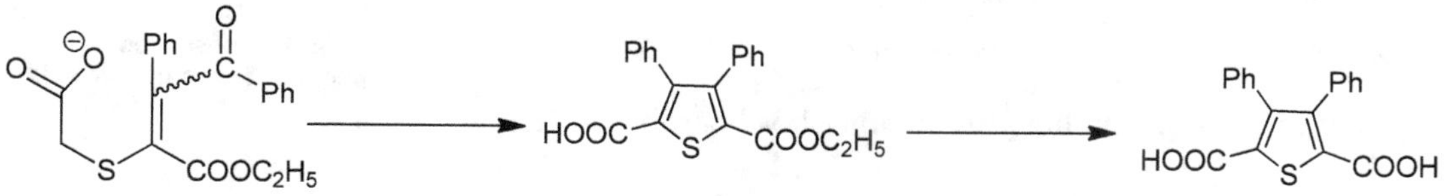

Applications

The reaction has been widely applicable in synthesis of many heterocyclic compounds mainly thiophene analogues as well as furan, selenophene, and pyrrole derivatives. Furan, selenophene, and pyrrole analogues are obtained by varying the ester from sulfur to oxygen, selenium, and nitrogen.

a) Synthesis of 5-cyano-3,4-dimethylthiophene-2-carboxamide.

Biacetyl   2-((isocyanomethyl)thio)acetonitrile   5-cyano-3,4-dimethylthiophene-2-carboxamide

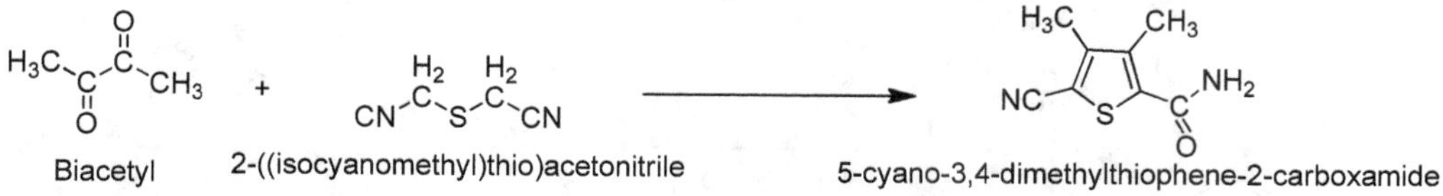

b) Synthesis of thiophene.

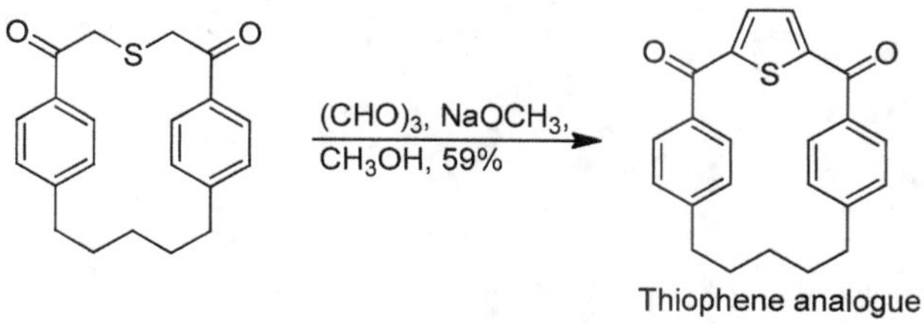

$(CHO)_3$, $NaOCH_3$, CH_3OH, 59%

Thiophene analogue

103. Hock Rearrangement

Principle

The rearrangement of hydroperoxides having unsaturated units connected to the carbon bearing hydroperoxide group into oxycarbonium ion, followed by nucleophilic attack by water, leading to the cleavage of C-C bond to form two carbonyl compounds. The protic acid (or Lewis acid) promoted rearrangement was reported by Hock and Schrader in 1936 is known as Hock rearrangement. The reaction occurs frequently in lipids and fatty acids, where allylic and dienylic hydroperoxides are generated upon radical hydrogen abstraction in presence of air or oxidation by $^1O_2$ (singlet oxygen).

General Reaction

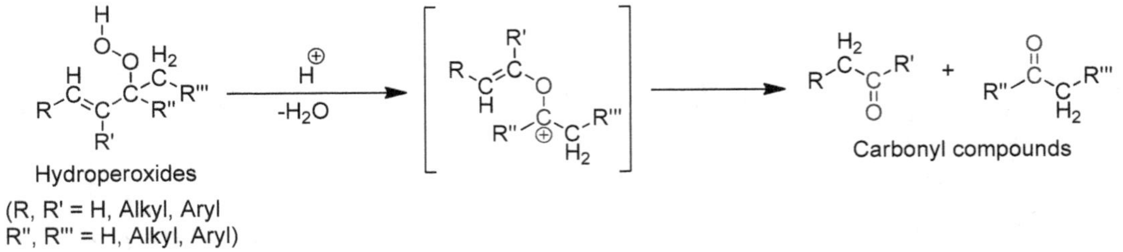

Hydroperoxides

(R, R' = H, Alkyl, Aryl
R", R''' = H, Alkyl, Aryl)

Carbonyl compounds

Mechanism

Step 1: Formation of oxycarbonium ion in presence of protic acid.

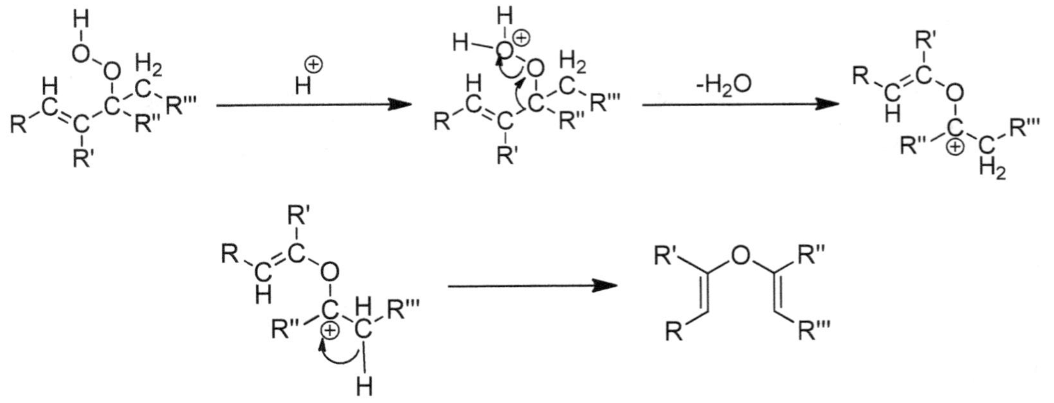

Step 2: Nucleophilic attack by water, leading to the cleavage of C-C bond to form two carbonyl compounds.

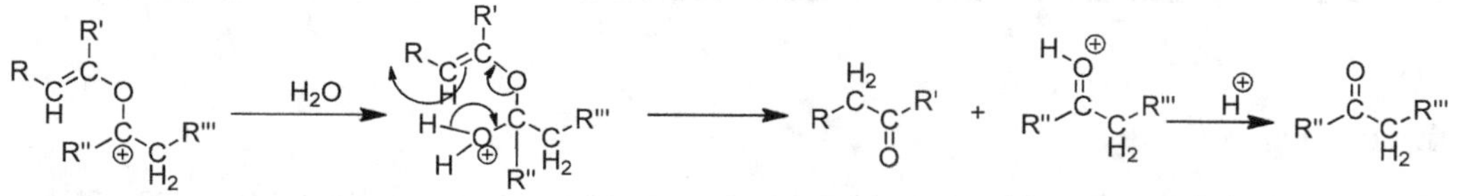

Applications

a) The degradation of long chain hydrocarbons with unsaturated unit(s) into carbonyl compounds has been carried out by using this reaction. For example: Degradation of (3-cyclopropylidenecyclobutyl) benzene into different carbonyl derivatives.

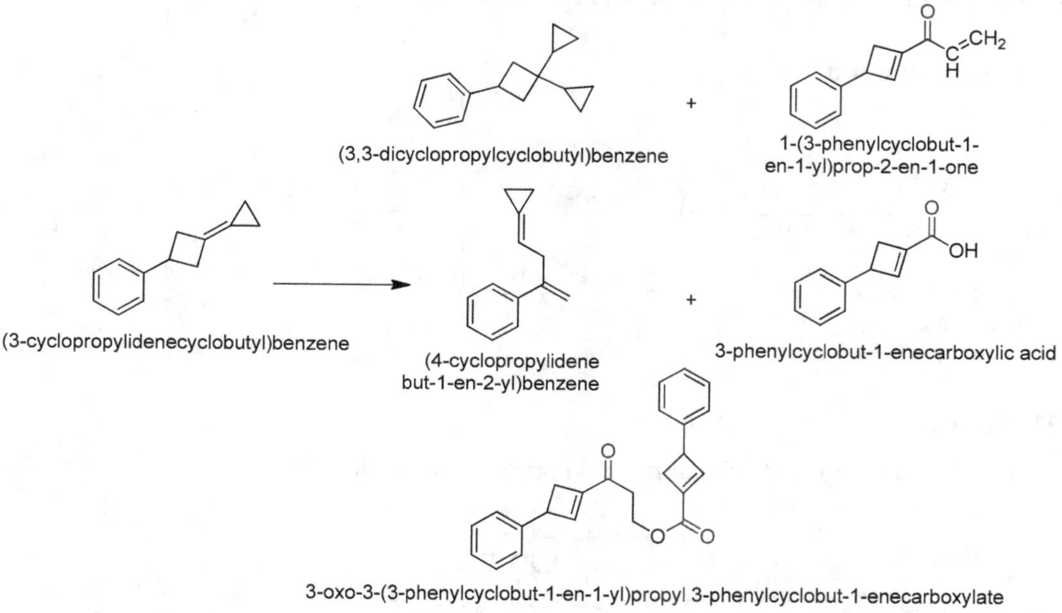

(3,3-dicyclopropylcyclobutyl)benzene

1-(3-phenylcyclobut-1-en-1-yl)prop-2-en-1-one

(3-cyclopropylidenecyclobutyl)benzene

(4-cyclopropylidene but-1-en-2-yl)benzene

3-phenylcyclobut-1-enecarboxylic acid

3-oxo-3-(3-phenylcyclobut-1-en-1-yl)propyl 3-phenylcyclobut-1-enecarboxylate

Principle

The reaction of amides with bromine in alkaline medium results in the formation of a primary amine containing one carbon atom less than the parent amide is known as Hoffmann degradation reaction. This degradation reaction was reported by Hofmann in 1881. It is widely used method for conversion of amides into primary amines. A unique feature of the outlined reaction is amines obtained in reaction have one carbon atom less than the initial reactant used. As this reaction involves migration of a group from carbonyl carbon to the adjacent electron deficient nitrogen atom, hence it is an example of molecular rearrangement reaction.

General Reaction

$$R-CONH_2 + Br_2 + KOH \longrightarrow R-NH_2 + K_2CO_3 + 2KBr + 2H_2O$$

An Amide   An amine

$$H_3C-CONH_2 + Br_2 + KOH \longrightarrow H_3C-NH_2 + K_2CO_3 + 2KBr + 2H_2O$$

Ethanamide   Methanamine

Mechanism

Hoffmann degradation proceeds *via* following steps:

Step 1: Halogenation of an amide to form *N*-bromoamide.

$$R-CONH_2 + Br_2 \longrightarrow R-CON(H)Br + HBr$$

Amide   *N*-bromoamide

Step 2: Abstraction of acidic hydrogen ion from *N*-bromoamide by hydroxide ion (base) to form *N*-bromoamide anion.

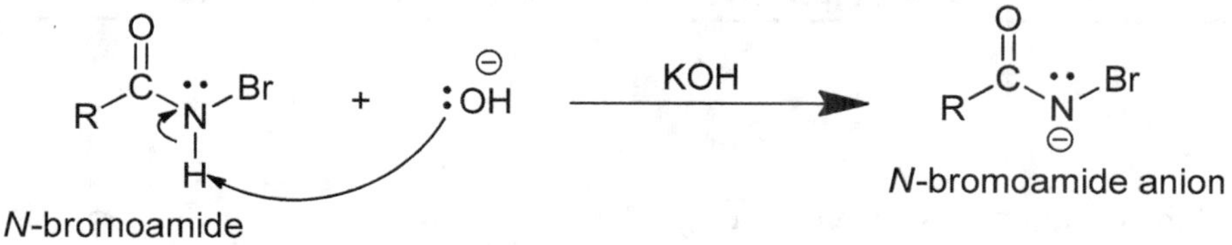

N-bromoamide

N-bromoamide anion

Step 3: Formation of electron deficient nitrogen atom by removal of electronegative bromide ion.

Step 4: 1, 2-migration of alkyl group is example of concerted mechanism. Step (**3**) and step (**4**) are believed to occur simultaneously, the attachment of R to nitrogen atom results in formation of isocyanate.

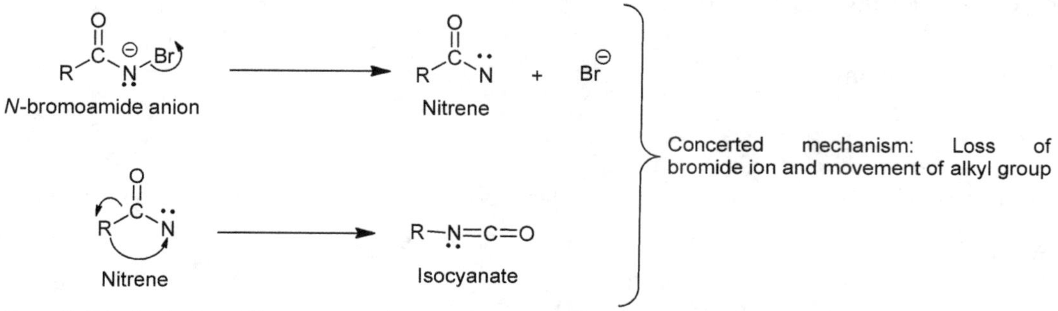

N-bromoamide anion

Nitrene

Nitrene

Isocyanate

$R{-}N{=}C{=}O$

Concerted mechanism: Loss of bromide ion and movement of alkyl group

Step 5: Hydrolysis of isocyanate intermediate to form an amine and carbonate ion.

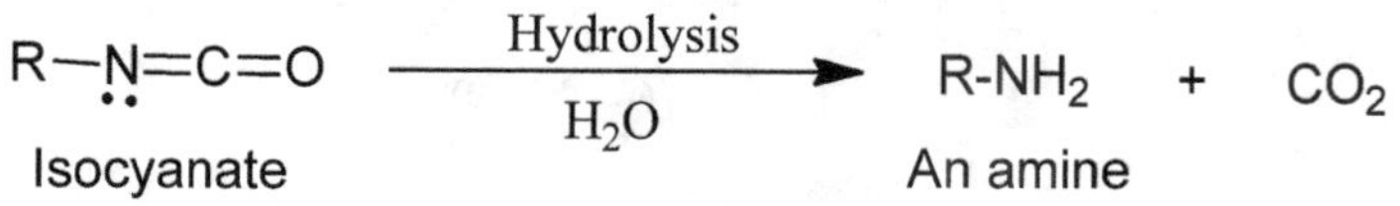

$$R{-}N{=}C{=}O \xrightarrow[\text{H}_2\text{O}]{\text{Hydrolysis}} R\text{-}NH_2 \; + \; CO_2$$

Isocyanate

An amine

Applications

a) *Preparation of primary amines*: *Hoffmann degradation reaction is applicable for conversion of carboxylic acid and its derivatives into primary amines containing one carbon atom less than the reactant used.*

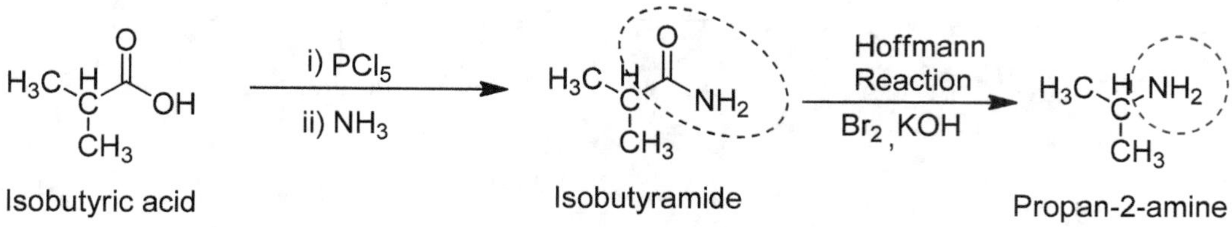

Isobutyric acid

Isobutyramide

Propan-2-amine

b)   *Preparation of β-aminopyridine*: *Nicotinamide is converted into β-aminopyridine by Hoffmann degradation reaction. Although β-aminopyridine is obtained from nitration of pyridine but by the Hoffmann degradation reaction it gives more yield.*

Nicotinamide   Pyridin-3-amine   (*β*-aminopyridine)

Principle

The thermal decomposition of quaternary ammonium hydroxide to form an alkene and quaternary amine is known as Hofmann elimination. As Hofmann invented this pyrolytic cleavage in 1851; it is popularly known as Hofmann decomposition or pyrolysis. A tetra-alkyl ammonium hydroxide $R(R')_3N^+$:OH on strong heating undergoes decomposition to produce water, a tertiary amine and an alkene (nature of alkene depends upon structure of alkyl group).

Statement: The Hofmann rule states that the major product will be an olefin having smallest number of attached alkyl groups; that is, the product of less thermodynamical stability. After extensive study, Hughes and Ingold proposed that the Hofmann rule was applicable to a bimolecular elimination (E2), which was controlled by polar factor from a positive charge in an onium ion; on the other hand the elimination that produces most substituted olefin follows Saytzeff rule, controlled by electronic factors. For most of aliphatic quaternary ammoni

um salts with a β-hydrogen atom *trans* to the amino group, the Hofmann elimination proceeds exclusively by anti-elimination, giving the olefins with less substituent's unless a substituent on the β-carbon enhances the acidity of that particular proton. The anti-elimination also predominates on six-membered rings. However, a few exceptions that precede E2 elimination *via* the *cis* transition state are also known, due to the enhanced acidity of proton, or conformational or steric hindrance. For example; Violation of Hofmann rule is the conversion of *trans*-2-phenylcyclohexyltrimethylammonium hydroxide into 1-phenylcyclohexene.

General Reaction

$$\left[\begin{array}{c} R\text{—}\overset{\overset{R}{|}}{\underset{\underset{R}{|}}{N^{\oplus}}}\text{—}\overset{\overset{H_2}{}}{C}\text{—}\overset{}{\underset{\underset{H}{}}{CH_2}} \end{array}\right] \overset{\ominus}{O}H \xrightarrow{\Delta} R\text{—}\overset{\overset{R}{|}}{\underset{\underset{R}{|}}{N}} \;+\; H_2C{=}CH_2 \;+\; H_2O$$

Quateranry ammonium hydroxides · · · · · · · Tertiary amine · · · Ethene

Quateranry ammonium hydroxides → Ethene (Major) + Prop-1-ene (Minor) + Tertiary amine

Mechanism

Heterolytic bond fission → Tertiary amine + $H_2C=CH_2$ (Ethene) + H_2O

Applications

a) *The rule is useful in predicting and rationalizing the structures of elimination products. For example: Degradation of quaternary ammonium hydroxides into methylenecyclohexane and 1-methylcyclohex-1-ene.*

Quaternary ammonium hydroxides → Methylenecyclohexane (Major) + 1-methylcyclohex-1-ene (Minor)

b) *Transformation of tert-pentyl acetate into 2-methylbut-1-ene and 2-methylbut-2-ene.*

tert-pentyl acetate → 2-methylbut-1-ene (Major) + 2-methylbut-2-ene (Minor)

106. Hoffmann-Martius Rearrangement

Principle

The pyrolysis of *N*-alkylated aromatic amines to *o*-alkylated and *p*-alkylated aromatic amines is reported by Hofmann and Martius in 1871. The *N*-alkylated aromatic amine hydrochlorides undergo pyrolysis followed by intermolecular alkylation, involving the migration of an alkyl group in the form of free radical, olefins or carbocation. The rearrangement reaction is known as Hoffmann-Martius rearrangement.

General Reaction

Mechanism

A cationic mechanism proposed for pyrolytic decomposiiton of *N*-methyl aniline hydrochloride, with the formation of methyl chloride as reactive intermediate.

Step 1: Formation of aniline in presence of aniline.

Step 2: Combination of chloromethane with aniline to give *o*-toluidine and *p*-toluidine.

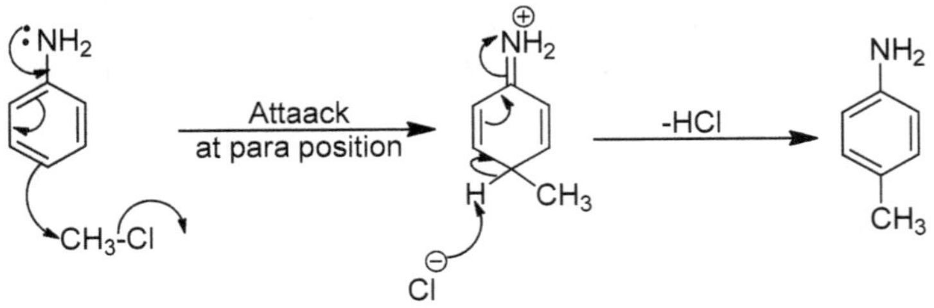

Reily-Hickinbottom rearrangement is a variation of the Hoffmann-Martius rearrangement in which Lewis acid is used instead of a protic acid. The reaction proceeds analogous to Hoffmann-Martius rearrangement.

For example: Decomposition of N-methyl aniline to p-toluidine and o-toluidine.

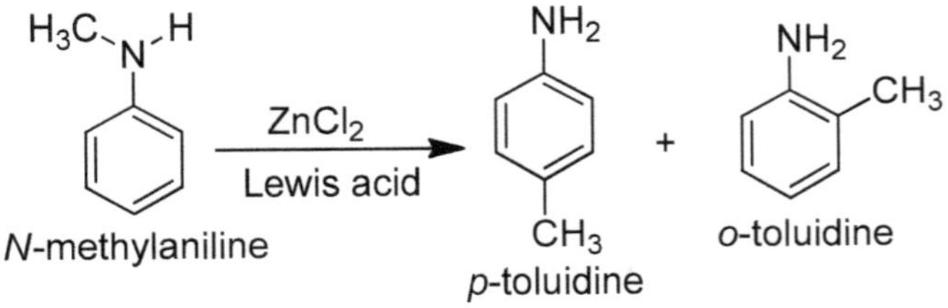

Applications

a) *The reaction is utilized for the synthesis of substituted aromatic amines. For example: Transformation of N-phenylbenzylamine into o-benzylaniline and p-benzylaniline.*

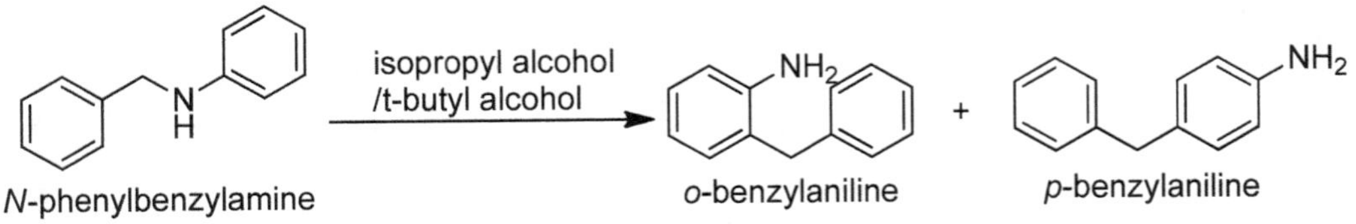

b) *Conversion of (E)-1-methyl-2-styrylpyridin-1-ium iodide into (E)-1-methyl-2-styrylpyridin-1-ium iodide and 2-((4-aminophenyl)(phenyl)methyl)aniline.*

(E)-1-methyl-2-styrylpyridin-1-ium iodide

2,4-diaminotriphenylmethane

+

2-((4-aminophenyl)(phenyl)methyl)aniline

107. Houben-Hoesch Reaction

Principle

Hoesch and Houben reported acylation of electron-rich aromatics such as multi-hydroxy phenols, their ethereal analogues, and heterocyclic compounds from organic nitriles *via* the intermediate of ketimines in presence of hydrochloric acid or Lewis acid or both is known as Houben-Hoesch reaction. When the same reaction proceeds intramolecularly; it is referred as Houben-Hoesch cyclization. Many different catalysts are useful for the reaction such as $ZnCl_2$, $AlCl_3$, the combination of BCl_3 and $AlCl_3$, super acids, and even the cation-exchange resin (Amberlite IR-120). Simple phenol does not undergo acylation and forms only a salt of iminomethyl phenyl ether; however, phenol can be *ortho-*acylated from nitriles by using combination of BCl_3 and $AlCl_3$ as catalyst. Houben-Hoesch reaction is Friedel-Crafts acylation with nitriles.

General Reaction

Resorcinol
(Phenol)

R–C≡N
Nitriles

HCl; AlCl$_3$
0 °C

$R-\overset{\oplus}{C}=NH_2$ HCl

H$_2$O

Acylated phenol

Synthesis of 1-(2,4-dihydroxyphenyl)ethanone from resorcinol and acetonitirle *via* ketimine chloride.

Resorcinol

H$_3$C–C≡N
Acetonitrile

HCl; AlCl$_3$
0 °C

$H_3C-\overset{\oplus}{C}=NH_2$ HCl

ketimine chloride

H$_2$O

1-(2,4-dihydroxyphenyl)ethanone

Mechanism

Step 1: Chelation with Lewis acid $AlCl_3$.

R–C≡N: + AlCl$_3$
Lewis acid
(Electron acceptor)

Chelation

$R-C≡\overset{\oplus}{N}-\overset{\ominus}{AlCl_3}$

Step 2: Electrophilic substitution reaction to form ketimines intermediate.

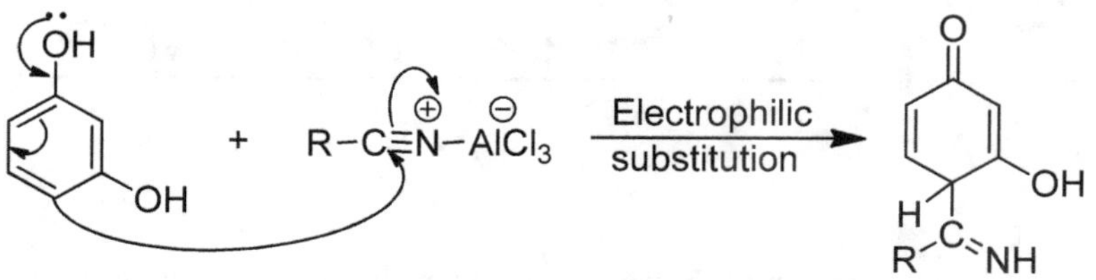

Step 3: Hydrolysis of ketimine hydrochloride and workup of reaction to form acylated product.

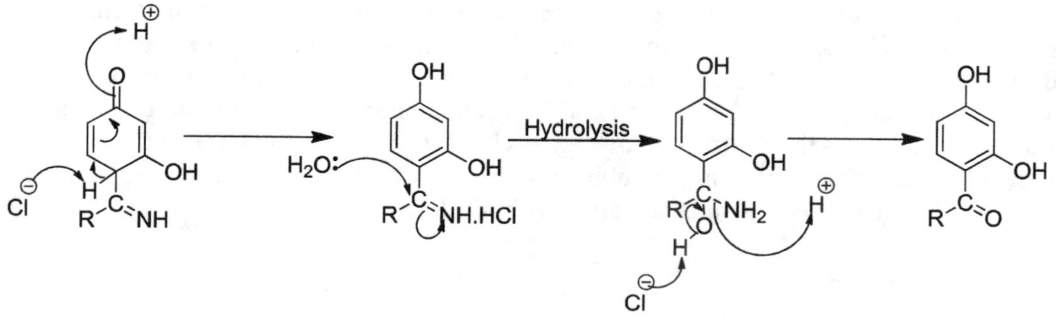

Applications

a) *The reaction is useful in synthesis of acylated phenols, their ethereal analogues, and some heterocycles.*

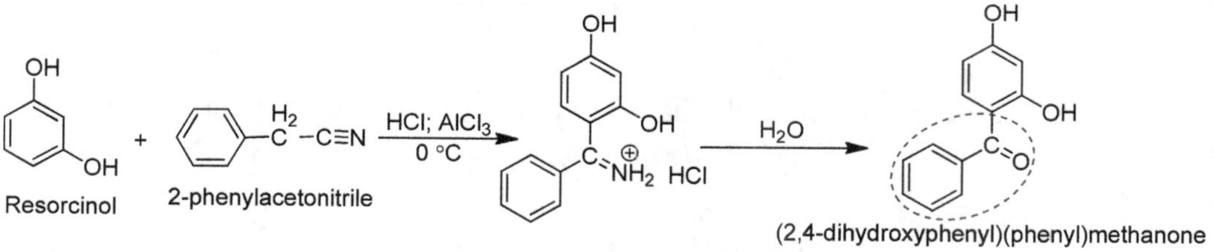

b) *Synthesis of coumarins. For example: The reaction of resorcinol with malonitrile in presence of catalyst to yield ketimines which on hydrolysis gives oxycoumarins.*

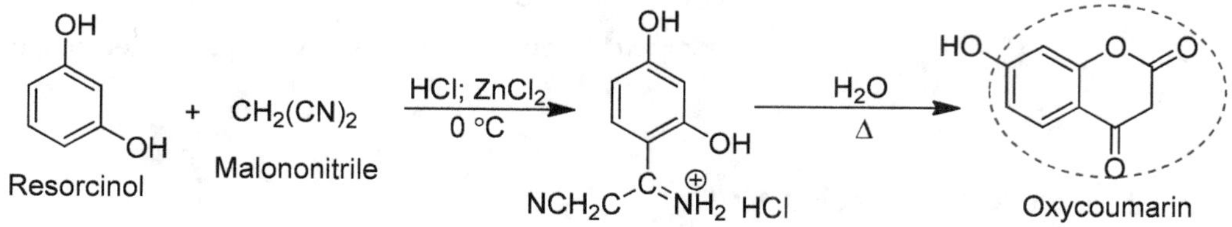

Principle

Hunsdiecker in 1942 reported an intramolecular condensation of 1,4-diketone (or γ-dicarbonyl compounds) in presence of base to yield 2,3-disubstituted cyclopentenones is known as Hunsdiecker condensation. When 2-acetyl ketone is treated with a base, two possible cyclopentenones (2-alkyl-3-methyl cyclopentenone and 3-alkyl cyclopentenone) were formed. But, only earlier has been reported as the isolated product because the intramolecular condensation occurs through reversible *Aldol Reaction*, and the transition state leading to the formation of 2-alkyl-3-methyl cyclopentenone is 2 Kcal/mol more stable than that needed to give 3-alkyl cyclopentenone.

General Reaction

NaOH, C_2H_5OH

2-acetyl ketone
(1,4-diketone)

2-alkyl-3-methyl cyclopentenone

3-alkyl cyclopentenone

Mechanism

Mechanism for intramolecular condensation of 1,4-diketone to form 2-alkyl-3-methyl cyclopentenone and 3-alkyl cyclopentenone in presence of sodium hydroxide is depicted below.

Application

a) *The reaction is widely applicable for the synthesis of substituted cyclopentenones. For example: Intramolecular condensation of (E)-undec-8-ene-2,5-dione to form (E)-3-methyl-2-(pent-2-en-1-yl)cyclopent-2-enone.*

(E)-undec-8-ene-2,5-dione $\xrightarrow{\text{NaOH, C}_2\text{H}_5\text{OH}}$ (E)-3-methyl-2-(pent-2-en-1-yl)cyclopent-2-enone

109. Hunsdiecker Reaction

Principle

The decomposition of silver salt of carboxylic acid on reaction with halogens (For example: bromine) into organic halides with one carbon atom less than the original acids is known as Hunsdiecker reaction. The reaction was reported by Borodine in 1861 for the synthesis of bromomethane from silver acetate and bromine, but Hunsdiecker extensively worked on this reaction to turn into practice. Hg (II), Ti (I), or Pb (IV) salt of the carboxylic acids also undergoes Hunsdiecker reaction. This reaction works for aliphatic acids with either long or short hydrocarbon chains, by which yields of haloalkanes decreases in the order of primary acid > secondary acid > tertiary acid.

General Reaction

$$R\text{-}COOAg + X_2 \xrightarrow{CCl_4} R\text{-}X + CO_2 \uparrow + AgX$$

Silver(I) salts of carboxylic acids   Halogens   Haloalkanes

$$H_3C\text{-}CH_2\text{-}COOAg + Br_2 \xrightarrow{CCl_4} H_3C\text{-}CH_2\text{-}Br + CO_2 \uparrow + AgBr$$

Silver(I) salt of propanoic acid   Bromine   Bromoethane

Mechanism

The reaction proceeds through free radical substitution reaction.

Step 1: Homolytic fission of halogen generates halogen free radicals.

$$X\text{-}X \xrightarrow[\text{fission}]{\text{Homolytic}} X^{\bullet} + X^{\bullet}$$

Free radicals

Step 2: Chain Initiation and generation of free radicals; acyl hypohalide undergoes homolysis.

$$R\text{-}COOAg + X^{\bullet} \longrightarrow R\text{-}CO\text{-}O\text{-}X + Ag\text{-}X$$

Step 3: Chain propagation and formation of haloalkanes; acyloxy radical loses carbon dioxide and resulting alkyl radical abstracts halogen to form haloalkane.

Applications

a) *The reaction widely used in preparation of aliphatic halides. For example: Synthesis of (E)-(2-bromovinyl)benzene from cinnamic acid.*

Cinnamic acid

1.05 eq. NBS, 0.2 eq. LiOAc
CH_3CN/H_2O (12:1)
MW, 200W, 1-2 min

(*E*)-(2-bromovinyl)benzene

b) *Synthesis of bromocyclopropane from cyclopropanecarboxylic acid.*

Cyclopropanecarboxylic acid

$2 \text{ cyclopropanecarboxylic acid} + 2 Br_2 \xrightarrow{HgO} \text{bromocyclopropane} + HgBr_2 + 2CO_2 + H_2O$

Bromocyclopropane

Principle

The transformation of polysubstituted aromatics into corresponding sulfonated aromatics with the migration of substituents in presence of concentrated sulfuric acid was reported by Jacobsen in 1886, hence this rearrangement is known as Jacobsen rearrangement. The halogenated polyalkylaromatics undergo disproportionation of halogen atoms under conditions of Jacobsen rearrangement. Only the tetraalkyl or pentaalkyl benzenes undergo this rearrangement, in which pentaalkyl benzenes produce 1,2,3,4-tetraalkyl benzenesulfonic acid and hexaalkyl benzenes, and the tetraalkyl benzenes intramolecularly rearrange to yield 1,2,3,4-tetralkyl benzenesulfonic acid as major product. It was found that tetra-ethylbenzenes rearrange more smoothly and quickly than tetramethylbenzenes; again penta-methylbenzene undergoes this reaction more readily than penta-ethylbenzene. In contrast, hexaalkyl benzenes, as the end point of the reaction, are stable toward concentrated sulfuric acid, even if they are in contact with concentrated sulfuric acid for several months. However, for less alkyl-substituted benzenes, such as trialkyl benzenes and dialkyl benzenes, no rearrangement occurs but sulfonation of aromatics was found. In addition, the Jacobsen rearrangement not only works for polyalkyl aromatics but also for halogenated aromatics as well as for mono-halogenated aromatics.

General Reaction

1,2,4,5-tetramethylbenzene   Conc. H_2SO_4 2,3,4,5,6-pentamethylbenzenesulfonic acid   Conc. H_2SO_4 1,2,3,5-tetramethylbenzene

Mechanism

The rearrangement involves the sulfonation and protonation of an aromatic ring.

Step 1: Formation of 2,3,5,6-tetramethylbenzenesulfonic acid in presence of concentrated sulfuric acid.

Step 2: Conversion of 2,3,5,6-tetramethylbenzenesulfonic acid into 2,3,5 trimethylbenzenesulfonic acid with the removal of methyl group.

Step 3: Formation of tertiary carbocation intermediate with elimination of methyl group which on reaction with concentrated sulfuric acid yield final product.

3°Carbocation intermediate

1,2,4,5-tetraethylbenzene

Conc. H_2SO_4

31.5%

1,2,3,4-tetraethylbenzene

Applications

The reaction is applicable in synthesis of polysubstituted alkyl aromatics with substituents in vicinal position, which is more difficult to synthesize from conventional Friedel-Crafts alkylation.

a) Rearrangement of 6-methyl-7-propyl-1,2,3,4-tetrahydronaphthalene.

6-methyl-7-propyl-1,2,3,4-
tetrahydronaphthalene

Conc. H_2SO_4

5-methyl-6-propyl-1,2,3,4-
tetrahydronaphthalene

+

6-methyl-5-propyl-1,2,3,4-
tetrahydronaphthalene

111. Jones Oxidation

Principle

Jones and his colleagues in 1946 reported oxidation of primary alcohols into carboxylic acids or transformation of secondary alcohols into ketones using chromic acid and the reaction is popularly known as Jones oxidation or Jones chromic acid oxidation. Chemically Jones reagent is CrO_3-H_2SO_4-H_2O. Reaction occurs smoothly in aqueous acetone and is fast with low selectivity. This reaction involves radical intermediate (conversion of Cr(IV) into Cr(II) species), but the predominant mechanism is transformation of Cr(VI) into Cr(IV) species. Furthermore, olefins in acetone can also be oxidized by the Jones reagent, especially in the presence of 20 mol% mercuric acetate or mercuric propionate, so that terminal olefins are oxidized to methyl ketones, and 1,2-disubstituted olefins are converted into ketones.

General Reaction

Primary alcohols $\xrightarrow{CrO_3,\ H_2SO_4}$ Carboxylic acids

Secondary alcohols $\xrightarrow{CrO_3,\ H_2SO_4}$ Ketones

R, R' = Alkyl, Aryl

Mechanism

Oxidation of a secondary alcohol in presence of chromic acid is depicted below.

Step 1: Generation of oxonium ion as intermediate by reaction of chromic acid with alcohol.

Step 2: Dissociation of intermediate to form oxidized product (carbonyl derivative).

Applications

a) Oxidation of secondary alcohol into ketone analogues.

CrO₃, H₂SO₄
Acetone, H₂O
73%

1-((R)-1-methylpyrrolidin-2-yl)-
3-((S)-1-methylpyrrolidin-2-yl)propan-2-ol

1-((R)-1-methylpyrrolidin-2-yl)-3-((S)-
1-methylpyrrolidin-2-yl)propan-2-one

112. Kharasch Addition Reaction

Principle

In presence of peroxides the addition of hydrogen bromide (polar reagent) to unsymmetrical alkenes results in the formation of product where bromide ion adds to a carbon of double bond having maximum number of hydrogen atoms to give the bromoalkane is popularly known as Kharasch effect or Kharasch addition reaction. It is also referred as *anti*-Markonikov's addition reaction. The peroxide effect is observed only in reaction with HBr and not with HI, HF, or HCl. The order of bond strength of haloacids is HF > HCl > HBr > HI. HF and HCl have high bond strength and polarity, so homolytic cleavage does not occur easily. In HBr and HI, homolytic cleavage occurs readily to produce bromide and iodine free radicals. However, iodine radical undergoes coupling reaction with other iodine radical to produce iodine. Thus, only HBr can provide bromine free radicals easily and is available for addition to olefinic carbon atoms.

General Reaction

For example: The reaction of HBr with an alkene in presence of peroxide, results in the formation of bromoalkane, where addition of HBr occurs anti to the Markonikov's rule. Reaction of HBr with propene gives 1-bromopropane, an anti-Markonikov's addition product in presence of peroxides.

Mechanism

The addition of HBr in presence of peroxide is a free radical addition. Peroxides are very good free radical generators. In peroxides, the oxygen-oxygen single bond undergoes homolytic fission to generate free radicals. The reaction proceeds *via* following steps:

Step 1: Generation of free radicals.

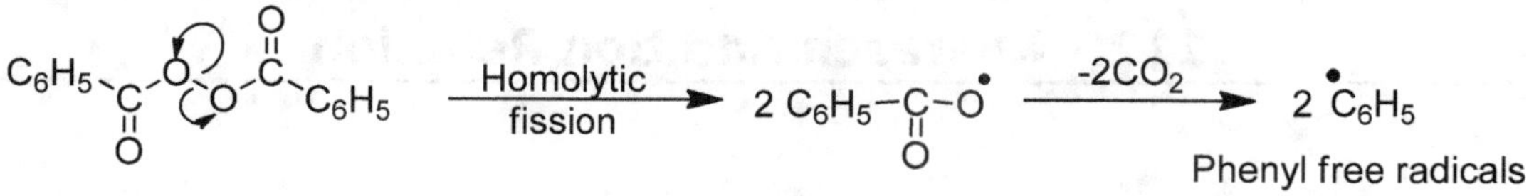

$$C_6H_5-\overset{O}{\underset{O}{C}}-O-O-\overset{O}{\underset{O}{C}}-C_6H_5 \xrightarrow[\text{fission}]{\text{Homolytic}} 2\ C_6H_5-\overset{\bullet}{\underset{O}{C}}-O^{\bullet} \xrightarrow{-2CO_2} 2\ \overset{\bullet}{C_6H_5}$$

Phenyl free radicals

Step 2: Generation of bromine free radical, The C_6H_5 abstracts H^+ from HBr to form benzene. This results in the generation of bromine free radical.

$$\overset{\bullet}{C_6H_5} \quad + \quad H-Br \longrightarrow C_6H_6 \quad + \quad \overset{\bullet}{Br}$$

Homolytic
fission

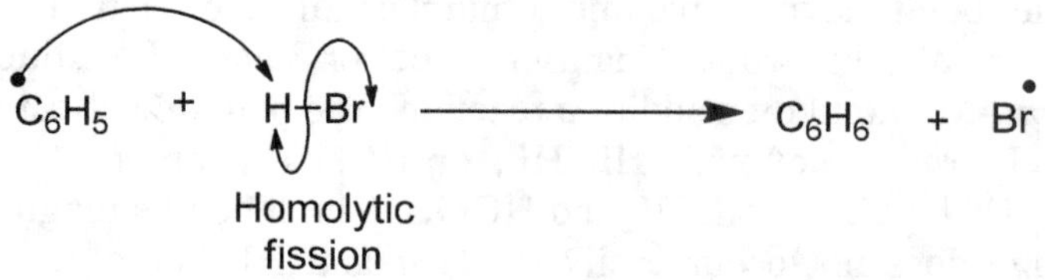

Step 3: Addition of bromine free radical into olefinic carbon atom. Here, more stable free radical (2°) is favored which further reacts with HBr to yield 1-bromopropane.

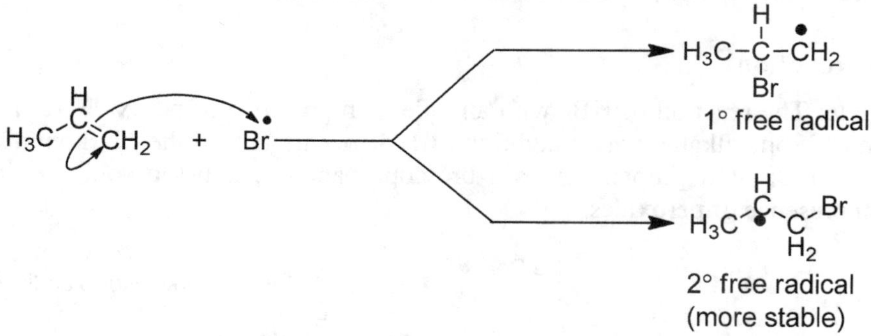

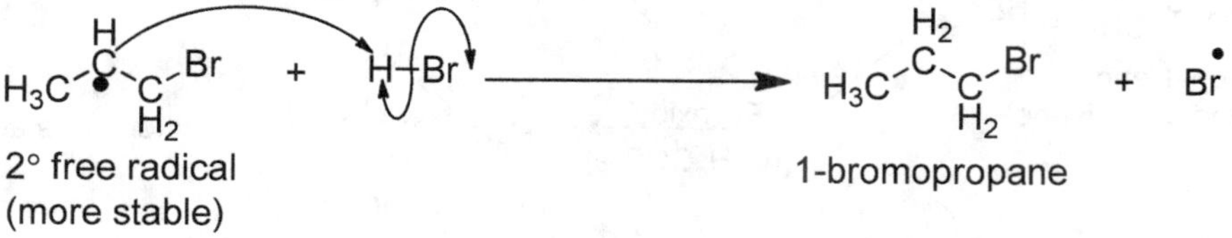

$$H_3C-\overset{H}{\underset{}{C}}=CH_2 \quad + \quad \overset{\bullet}{Br}$$

$$H_3C-\overset{H}{\underset{Br}{C}}-\overset{\bullet}{C}H_2$$

1° free radical

$$H_3C-\overset{\bullet}{C}-\overset{H}{\underset{H_2}{C}}-Br$$

2° free radical
(more stable)

$$H_3C-\overset{\bullet}{C}-\overset{H}{\underset{H_2}{C}}-Br \quad + \quad H-Br \longrightarrow H_3C-\overset{H_2}{\underset{H_2}{C}}-C-Br \quad + \quad \overset{\bullet}{Br}$$

2° free radical
(more stable)

1-bromopropane

Principle

An aldose can be easily converted into another aldose with one carbon atom more than the parent aldose using Killaini-Fisher synthesis. The following example illustrates the conversion of an aldopentose to aldohexose.

Conversion of D-arabinose to D-glucose and D-mannose; the synthesis involves following steps:

a)   Addition of HCN (formation of diastereomeric cyanohydrin): Addition of cyanide to aldehyde group lengthens the chain by one carbon and generates a new stereocentre around which two configurations are possible. As a result, two diastereomeric cyanohydrins are formed.

b)   Hydrolysis (formation of two diastereomeric aldonic acids).

c)   Dehydration (formation of lactone).

d)   Reduction of lactone with Sodium amalgam (Na-Hg) (formation of D-glucose and D-mannose): The acids on dehydration followed by reduction with sodium amalgam result in the formation of two aldoses with one carbon more than the parent aldose. The two new aldoses formed differ only in their configuration at second carbon and are termed as epimers.

General Reaction

Kiliani-Fisher synthesis

Note: In Kiliani-Fischer synthesis yield of sugar is poor. The modified Kiliani-Fischer synthesis involves the direct reduction of cyanohydrins in presence of $Pd/BaSO_4$ as catalyst and water as solvent resulted in good to better yield of monosaccharides.

114. Knoevenagel Condensation

Principle

Nucleophilic addition of a compound with an active methylene group to the carbonyl compounds, followed by the elimination of water molecule to yield olefins is known as Knoevenagel condensation. The reaction was initially given by Emil Knoevenagel in 1894. In Knoevenagel polycondensation reaction, the condensation between compounds with two active methylene groups and two carbonyl groups generate polymers. Reaction is popular for synthesis of α,β-unsaturated compounds. The reaction is exploration of aldol condensation and is reversible. The yield depends not only on the relative reactivity of the reactants, the strength of the bases employed, and the nature of solvents, but also on the removal of water molecule during the reaction. The activation of methylene group is carried out by substituting at least one electron-withdrawing group, such as nitro, cyano, carbonyl, or sulfonyl group. Hence compounds with combination of these electron-withdrawing components adjacent to a methylene group can function as nucleophiles, for example: malonic acid, malonic esters, malonic amides, cyclohexa-1,3-dione, β-ketoacid, β-ketoester, β-ketoamide, ethyl cyanoacetate, malononitrile, phenylcinnamonitriles, o-nitroarylacetonitriles, cyanoacetamides, oxindole, isatin, α-keto aryl sulfones, α-keto sulfones, and keto-thioester.

General Reaction

Mechanism

Step 1: The base abstracts proton from active methylene group to form carbanion intermediate.

Step 2: Attack of carbanion on carbonyl carbon of aldehyde to generation anion.

Aldehyde + Carbanion ⟶

Step 3: Protonated base gives its proton to form stable intermediate and regeneration of base.

Step 4: Dehydration followed by decarboxylation gives α,β-unsaturated carboxylic acid.

$$R-\overset{|}{\underset{H}{C}}-HC(COOC_2H_5)_2 \xrightarrow[-H_2O]{\text{Dehydration}} R-C=C(COOC_2H_5)_2$$

$$R-C=C(COOC_2H_5)_2 \xrightarrow[\text{2. } -CO_2]{\text{1. } H_2O} R-C=C-COOH$$

α,β–unsaturated carboxylic acid

Applications

a) *Generally useful for the synthesis of most of unsaturated carboxylic acids like fumaric acid, maleic acid, β-piperonyl acrylic acid etc. For example: Synthesis of but-2-enedioic acid.*

$$HOOC-CHO + H_2C(COOC_2H_5)_2 \xrightarrow{\text{Pyridine}} HOOC-C=C(COOC_2H_5)_2 \xrightarrow[\text{2. } -CO_2]{\text{1. } H_2O} HOOC-C=C-COOH$$

2-oxoacetic acid   Diethyl malonate

4-ethoxy-3-(ethoxycarbonyl)-4-oxobut-2-enoic acid

But-2-enedioic acid
Maleic acid

b) Synthesis of β-piperonyl acrylic acid.

Benzo[*d*][1,3]dioxole-5-carbaldehyde
(Piperonal)

Malonic acid

1. Pyridine
2. H_2O, $-CO_2$

3-(benzo[*d*][1,3]dioxol-5-yl)acrylic acid
(β–Piperonyl acrylic acid)

c) Synthesis of (E)-1-(4-methoxyphenyl)-3-(p-tolyl)-2-tosylprop-2-en-1-one.

2-((4-methoxyphenyl)sulfonyl)-1-(*p*-tolyl)ethanone

4-methylbenzaldehyde

Piperidine, AcOH
Toluene, Δ

(*E*)-1-(4-methoxyphenyl)-3-(*p*-tolyl)-2-tosylprop-2-en-1-one

d) Synthesis of 4-((4-chloro-1,1,2,2,3,3,4,4-octafluorobutyl)sulfonyl)-3,5-diphenyl-2,3-dihydrofuran-2-yl)(phenyl)methanone.

2-((4-chloro-1,1,2,2,3,3,4,4-octafluorobutyl)sulfonyl)-1-phenylethanone

4-methylbenzaldehyde

Piperidine, AcOH
Toluene, Δ

(4-((4-chloro-1,1,2,2,3,3,4,4-octafluorobutyl)sulfonyl)-3,5-diphenyl-2,3-dihydrofuran-2-yl)(phenyl)methanone

Principle

The preparation of tertiary carboxylic acids or esters by reaction of alcohols or alkenes with anhydrous formic acid is known as Koch-Haaf carbonylation reaction. In this reaction, formic acid acts as both an acidic catalyst and a source of carbon monoxide.

The reaction was first reported by Koch in 1955 and subsequently modified by Koch and Haaf. It has been found that a variety of Bronsted and Lewis acids can catalyze this reaction, among these acids; boron trifluride is most effective catalyst. However, in presence of a super acid, it is possible to formylate isoalkanes to aldehydes that subsequently rearrange to the corresponding branched ketones.

General Reaction

Alcohol   Carbon monoxide

Carboxylic acid

Mechanism

Step 1: The formation of an acyl cation from carbocation generated from acidic dehydration.

Protonation

E1

Alkyl migration

Tertiary Carbocation

Step 2: Reaction between carbon monoxide and tertiary carbocation (3°) to form oxonium ion.

Step 3: Coupling between the oxonium ion and water or alcohol to form the acid or ester derivative.

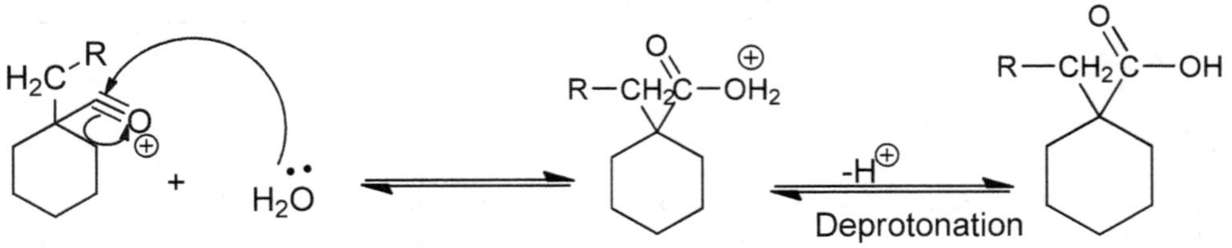

Applications

a) *The reaction has wide application in industry for synthesis of different substituted carboxylic acids. For example: Synthesis of (3r,5r,7r)-adamantane-1-carboxylic acid.*

Adamantane

(3*r*,5*r*,7*r*)-adamantane-1-carboxylic acid

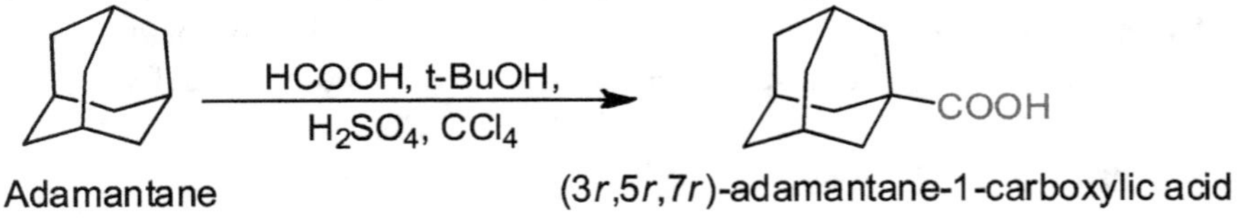

116. Kochi Reaction

Principle

Kochi in 1965 reported one-carbon oxidative degradation of carboxylic acids, and is a valuable alternative to the Hunsdiecker reaction. A lead (IV) acetate [Pb (IV)] reagent is oxidant, and this reaction is suitable for synthesis of secondary and tertiary chlorides. It is the conversion of carboxylic acids into corresponding halides through oxidative degradation of carboxylic acid by lead tetraacetate accompanied by a simultaneous replacement with a halogen under free-radical conditions in presence of a stoichiometric amount of metal halide. It is also referred as Kochi oxidative decarboxylation. In the absence of metal halide, the carboxylic acids will be oxidized into various compounds, depending on experimental conditions, either thermally or photochemically.

General Reaction

$$R = H, Alkyl, Ar$$
$$R = Alkyl, Ar$$
$$R = Alkyl, Ar$$

Carboxylic acid $\xrightarrow{\text{Pb(OAc)}_4,\ \text{LiCl, Benzene, }\Delta}$ Haloalkane $+\ CO_2\ +\ LiPb(OAc)_3\ +\ CH_3COOH$

Mechanism

Applications

a) *The reaction is widely applicable for preparation of secondary and tertiary halides. For example: Preparation of 2-chlorobutane.*

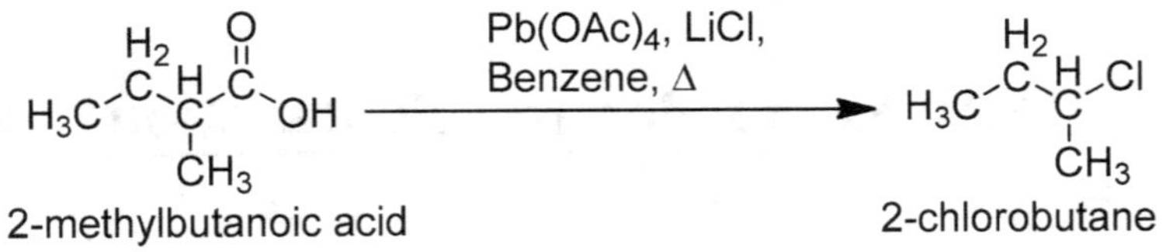

2-methylbutanoic acid $\xrightarrow{\text{Pb(OAc)}_4,\ \text{LiCl,}\ \text{Benzene},\ \Delta}$ 2-chlorobutane

b) Preparation of (E)-(2-bromovinyl)benzene.

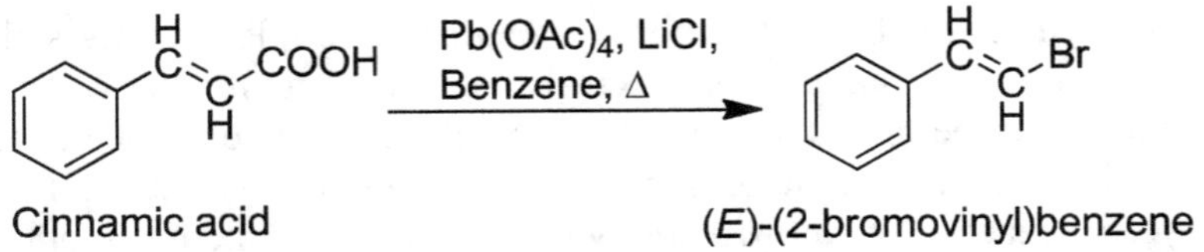

Cinnamic acid $\xrightarrow{\text{Pb(OAc)}_4,\ \text{LiCl,}\ \text{Benzene},\ \Delta}$ (E)-(2-bromovinyl)benzene

117. Kolbe Electrolysis Reaction

Principle

The synthesis of symmetrical hydrocarbons through the coupling of radicals generated from carboxylic acid at an anode *via* electrolysis is commonly known as the Kolbe electrolysis reaction. Kolbe in 1849 reported this reaction. In the presence of free radicals, electrolysis reaction becomes intiator for polymerization to synthesize low molecular weight polystyrene; reaction also proceeds with vinyl acetate, methyl methacrylate, and vinyl chloride. However, free radicals also disproportionate or lose one more electron to form carbocations; therefore olefins, esters, ethers and alcohols are also obtained by this reaction.

General Reaction

$$2\ R-CO_2^{\ominus}Na^{\oplus} \xrightarrow{\text{Oxidative decarboxylation}} \underbrace{R-R + CO_2}_{\text{At anode}} + \underbrace{NaOH + H_2}_{\text{At cathode}}$$

Sodium salt of carboxylic acid → R–R (Hydrocarbons) + CO_2 + NaOH + H_2

Mechanism

Step 1: One-electron transfer process at the anode from the carboxylate to form an acyloxy radical.

$$2\ R-CO_2^{\ominus}Na^{\oplus} \xrightarrow{\text{Ionic mechanism}} 2\ R-CO_2^{\ominus} + Na^{\oplus}$$

Acyloxy radical

Step 2: Decarboxylation of an acyloxy radical to form an alkyl free radical.

$$2\ R-CO_2^{\ominus} \xrightarrow{-2e^{\ominus}} [R-CO_2^{\oplus}] \xrightarrow[\substack{-CO_2}]{\text{Oxidative decarboxylation}} R^{\bullet} + R^{\bullet}$$

Alkyl free radicals

Step 3: Dimerization of alkyl free radicals to produce hydrocarbon.

$$R^{\bullet} + R^{\bullet} \xrightarrow{\text{Dimerization}} R-R$$

Free radicals → R–R (Hydrocarbons)

Applications

a) *Hydrocarbons have been synthesized quantitatively by this reaction; hence practically useful in industry for production of special hydrocarbons. For example: The sodium or potassium salt of monocarboxylic acid on electrolysis gives alkanes. The reaction results symmetrical alkanes which have higher number of carbon atoms than the parent carboxylic acid salt. Methane cannot be synthesized by this method.*

$$2\ H_3C-\underset{\substack{||\\O}}{C}-\overset{\ominus}{O}\ \overset{\oplus}{Na} \xrightarrow[\text{decarboxylation}]{\text{Oxidative}} H_3C-CH_3 \ + \ CO_2$$

Sodium ethanoate   ethane

b) *The concentrated aqueous solution of alkali salts of dicarboxylic acids on electrolysis results in the formation of an alkene at anode.*

$$\xrightarrow[\text{decarboxylation}]{\text{Oxidative}} \underbrace{H_2C{=}CH_2 \ + \ CO_2}_{\text{At anode}} + \ \underbrace{NaOH \ + \ H_2}_{\text{At cathode}}$$

Sodium succinate — Ethene — At anode; At cathode

c) *The concentrated aqueous solution of alkali salts of an unsaturated dicarboxylic acid results in the formation alkynes, at anode.*

$$\xrightarrow{\text{Electrolysis}} \underbrace{HC{\equiv}CH}_{\substack{\text{Ethyne}\\\text{At anode}}} + \ \underbrace{NaOH \ + \ H_2}_{\text{At cathode}}$$

Disodium but-2-endioate

Principle

Kolbe and Schmidt reported the synthesis of *ortho* or *para* hydroxy benzoic acid from carbon dioxide and sodium or potassium phenolate is known as the Kolbe-Schmidt reaction. It is interesting that the sodium phenolate always leads to formation of *o*-hydroxy benzoic acid (salicylic acid), whereas potassium phenolate often results in the production of *p*-hydroxy benzoic acid.

General Reaction

The sodium salt of phenol on reaction with carbon dioxide at 120-140°C at high pressure (5-6 bars) gives *o*-hydroxy benzoic acid.

Sodium phenolate + $O=C=O$ $\xrightarrow[\text{5-6 bar}]{\text{120-140°C}}$ Sodium 2-hydroxybenzoate $\xrightarrow{H^{\oplus}}$ 2-hydroxybenzoic acid (Salicylic acid)

Phenols are transformed into 4-hydroxybenzoic acid.

Phenol + $O=C=O$ $\xrightarrow[\text{5-6 bar}]{\text{KHCO}_3 \ \text{120-140°C}}$ 4-hydroxybenzoic acid

Mechanism

It is an example of electrophilic substitution reaction, where CO_2 acts as an electrophile. In phenoxide ion the aromatic ring is strongly activated and easily undergoes electrophilic substitution at *ortho* position, even with a weak electrophile such as carbon dioxide. The electrophile gets added to phenoxide ion at *ortho* position. This results in the formation of intermediate, stabilization by chelation and results in the formation of a substituted *ortho* product. Salicylic acid is major product formed in the reaction with formation of small amount of *p*-hydroxy benzoic acid.

Applications

The reaction is broadly applicable in preparation of aromatic carboxylic acids containing strong electron-donating groups.

a) Transformation of resorcinol into 2,4-dihydroxybenzoic acid.

Resorcinol

Potassium 2,4-dihydroxybenzoate

2,4-dihydroxybenzoic acid

b) Conversion of 2-methylpyridin-3-ol into 5-hydroxy-6-methylpicolinic acid.

2-methylpyridin-3-ol

5-hydroxy-6-methylpicolinic acid

Principle

Lander in 1900 reported thermal conversion of alkyl imidates (imino ethers) into *N*-alkyl amides in presence of an alkylting agent is known as Lander rearrangement. The rearrangement proceeds through 1,3-shift with following two possibilities: a) an alkyl group of alkylating agent is different from that of the substrate, mixed alkyl amides are obtained; b) an alkylating agent containing lower alkyl group is used for pyrolysis of alkyl imidates, the *N*-alkyl amides more likely form, with alkyl groups substituted by lower one.

General Reaction

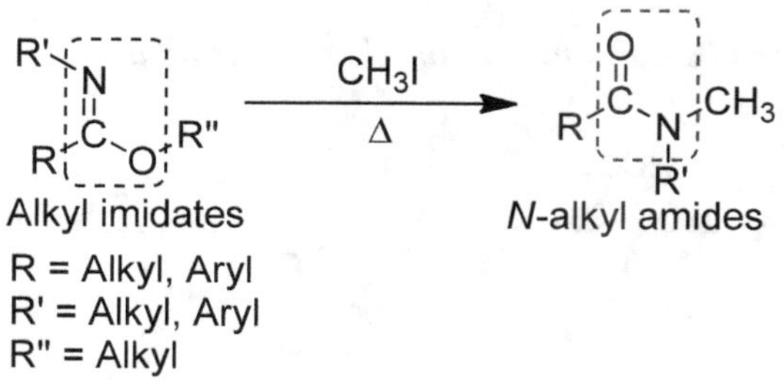

Alkyl imidates

R = Alkyl, Aryl
R' = Alkyl, Aryl
R" = Alkyl

N-alkyl amides

Mechanism

Lander rearrangement involves intramolecular process and mechanism proceeds *via* two different schemes as illustrated below:

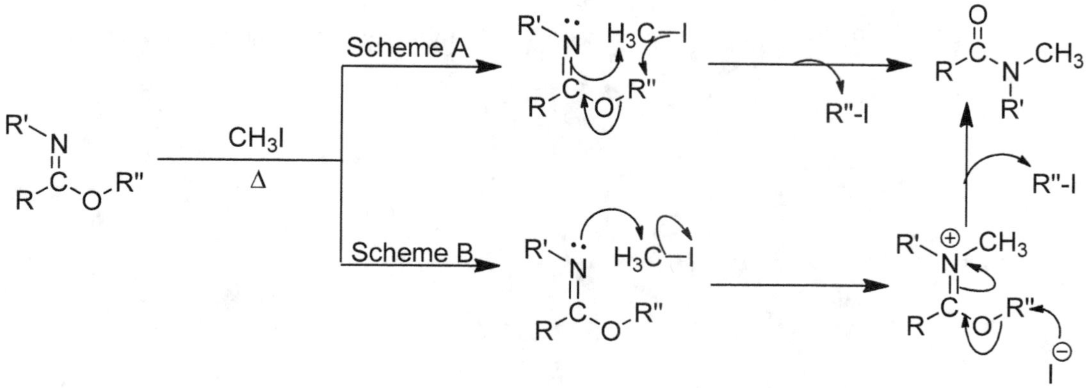

Scheme A

Scheme B

Applications

a) *The reaction is useful for transformation of methoxypyridines into N-methylpyridones. For example: 1,5-dimethyl-1,5-naphthyridine-4,8(1H,5H)-dione convert into 3,7-dimethyl-1,5-naphthyridine-4,8(1H,5H)-dione.*

1,5-dimethyl-1,5-naphthyridine-4,8(1H,5H)-dione   4,8-dimethoxy-1,5-naphthyridine   3,7-dimethyl-1,5-naphthyridine-4,8(1H,5H)-dione

120. Larock Indole Synthesis

Principle

Larock and his collegues in 1991 reported palladium catalyzed chemoselective and ligandless heteroannulation reaction between *o*-iodoanilines and disubstituted alkynes to produce substituted indoles is popularly known as Larock indole synthesis. The reaction is well tolerated by varied substituents on the alkynes and nitrogen atom of anilines. For example, the substituents on nitrogen atom of anilines are methyl, acetyl, etc. and alkynes include alkyl, aryl, alkenyl, hydroxyl, and silyl substituents. Even bulky tertiary alkyl or trimethyl silyl groups are readily accommodated by Larock reaction. Larock reaction is highly regioselective, whereas the more sterically bulky group ends up at 2-position.

General Reaction

O-iodoanilines + $R_1-C{\equiv}C-R_2$ (Disubstituted alkynes) $\xrightarrow{\text{Pd, Base, Cl}^{\ominus}}$ Substituted indole analogues

Mechanism

The reaction mechanism for Larock indole synthesis is illustrated by following steps:

(1) Coordination of chloride to palladium to initiate the catalytic cycle;

(2) Oxidative addition of aryl iodide to palladium;

(3) Coordination of substituted alkyne to palladium atom and subsequent regioselective syn-insertion into the aryl palladium bond;

(4) Nitrogen displacement of the halide in the resulting vinylic palladium intermediate to form a palladacycle, and;

(5) Reductive elimination to form indole and the regenerated palladium re-enters the catalytic cycle.

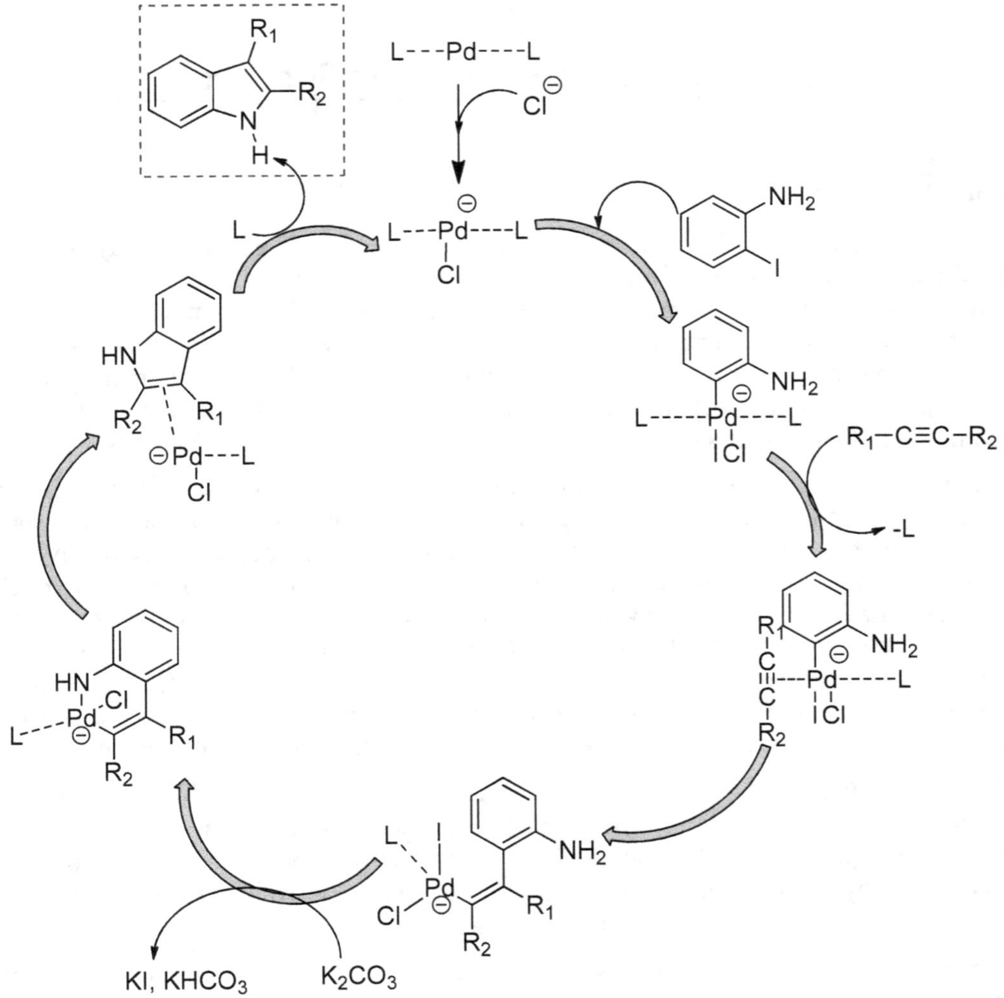

Applications

a) *The reaction is helpful to synthesize wide varieties of heterocyclic compounds, but it is more popular for synthesis of substituted indole analogues.*

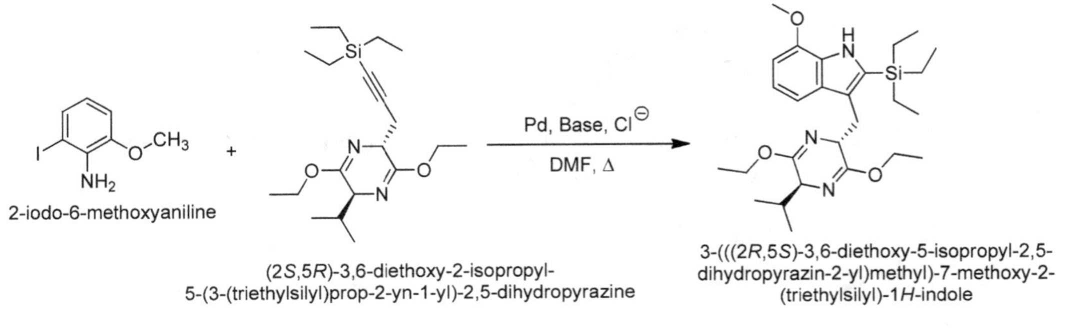

2-iodo-6-methoxyaniline

(2S,5R)-3,6-diethoxy-2-isopropyl-5-(3-(triethylsilyl)prop-2-yn-1-yl)-2,5-dihydropyrazine

Pd, Base, $Cl^{\ominus}$

DMF, Δ

3-(((2R,5S)-3,6-diethoxy-5-isopropyl-2,5-dihydropyrazin-2-yl)methyl)-7-methoxy-2-(triethylsilyl)-1H-indole

Principle

Lawesson's reagent chemically consists of 2,4-bis(4-methoxyphenyl)-1,3,2,4-dithiadiphosphetane 2,4-disulfide, required for transformation of carbonyl groups of ketones, amides and esters into corresponding thiocarbonyl compounds. Lawesson's reagent is also referred as 2,4-bis(4-methoxyphenyl)-2,4-dithioxo-1,3,2,4-dithiadiphosphetane, 4-methoxy-phenylthionophosphine sulphide dimmer, or (*p*-methoxyphenyl) thionophosphine sulfide. Lawesson's reagent is a powerful, mild, and versatile agent that converts oxygen functionalities into respective *thio*-analogs.

First, it converts carbonyl group into thiocarbonyl group, as shown in the transformation of ketone into thioketone, amide to thioamide, ester to thioester, urea to thiourea, trimethylsilyl ferrocenoketone to analgous thioketone, enaminoaldehyde to enaminothioaldehyde, enaminoketone to enaminothioketone, oxalamide to oxala-thioamide, flavones to thioflavones, isoflavones to thioisoflavones, lactones to thiolactone, and coumarin to 2*H*-chromene-2-thione. Second, the Lawesson reagent also converts the carbon-oxygen single bond into a carbon-sulfur single bond, as indicated in the transformation of benzylic and related alcohols into corresponding thiols such as 17β-lactone to β-thio lactone; and orthoesters, acetals, or epoxides to their thio-analogues.

Third, because of the versatility in replacing oxygen functionalities, Lawesson's reagent has been used for the cyclization of compounds containing at least two oxygen functional groups. The transformation of 1,4-dicarbonyl compounds into 2,5-disubstituted thiophenes; 4-hydroxy enone or 4-keto enol into thiophene derivatives; α-keto amide into 1,3-thiazoles can be carried out by using this reagent.

Lawesson's reagent also faces some disadvantages during functional group transformation. For example, due to its relatively high molecular weight, reagent is normally used for small-scale reactions. In addition, the by-products derived from this reagent cannot be separated by simple extraction but usually by column chromatography.

General Reaction

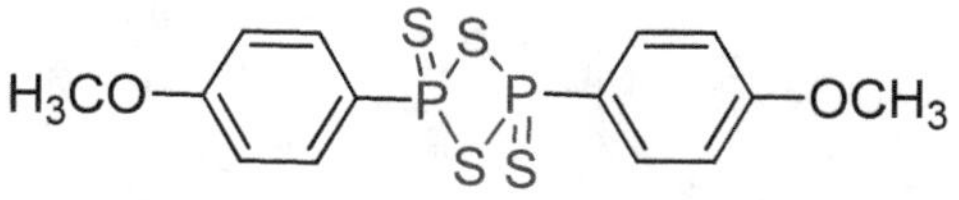

2,4-bis(4-methoxyphenyl)-1,3,2,4-dithiadiphosphetane 2,4-disulfide
(Lawesson's reagent)

Ketones or Esters or Amides
R' = R", OR", NHR"

2,4-bis(4-methoxyphenyl)-1,3,2,4-
dithiadiphosphetane 2,4-disulfide

(Lawesson's reagent)

Thioketones or Thioesters or Thioamides

(4-methoxyphenyl)(thioxo)phosphine oxide

Mechanism

The mechanism for the formation of thioketone from ketone is illustrated here:

Thia-analogues (4-methoxyphenyl)(thioxo)phosphine oxide

Applications

a) Synthesis of pyrrolidine-2-thione from pyrrolidin-2-one and Lawesson's reagent.

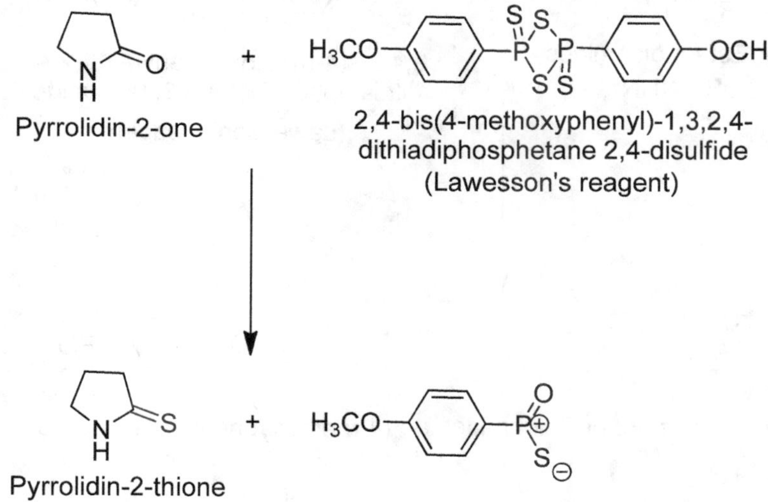

122. Leuckart's Reaction

Principle

Leuckart in 1885 reported reductive amination of carbonyl compounds with ammonium formate (or the combination of formic acid and formamides) to produce primary, secondary, or tertiary amines is known as the Leuckart reaction. The reaction occurs between carbonyl compound and formamide, or ammonium formate, or a combination of formic acid and formamide, from which the primary and secondary amines are produced as formyl derivatives; tertiary amines are obtained as formates. As Leuckart reaction is not moisture-sensitive reaction, it has been proved that if water is removed from the reaction, the yields of corresponding amines will be increased for any reagents other than formamide, in following sequence: formamide < ammonium formate < combination of formamide and formic acid.

General Reaction

$$R\text{-}CO\text{-}R' + 2\ HCOONH_4 \xrightarrow[\Delta]{180\text{-}200°C} \text{Formyl derivatives of primary amine} + 2H_2O + CO_2 + NH_3$$

Carbonyl compounds   Ammonium formate

Formyl derivatives of primary amine $\xrightarrow{H^{\oplus}}$ Primary amine

$$R\text{-}CO\text{-}R' + 2\ HCONH_2 \xrightarrow[\Delta]{180\text{-}200°C} \text{Formyl derivatives of primary amine} + H_2O + CO_2 + NH_3$$

Carbonyl compounds   formamide

Formyl derivatives of primary amine $\xrightarrow{H^{\oplus}}$ Primary amine

Mechanism

Step 1: Formation of formyl derivative of primary amine.

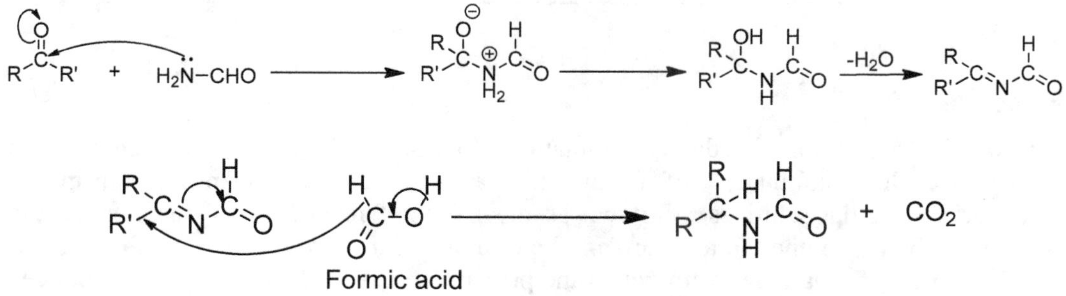

Formic acid

Step 2: Hydrolysis of formyl derivative to give primary amine.

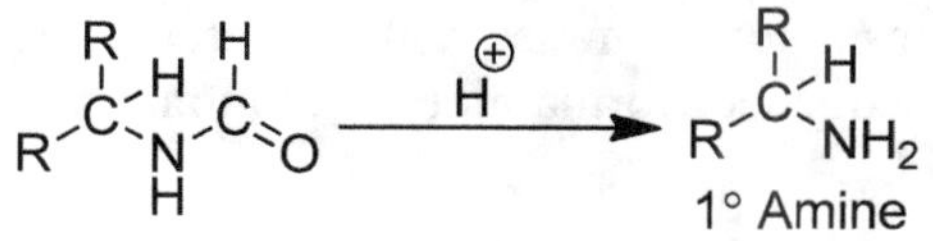

1° Amine

Applications

a) *Leuckart's reaction has wide application in preparation of primary, secondary, and tertiary amines. For example: synthesis of 1-phenylethanamine from acetophenone and ammonium formate.*

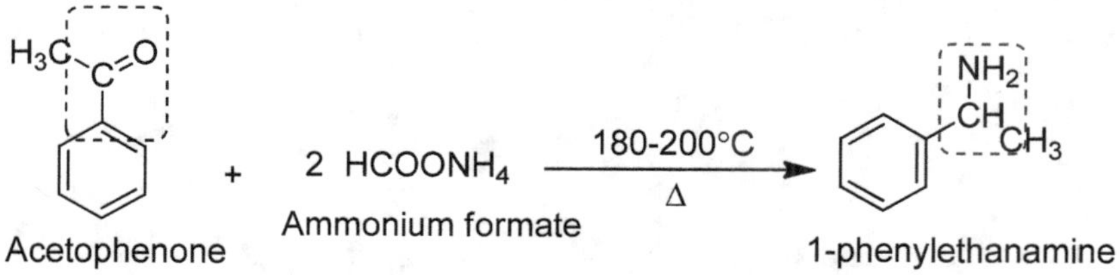

Acetophenone   +   2 HCOONH$_4$ $\xrightarrow[\Delta]{180\text{-}200°C}$ 1-phenylethanamine

Ammonium formate

123. Leucart-Wallach Reaction

Principle

The Leuckart reaction is explored and extended by Rudolf Leuckart. The conversion of carbonyl compounds (aldehydes and ketones) into amines by reductive amination in presence of heat is known as Leucart-Wallach reaction. The reaction proceeds through two different mechanisms: i) using ammonium formate; ii) using formamide as reducing agent. It requires high temperatures, usually between 120-130°C; although under the presence of formamide, temperature should be more than 165°C.

General Reaction

Ketones + Amines $\xrightarrow{\text{H-COOH}}$ Higher amines + $CO_2 \uparrow$ + H_2O

Mechanism

Step 1: Amine first nucleophilically attacks at the carbonyl carbon. The oxygen is protonated by abstracting hydrogen from nitrogen atom, subsequently forming a water molecule that leaves, forming iminium ion.

α-aminoalcohol $\xrightarrow{-H_2O}$ Iminium ion

Step 2: Hydride shift occurs, resulting in loss of carbon dioxide. The resulting intermediate with loss of proton gives higher amine.

$\xrightarrow[-CO_2]{\text{Reduction}}$ $\xrightarrow{-H^{\oplus}}$

Applications

Reaction is widely applicable in synthesis of different substituted amine analogues.

a) Synthesis of 4-isobutylmorpholine.

Isobutyraldehyde Morpholine $\xrightarrow[60°C, 57\%]{\text{H-COOH}}$ 4-isobutylmorpholine

b) Synthesis of N,N-dimethylcyclooctanamine.

Cyclooctanone *N,N*-dimethylformamide $\xrightarrow[190°C, 16H, 75\%]{\text{H-COOH, H}_2\text{O}}$ *N,N*-dimethylcyclooctanamine

124. Lindlar Hydrogenation

Principle

Lindlar in 1952 optimized and reported stereoselective reduction of alkynes into *cis*-alkenes by hydrogen in presence of $Pd/CaCO_3$ as catalyst and then deactivated by lead acetate and quinoline is known as Lindlar reduction or Lindlar hydrogenation. The palladium on calcium carbonate poisoned with lead acetate and quinoline is commonly known as Lindlar catalyst. It is believed that the stereoselective reduction of alkynes into only *cis*-alkenes is possible; because one face of the triple bond is shielded by the Lindlar catalyst so that hydrogen is restricted to approach the triple bond from the other side.

General Reaction

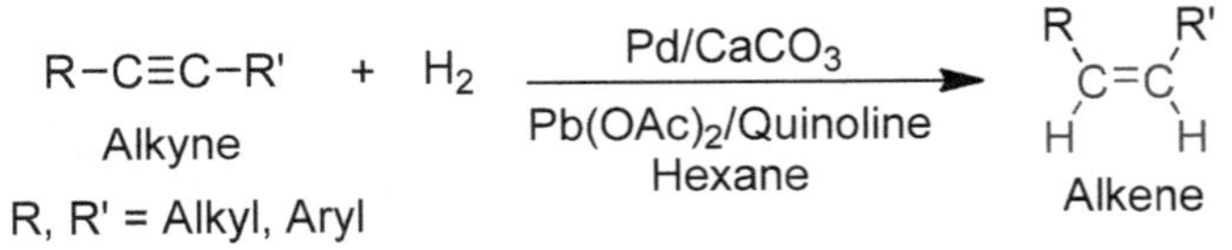

$$R-C{\equiv}C-R' \; + \; H_2 \xrightarrow[\substack{Pb(OAc)_2/Quinoline \\ Hexane}]{Pd/CaCO_3} \; \underset{\substack{H \quad\quad H \\ Alkene}}{\overset{R \quad\quad R'}{C=C}}$$

Alkyne

R, R' = Alkyl, Aryl

Mechanism

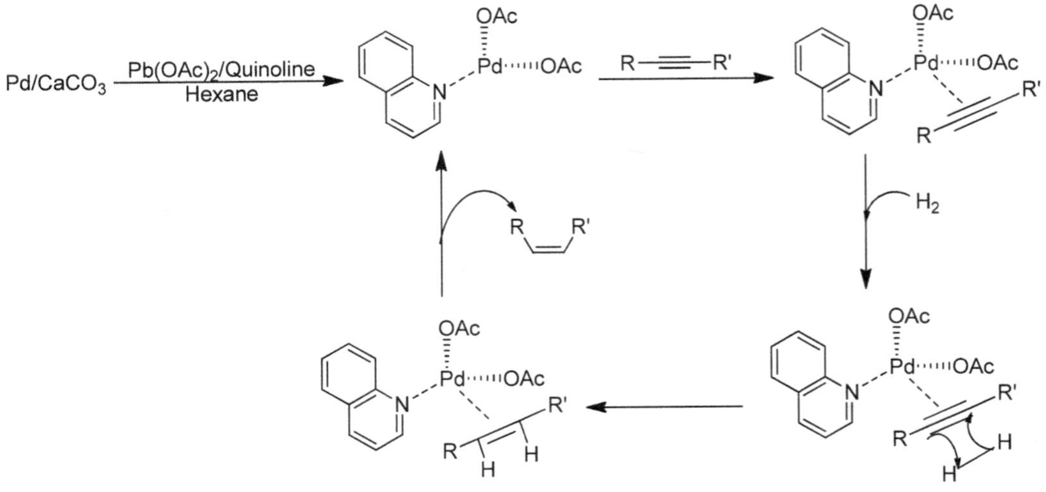

Applications

a) *The reduction reaction is widely useful in synthesis of natural product.*

7-phenylhept-4-yn-1-amine →[H₂/Lindlar Catalyst / DME/DMF] 7-phenylhept-4-en-1-amine

125. Lossen Rearrangement

Principle

Lossen in 1872 carried out thermal or alkaline decomposition of derivatives of hydroxamic acid into an isocyanate which on hydrolysis gives amine is known as Lossen rearrangement. In presence of water, amine, or alcohol, the isocyanate is converted into corresponding amine, urea or urethane respectively. Initially, this reaction was thought to occur *via* a similar mechanism of *Curtius rearrangement* or *Hofmann rearrangement* by means of the formation of univalent nitrogen intermediate (nitrene), from which alkyl group migrates from a carbon to a nitrogen atom to form isocyanate. However, the explored and accepted mechanism involves deprotonation of N-H from a base, followed by migration of an alkyl group and release of carboxylate (or sulfonate or phosphonate) if the hydroxamic acid is activated by an acyl group (or correspondingly by a sulfonyl or phosphoryl group). The rearrangement rate is directly proportional to the acidity of leaving group or the conjugate acid of leaving group if the leaving group is an anion; in addition, in case of *o*-benzoyl phenylhydroxamic acid, the electron-donating group on a phenyl and electron-withdrawing group on a benzoyl moiety will facilitate the rearrangement, especially when these substituents are in *meta* or *para*-positions.

General Reaction

$$R-\overset{\overset{\text{O}}{\|}}{C}-\underset{\underset{\text{H}}{|}}{N}-OH \xrightarrow{\text{KOH}} R-NH_2 \ (\text{An amine}) \ + \ CO_2$$

Hydroxamic acid

Mechanism

Step 1: The hydroxamic acid on heating in the inert solvent in presence of acetic anhydride undergoes loss of water to form an acetyl derivative of hydroxamic acid.

$$R-\overset{\overset{\text{O}}{\|}}{C}-\underset{\underset{\text{H}}{|}}{N}-OH \ + \ H_3C-\overset{\overset{\text{O}}{\|}}{C}-O-\overset{\overset{\text{O}}{\|}}{C}-CH_3 \xrightarrow{-H_2O} R-\overset{\overset{\text{O}}{\|}}{C}-\overset{\ominus}{N}-O-\overset{\overset{\text{O}}{\|}}{C}-CH_3$$

Hydroxamic acid   Acetic anhydride   Acetyl derivative of hydroxamic acid

Step 2: 1,2-migration of alkyl group in alkaline medium to form isocyanate ion and loss of acetate ion.

R−C(=O)−N(−)−O−C(=O)−CH₃ ⟶ R−N=C=O

Migration of alkyl group   Isocyanate

Step 3: Hydrolysis of isocyanate intermediate to form an amine.

R−N=C=O $\xrightarrow[\text{H}_2\text{O}]{\text{Hydrolysis}}$ R-NH₂ + CO₂

Isocyanate   An amine

Applications

The reaction has general application in the synthesis of amine, urea, and urethane analogues.

a) *Synthesis of urea derivatives.*

R−N=C=O $\xrightarrow{\text{R'—NH}_2}$ R−NH−C(=O)−NH−R'

Isocyanate   Urea derivatives

b) *Synthesis of urethane analogues.*

R−N=C=O $\xrightarrow{\text{R'—OH}}$ R−NH−C(=O)−O−R'

Isocyanate   Urethane analogues

c) *Synthesis of Benzyl cyclopent-3-en-1-ylcarbamate.*

10 mol% Zn(OTf)₂, 2,6-DTBP, CH₃CN, Δ

tert-butyl cyclopent-3-enecarbonyl((methylsulfonyl)oxy)carbamate   Benzyl cyclopent-3-en-1-ylcarbamate

126. Malaprade Reaction

Principle

Malaprade in 1928 carried out an oxidation of vicinal diols with periodic acid or its salt in aqueous medium is known as Malaprade reaction or Malaprade oxidation. To enhance the solubility of organic substrates solvents such as methanol, ethanol, and acetic acid is used in this oxidation reaction. This reaction has been extended to the cleavage of α-hydroxy carbonyl compounds, 1,2-dicarbonyl compounds, α-amino alcohols, α-amino acids, and polyhydroxy alcohols. In these oxidations, the hydroxyl group is oxidized to an aldehyde or a ketone group, the carbonyl group is oxidized to a carboxyl group, and the amino group is transformed into an aldehyde group and ammonia. In the case of an adjacent polyhydroxyl alcohol, the reaction occurs at the terminal hydroxyl group, and middle hydroxyl groups are converted into formic acid. The reaction proceeds faster under acidic conditions than that in neutral or basic solutions.

General Reaction

$$\text{HIO}_4 \text{ OR NaIO}_4$$

Vicinal diols

Carbonyl compounds

R = H, Alkyl, Aryl; R' = H, Alkyl, Aryl
R" = H, Alkyl, Aryl; R''' = H, Alkyl, Aryl

Mechanism

The complete mechanism for Malaprade reaction is illustrated below:

Applications

a) *The reaction has broad application in the oxidation of vicinal diols, amino alcohols, and similar functionalities. The reaction has been successfully used for degradation of carbohydrates and in structural analysis.*

127. Malonic Ester Synthesis

Principle

Ethyl acetoacetate (EAA) and Diethyl malonate (DEM) are two active methylene group containing compounds widely used in organic synthesis. The synthetic utility of ethylacetoacetate depends on following factors:

a) One of the two active hydrogens is easily replaced by sodium ethoxide to give sodio ethylacetoacetate. Sodio ethyl acetoacetate is resonance stabilized carbanion.

b) Nucleophilic sodio ethyl acetate easily attacks the electrophilic carbon (primary or secondary), tertiary, vinylic, or aromatic carbons are not attacked by the sodio ethyl acetoacetate.

c) The monoalkylated product still contains the other active hydrogen for which the whole sequence of reactions can be repeated to form dialkyl ethylacetoacetate.

Ethylacetoacetate or diethyl malonate easily converted into sodio diethyl malonate with sodium ethoxide. Sodio diethyl malonate can be alkylated with RX to give alkyl diethyl malonate that still contains the other active hydrogen atom. For this, the whole sequence of reactions can be repeated to obtain dialkyl diethyl malonate. Mono and dialkyl diethyl malonates on hydrolysis affords numerous monocarboxylic acids, the whole sequence of reactions is collectively known as Malonic ester synthesis.

General Reaction

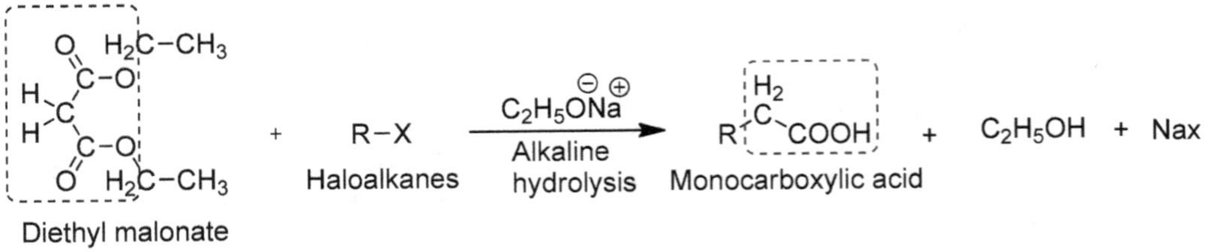

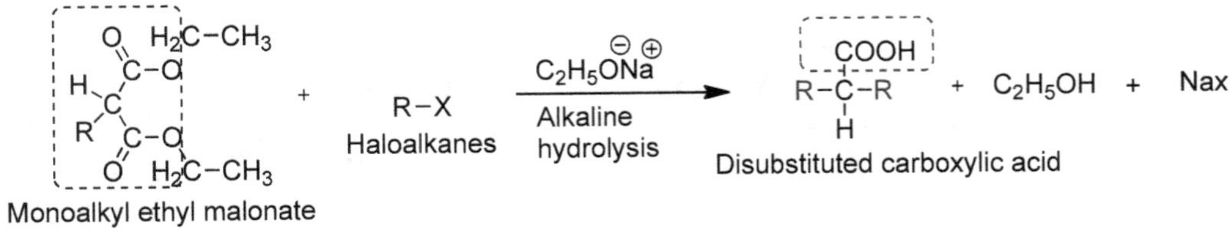

Mechanism

Step 1: Formation of sodio diethyl malonate by abstraction of acidic hydrogen by ethoxide ion.

Step 2: Coupling of sodio diethyl malonate with haloalkane to give mono-alkyl diethyl malonate.

Step 3: Mono-alkyl diethyl malonate on alkaline hydrolysis followed by decarboxylation at 140°C results in the formation of mono substituted carboxylic acid. Di-alkyl diethyl malonate on alkaline hydrolysis followed by decarboxylation gives di-substituted carboxylic acid.

Applications

a) Synthesis of diethyl 2-(3,4-dihydronaphthalen-2-yl)malonate.

(cyclopropylidenemethyl)benzene   Diethyl malonate

Mn(OAc)$_3$,
AcOH/Ac$_2$O
68%

Diethyl 2-(3,4-dihydronaphthalen-2-yl)malonate

128. Mannich Reaction

Principle

The aldehydes and ketones containing α-hydrogens undergo condensation with formaldehyde and ammonia (or primary and secondary amines) to produce β-amino carbonyl compounds. This condensation reaction is known as Mannich reaction. The reaction may be acid or base catalysed.

It is a multicomponent condensation between an amine (1°, 2° amine or ammonia), an enolizable carbonyl compound (donor), and a nonenolizable carbonyl compound (acceptor) to form a β-amino carbonyl compound, with the concomitant formation of both carbon-carbon and carbon-nitrogen bonds, known as the Mannich base. The iminium derivative of aldehyde is acceptor in reaction.

The reaction was first reported by Mannich and Kroesche in 1912. The reaction is extensively optimized for different substrates: (*a*) aldimines, such as the reaction between aldimine and silyl enol ether or silicon enolate, aldimine and fluorovinyl ether, chiral aldimines and silyloxydienes, and fluorinated aldimines with aliphatic aldehydes; (*b*) iminium ions, such as the reaction between cyclic iminium ions and enol derivatives (enol silane, vinyloxytrimethylsilane), and cyclic *N*-alkoxycarbonyl pyrrolinium ion and 2-trimethylsilyloxyfuran; (*c*) hydrozones, such as the coupling between chiral *N*-acylhydrazones and silyl enolates, hydrazones with ketene silyl acetal, and hydrazone ester and silicon enolates; (*d*) imino esters, such as the reaction between α-imino ester and aldehydes, *N*-PMP-protected α-imino glyoxylate and α,α-disubstituted aldehyde, α-imino ester and a glycinated Schiff base, and *N*-acylimino ester and silyl enol ethers or vinyl ethers.

General Reaction

Mechanism

Base catalyzed mechanism

Step 1: Formation of carbanion by abstraction of α-hydrogen by base.

Ketone → Base → Carbanion → Resonating structures

Step 2: Nucleophilic addition of amines into formaldehyde to form adduct.

Methanal + 2° amine → → N,N'-dialkyl amino methanol

Step 3: Nucleophilic substitution reaction to form β-amino carbonyl compounds.

Carbanion + N,N'-dialkyl amino methanol → Nucleophilic substitution → β-amino carbonyl compound + $:\overset{\ominus}{O}H$

Acid Catalyzed Mechanism

Step 1: Acid catalyzed Keto-enol tautomerism.

Keto → Enol

Step 2: Addition of formaldehyde into amines to form Immonium ion.

Methanal + 2° amine → → → N,N'-dialkyl amino methanol

Step 3: Addition of enol into immonium ion to form β-amino carbonyl compounds.

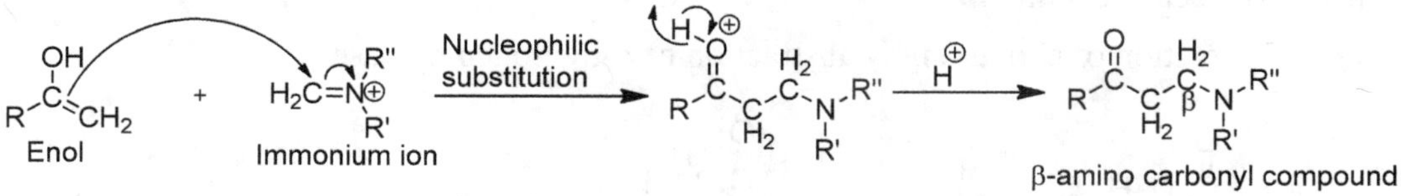

Applications

The involvement of the Mannich reaction has been proposed in many biosynthetic pathways, especially in the synthesis of alkaloids.

a) *Many important natural products, like alkaloids have been easily synthesized by this reaction. For example: Robinson's synthesis of tropinone by a double Mannich condensation which is further utilized for synthesis of atropine.*

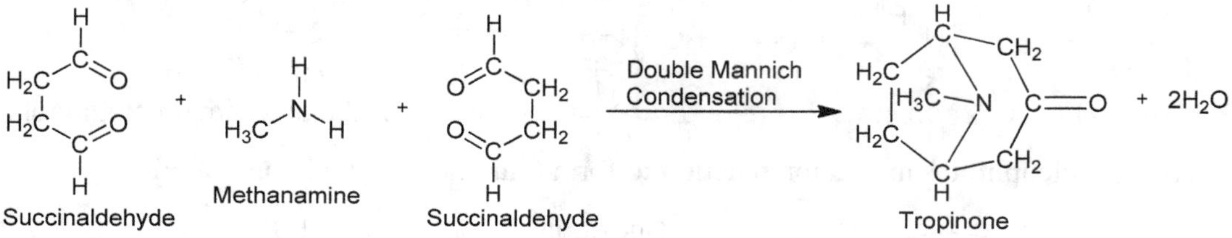

Principle

The Markovnikov's rule was first formulated by Markownik off in 1870. The rule states that, in the addition of polar reagents to unsymmetrical alkenes, the negative part of the polar reagent adds to that carbon of the double bond which contains minimum number of hydrogen atoms. It is a regioselective addition of polar reagent. This rule has been extended to the heterolytic addition of a polar molecule to an alkene or alkyne, of which the more electronegative atom (or group) of the polar molecule attaches to the carbon with more substituents.

General Reaction

Prop-1-ene (Unsymmetrical alkene) $H_3C-CH=CH_2$ reacts with HCl (Polar reagent) to give:

- $H_3C-CHCl-CH_3$ — Markonikov's addition, 2-chloropropane (major)
- $H_3C-CH_2-CH_2Cl$ — 1-chloropropane (minor)

Mechanism

Consider the mechanism of addition of HCl to propene. The attack of the electrophile H^+ on either of the olefinic carbons may result in the formation of a 1° and 2° carbocation. In general, the attack of the electrophile H^+ is always favored in a direction which results in the formation of more stable carbocation. In
predominantly through the formation of a more stable 2

At C-1: $H_3C-\overset{+}{C}H-CH_3$ (2° carbocation, more stable) + Cl^- → $H_3C-CHCl-CH_3$, 2-chloropropane (major), Markonikov's product

At C-2: $H_3C-CH_2-\overset{+}{C}H_2$ (1° carbocation, less stable) + Cl^- → $H_3C-CH_2-CH_2Cl$, 1-chloropropane (minor)

Applications: *The rule is used to rationalize the addition products.*

130. Mclafferty Rearrangement

Principle

Happ and Stewart in 1952 reported the rearrangement from mass spectroscopy of butyric acid; but McLafferty studied this reaction extensively and stated the importance of fragmentation pattern for ionized organic molecules in mass spectroscopy in 1956. McLafferty also reported cyclic transition state; hence this rearrangement is known as McLafferty rearrangement. This is the rearrangement of monounsaturated molecular ion or radical ion with cleavage of α,β-bond of unsaturated chain along with movement of γ-hydrogen *via* a six-membered transition state by formation of a pair of unsaturated fragments, regardless of which fragment holds the charge.

General Reaction

Monounsaturated
molecularion or radical ion
R = H, Alkyl, Aryl, OH, Alkoxy, Amino
R' = H, Alkyl, Aryl

Mechanism

The detailed mechanism of McLafferty rearrangement for carbonyl compounds includes hydrogen transfer to oxygen and the subsequent β-cleavage of α,β-bond. In this mechanism, n-bonding carbonyl group is highest occupied molecular orbital in carbonyl compounds, excited to lose one electron to form a radical cation, which undergoes β-cleavage to afford the two fragments.

Applications

a) The reaction is helpful to understand most common fragmentations and studied most extensively in the field of mass spectroscopy. The rearrangement has been used in structural elucidation using mass spectroscopy as the tool.

131. Meerwein-Ponndorf-Verley Reduction

Principle

An aluminum alkoxide-catalyzed reduction of carbonyl compounds (ketones and aldehydes) to corresponding alcohols using alcohol (2° alcohol) as a reducing agent or hydride source is known as the Meerwein-Ponndorf-Verley reduction (MPV). The reaction was given by Meerwein, Schmidt and Verley in 1925, followed by Ponndorf in 1926. After 12 years, Oppenauer reported the reversal of alcohols into carbonyl compounds by oxidation reaction. Since then, the reversible reaction between carbonyl compounds and alcohols in presence of aluminum alkoxide are called Meerwein-Ponndorf-Oppenauer-Verley reduction. This reduction has advantages of chemoselectivity, mild reaction conditions, ease of operation, cost feasibility and scalability. In addition, this reaction does not affect other double bonds or triple bonds and other enolizable carbonyl compounds (β-keto ester and β-diketone). The catalysts suitable for reaction are aluminum alkoxide and some transition-metal alkoxides (For example: zirconium, iron, titanium, and lanthanide), used as homogenous catalysts.

General Reaction

$$\underset{\text{Ketone}}{\overset{R}{\underset{R'}{>}}C=O} + \underset{\text{Aluminiium isopropoxide}}{\frac{(CH_3)_2CHO}{3}Al} \xrightarrow{\text{Isopropanol}} \overset{R}{\underset{R'}{>}}\frac{CHO}{3}Al + (CH_3)_2CO \xrightarrow{\text{Dil. }H_2SO_4} \underset{\text{Alcohol}}{\overset{R}{\underset{R'}{>}}CHOH}$$

a) Transformation of acetophenone to 1-phenylethanol.

Acetophenone $\xrightarrow[\text{(CH}_3)_2\text{CHOH}]{[(CH_3)_2CHO]_3Al}$ 1-phenylethanol

b) Conversion of benzophenone to benzhydrol.

Benzophenone $\xrightarrow[\text{(CH}_3)_2\text{CHOH}]{[(CH_3)_2CHO]_3Al}$ Benzhydrol

Mechanism

The reaction involves following steps:

Step 1:

(a) Co-ordination of a carbonyl compound to the metal atom of the alkoxide catalyst, typically aluminium isopropoxide, to increase the positive character of the carbonyl group;

(b) Internal hydride transfer from an alkoxide ligand to the carbonyl group coordinating to the metal atom through a six-membered transition state, to generate a new alkoxide ligand and carbonyl compound.

Step 2: Dissociation of the new carbonyl compound and alcoholysis of the mixed metal alkoxide to release the alcohol and regenerate the catalyst.

Applications

The reaction is useful in reduction of carbonyl compounds.

a) *To reduce α,β-unsaturated aldehydes to α,β-unsaturated alcohols.*

(E)-but-2-enal → (E)-but-2-en-1-ol

b) *Reduction of alicyclic ketones; for example: cyclohex-2-enone is reduced to cyclohex-2-enol.*

Cyclohex-2-enone → Cyclohex-2-enol

132. Michael Addition Reaction

Principle

Michael addition was reported 1887. A 1,4-addition (or conjugated addition) of a nucleophile to an electron-deficient alkene or alkyne gives Michael addition product. The nucleophiles are Michael donors, while the electron-deficient alkenes or alkynes are Michael acceptors, and the product formed is known as Michael adducts. In presence of nucleophile and electron-deficient acceptors, Michael addition occurs more easily. Because both nucleophile and Michael acceptor become a part of the Michael adduct, this reaction is one of the most atom-efficient reactions and has been widely explored in many aspects.

Michael donors: These are the nucleophiles acts as Michael donors; For examples: heteroatoms containing lone-pair electrons, such as primary and secondary amines, thiols, and fullerenol; neutral carbon-nucleophiles such as indoles, (*tert*butyldimethylsiloxy) pyrrole, silyl ketene acetals, and enolsilanes; carbon-nucleophiles with α-active protons, such as malonates, β-ketoesters, silyl nitronates, and nitroalkanes; and anionic carbon-nucleophiles, such as enolates, chiral Fischer aminocarbene complexes, α-lithiated vinylic sulfoxide, and enantio-enriched allenyltitaniums; lithium eneselenolates generated from selenoamides; and acylperoxy radicals.

Michael acceptors: α,β-unsaturated carbonyl compounds (For examples: acrylates and ethyl vinyl ketone), have been extended to a variety of electron-deficient alkenes or alkynes, such as 2-*oxo*-3-butynoate, 2-*oxo*-3-pentynoate, cyclic enones (cyclohexenone), chalcone, acrynitrile, alkylidene malonates, α, β-unsaturated imides, nitroalkenes, nitrostyrene, α-methylene lactones, 1,4-benzoquinones, 3-alkylidene-3*H*-indole-1-oxide, 3-(*E*-enoyl)-4-phenyl-1,3-oxazolidin-2-ones, and α, β-unsaturated iodonium salt.

General Reaction

The addition of active methylene group containing compound (diethyl malonate) to α,β-unsaturated compound (chalcone) in presence of basic catalyst (Sodium ethoxide or piperdine) to give Michael addition product.

Benzylideneacetophenone (chalcone) + Diethyl malonate $\xrightarrow{\text{C}_2\text{H}_5\text{ONa/ Piperidine}}$ (S)-diethyl 2-(3-oxo-1,3-diphenylpropyl)malonate

Mechanism

Step 1: Abstraction of active hydrogen from diethyl malonate by ethoxide ion (base) to generate carbanion.

diethyl malonate + C_2H_5ONa ⇌ Carbanion + C_2H_5OH

Step 2: Addition of carbanion to the β-carbon atom to form intermediate.

+ Carbanion → (addition) → $CH(COOC_2H_5)_2$

Step 3: Workup of reaction to give addition product.

$CH(COOC_2H_5)_2$ + C_2H_5OH / $-C_2H_5O^{\ominus}$ → $CH(COOC_2H_5)_2$ (1,4-addition product) Enol ← tautomerism ← $CH(COOC_2H_5)_2$ (1,2-addition product) (S)-diethyl 2-(3-oxo-1,3-diphenylpropyl)malonate

Applications

a) *The reaction is applicable in organic synthesis for the synthesis of different important intermediates. For example: synthesis of diethyl 2-(2-cyanoethyl)malonate from acrylnitrile and diethyl malonate.*

$N\equiv C-CH=CH_2$ Acrylonitrile + Diethyl malonate → (C_2H_5ONa/ Piperidine) → $N\equiv C-CH_2-CH_2-CH(COOC_2H_5)_2$ Diethyl 2-(2-cyanoethyl)malonate

b) *Synthesis of propane-1,2,3-tricarboxylic acid from diethyl maleate and diethyl malonate.*

Diethyl maleate + Diethyl malonate → (C_2H_5ONa) → $H_2C-COOC_2H_5$ / $H-C-COOC_2H_5$ / $CH(COOC_2H_5)_2$ → (Hydrolysis, $-C_2H_5OH$, $-CO_2$) → CH_2COOH / $CHCOOH$ / CH_2COOH Propane-1,2,3-tricarboxylic acid

133. Mitsunobu Reaction

Principle

Mitsunobu in 1967 reported an alkylation of organic compounds with active protons by using primary or secondary alcohols as alkylating agents in combination with triphenylphosphine and DEAD or DIAD to yield functionalities such as esters, ethers, thioethers, and amines is known as Mitsunobu reaction. The formation of esters by reacting DEAD (or DIAD) and triphenyl phosphene is Mitsunobu esterification. Analogously, the transformation of alcohol into amines is Mitsunobu transformation; whereas synthesis of cyclic compounds is Mitsunobu cyclization.

General Reaction

a) Formation of esters

Primary or Secondary alcohols   Acids   Esters

b) Formation of azides

Primary or Secondary alcohols   HN$_3$ Hydrogen azide

Mechanism

A general mechanism containing the formation of phosphonium salt and phosphorane intermediates.

Step 1:

Step 2:

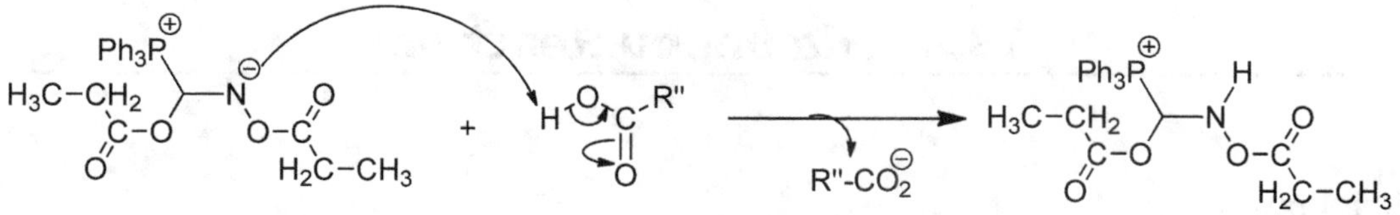

Step 3:

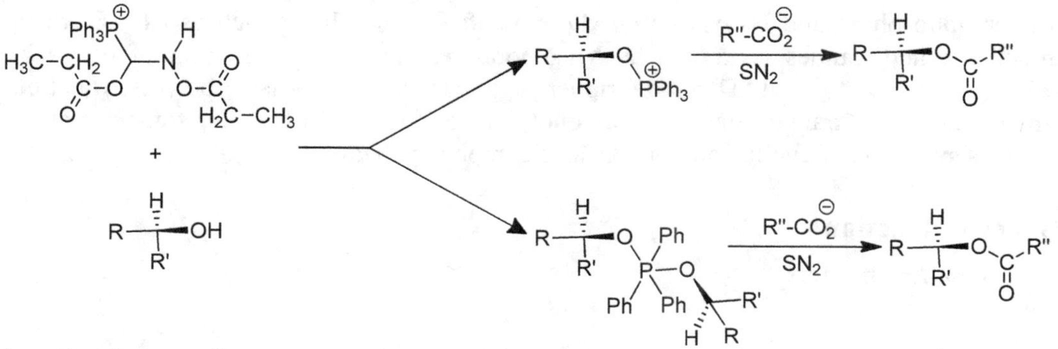

Applications

a) *The reaction has very wide application in organic synthesis, including the synthesis of esters, ethers, thioethers, amines, and azides.*

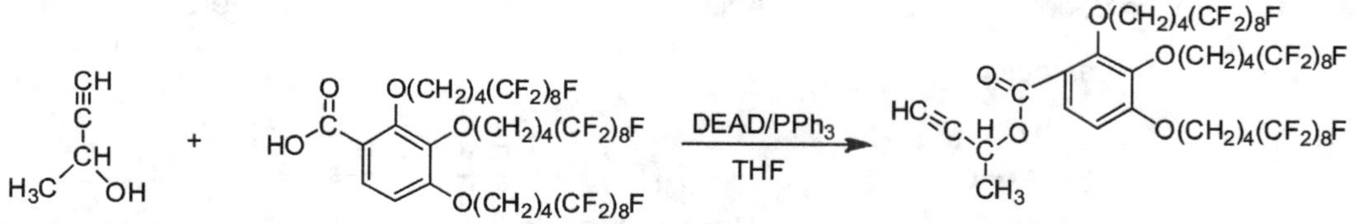

But-3-yn-2-ol

2,3,4-tris(heptadecafluorododecyl) oxy)benzoic acid

DEAD/PPh$_3$

THF

But-3-yn-2-yl 2,3,4-tris(heptadecafluorododecyl)oxy)benzoate

134. Morin Rearrangement

Principle

An acid-promoted stereospecific conversion of penicillin sulfoxides into corresponding desacetoxycephalosporins was reported by Morin in 1963 is commonly referred as Morin rearrangement.

The reaction involves following steps:

1) Activation of sulfoxide by an acid,

2) Generation of azetidinone sulfenic acid via stereospecifically sigmatropic opening of the thiazolidine ring without concomitant rupture of the β-lactam by abstracting a proton from the C-2 methyl group cis to the sulfoxide S-O bond,

3) Protonation of sulfenic acid, and

4) Cyclization of protonated sulfenic acid with the adjacent π-bond and the final deprotonation. The reaction proved that there was a thermal equilibrium between sulfoxide and sulfenic acid.

The reaction also proved chemical correlation between penicillins and cephalosporins that was helpful for synthesis of important antibiotics containing deacetoxycephalosporin moiety.

General Reaction

Penicillin sulfoxides

Desacetoxycephalosporins

Mechanism

Step 1: Protonation

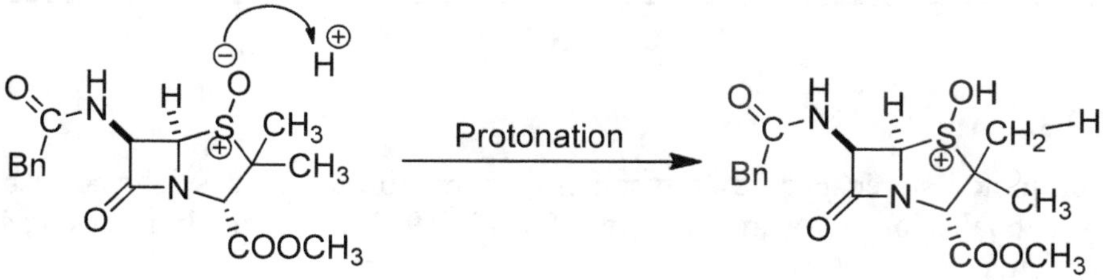

Step 2: Ring opening to yield sulfenic acid

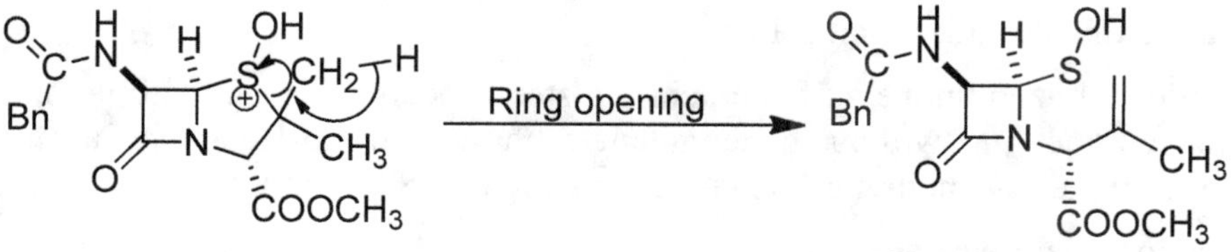

Sulfenic acid

Step 3: Ring closure

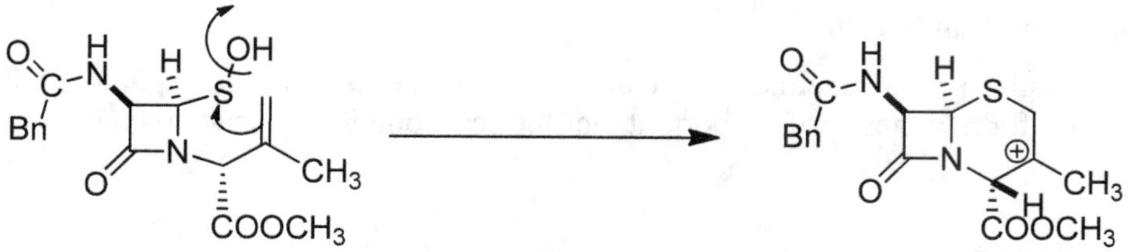

Step 4: Work-up

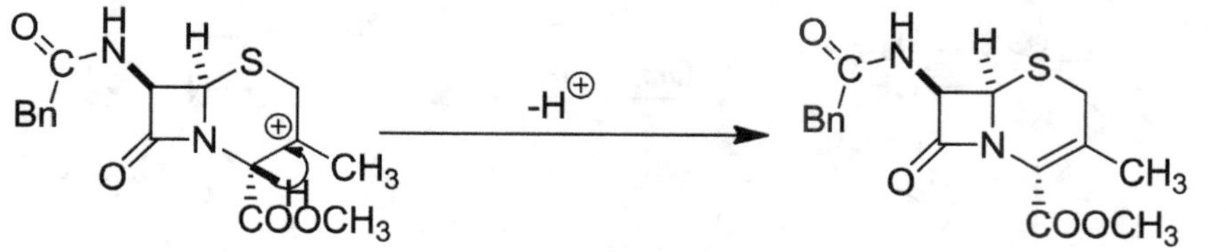

Applications

a) *The rearrangement is useful in synthesis of penicillin and hetacillin sulfoxide analogues. For example: Synthesis of β-lactum antibiotics.*

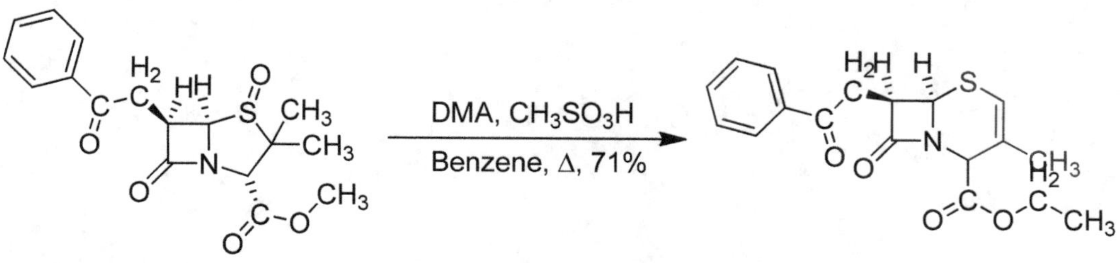

135. Neber Rearrangement

Principle

Neber in 1926 reported conversion of ketoxime o-arylsulfonate into α-amino ketone through a deprotonation of α-methylene hydrogen by a base followed by acidic hydrolysis of azirine intermediate is known as Neber rearrangement. In this rearrangement, amino group is introduced at the α-position with higher acidic hydrogen in ketoxime o-arylsulfonate. The rearrangement is performed in ethanol by reacting ketoxime tosylate with sodium ethoxide, subsequently by acidic hydrolysis. For cyclic ketones, the amino group is added at equatorial positions at α-carbon. By Neber rearrangement, aldoxime tosylate gives corresponding nitrile or isonitrile under similar conditions. The starting material for Neber rearrangement is synthesized by treatment of oxime with p-toluenesulfonyl chloride in pyridine.

General Reaction

Preparation of ketoxime o-arylsulfonate (Ketoxime tosylate)

Ketone   Hydroxyl amine   Ketoxime   4-methylbenzene-1-sulfonyl chloride

Pyridine
-HCl

Ketoxime tosylate

Preparation of α-amino ketone

Ketoxime tosylate   NaOEt, Pyridine   α-aminoketone   4-methylbenzenesulfonic acid

Mechanism

Step 1: The mechanism involves initial deprotonation from ketoxime tosylate to from carbanion. For this reason, α-carbon of the reactant should have atleast one acidic hydrogen. The reaction is more efficient if R is a phenyl group.

Ketoxime tosylate

Step 2: The carbanion undergoes internal displacement of the leaving group to form an azirine intermediate.

Azirine   4-methylbenzenesulfonic acid

Step 3: Azirine on hydrolysis affords α-amine ketone.

Azirine intermediate α–aminoketone

Applications

a) *The reaction is applicable in synthesis of different α-amino ketones. For example: Synthesis of 2-amino-1-phenylethanone and 4-methylbenzenesulfonic acid.*

Acetophenone *O*-tosyl oxime   NaOEt, Pyridine   2-amino-1-phenylethanone   4-methylbenzenesulfonic acid

b) *Transformation of Propan-2-one o-tosyloxime into 1-aminopropan-2-one and 4-methylbenzenesulfonic acid.*

Propan-2-one *o*-tosyl oxime   NaOEt, Pyridine   1-aminopropan-2-one   4-methylbenzenesulfonic acid

Principle

Nef in 1894 reported and explained the conversion of primary or secondary nitroalkane into appropriate carbonyl compounds (aldehyde or ketone) by means of deprotonation of nitroalkane to an *aci*-nitro compound followed by hydrolysis with a strong acid is known as Nef reaction. The primary nitroalkanes also converted into carboxylic acids under similar conditions, depending on the proton donating capability of the acid. Again, it should be pointed out that the yield of a ketone is generally low if the alkaline solution of a nitro-alkane is added to acid. Furthermore, if there is considerable resonance stabilization in nitroalkane, for example: conjugation with an alkene, or a steric hindrance, the Nef reaction will proceed with difficulty. It is very easy to monitor the reaction because of its characteristic transitory blue color.

General Reaction

$$O_2N-\underset{\underset{H}{|}}{\overset{\overset{R'}{|}}{C}}-R \quad \xrightarrow[\text{ii) H}_2\text{SO}_4]{\text{i) NaOH}} \quad R-\overset{\overset{O}{\|}}{C}-R'$$

Nitroalkane → Carbonyl compounds

Mechanism

Step 1: Formation of nitronic acid.

$$\underset{\text{deprotonation}}{\xrightarrow{\quad H^{\oplus} \quad}} \quad \text{nitronate} \quad \xrightarrow{\;+H^{\oplus}\;} \quad \text{nitronic acid}$$

Step 2: Hydrolysis

$$\xrightarrow{\text{hydrolysis}} \quad \xrightarrow{\;-H_2O\;}$$

Step 3: Formation of aldehyde or ketone (carbonyl compounds).

$R' = H$, alkyl group
$R = H$, alkyl group

$1/2 N_2O + 1/2 H_2O$

Applications

a) *The reaction has been broadly applicable in preparation of aldehydes and ketones, such as 1,4-diketones and dihydrofurans.*

Nitroethane $\xrightarrow{\text{NaOH}}$ Sodium salt of nitroethane $\xrightarrow{H_2SO_4/H^{\oplus}}$ H_3C–CHO (Ethanal)

2-nitropropane $\xrightarrow{\text{NaOH}}$ Sodium salt of nitropropane $\xrightarrow{H_2SO_4/H^{\oplus}}$ propan-2-one

b) *The 1,4-diketones has been synthesized by Michael reaction between a nitro-alkane and an α,β-unsaturated ketone followed by Nef reaction, whereas dihydrofurans were prepared by Nef reaction followed by cyclohydration.*

$\xrightarrow{\text{1. CTABr, NaOH, } H_2O \quad \text{2. 3.75 M } H_2SO_4}$

(6a*R*,10a*S*)-8,9-dimethyl-6a-nitro-6a,7,10,10a-tetrahydro-6*H*-benzo[*c*]chromen-6-one

2,3-dimethyl-1,4-dihydrodibenzo[*b*,*d*]furan

Principle

In haloalkanes, the C-X bond is more polarized due to high electronegativity of halogen, and thus carbon is electrophilic in nature. The substitution reactions in haloalkanes involve the replacement of halogen (leaving group) by a nucleophile are known as nucleophilic substitution reactions. Halide ions are acting as good leaving group and hence easily replaced by nucleophilic attack.

General Reaction

$$R-X \ + \ ^{\ominus}\ddot{N}u \ \longrightarrow \ R-Nu \ + \ \overset{\ominus}{\ddot{X}:}$$

Haloalkane (Substrate)   Nucleophile (Entering group)   Nucleophilic substituted product   Leaving group (Halide ion)

Depending upon nature of substrate, nature of reagent and reaction conditions; the nucleophilic reactions proceeds:

a) Substitutions nucleophilic unimolecular mechanism

b) Substitutions nucleophilic bimolecular mechanism

c) Substitutions nucleophilic internal mechanism

Unimolecular nucleophilic substitution reaction

Haloalkane   Carbocation   Nucleophilic substituted product

Mechanism

Unimolecular nucleophilic substitution mechanism

The nucleophilic substitution in a unimolecular reaction, *i.e.* S_N1 mechanism is favored in presence of weak bases or poor nucleophiles.

It involves two steps: The first step involves heterolytic cleavage of C-X bond that results in removal of halide ion and formation of a carbocation. The second step involves the attack of nucleophile on carbocation to form the substituted product.

Step 1: Heterolytic cleavage of C-X bond results in formation of carbocation. In haloalkanes, the C-X carbon is SP^3 hybridized and results in formation of carbocation as reactive intermediate.

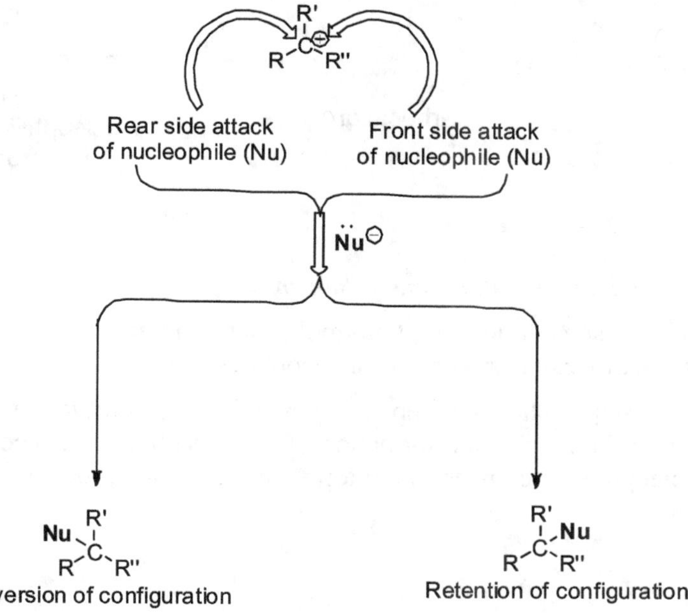

The formation of carbocation is slow and rate determining step. The rate of reaction depends only on the concentration of haloalkanes. After formation of carbocation, it is readily reacts with nucleophile to give product.

As the rate of reaction depends upon only one reactant, it is known as unimolecular reaction following first order kinetics.

The ease of formation of carbocation is directly related to the stability of carbocation. More stable carbocation is formed more readily and order of stability of carbocation is $3° > 2° > 1° > {}^+CH_3$. The order of reactivity for haloalkanes is $3° > 2° > 1° > CH_3\text{-}X$.

Step 2: Fast step of attack of nucleophile on formed carbocation.

The SP^2 hybridized carbocation has planar structure and undergo attack by nucleophile from front side or rear side results in formation of product.

The retention of configuration results when the nucleophile attacks from the front side. The back side attack of nucleophile results in product having inversion of configuration to that of haloalkanes.

The S_N1 mechanism involves retention as well as inversion of configuration as follows:

The (*S*)-3-bromo-3-methylhexane undergo S_N1 reaction in presence of sodium hydroxide to produce both inversed and retention product as follows:

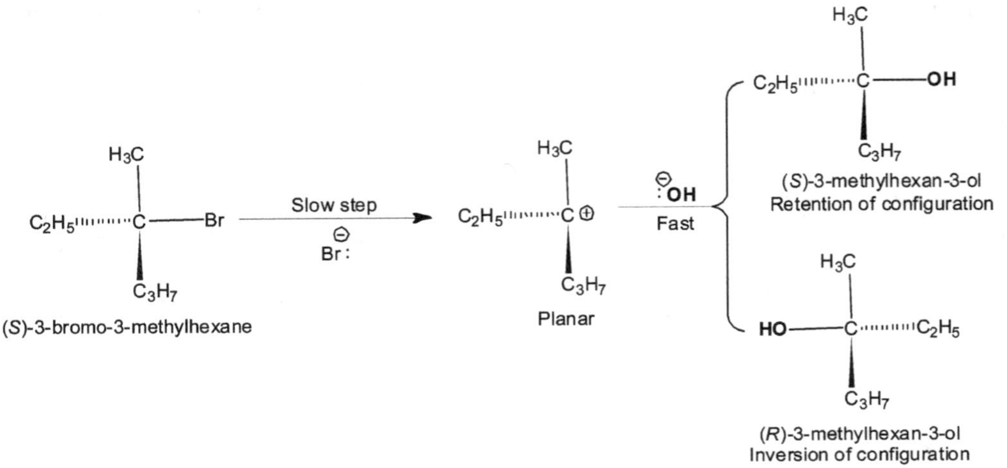

(*S*)-3-methylhexan-3-ol
Retention of configuration

(*S*)-3-bromo-3-methylhexane

Slow step

Planar

(*R*)-3-methylhexan-3-ol
Inversion of configuration

Bimolecular nucleophilic substitution reactions

In this reaction, the simultaneous breaking and making of bond takes place to give the product in a single step. The attack of nucleophile at SP^3 hybridized carbon of haloalkane and removal of halide ion as a leaving group occurs simultaneously to give substituted product. Hence in this reaction, the attack of nucleophile always occurs through back side of halogen. The reaction has pentavalent transition state; where nucleophile and halogen atom attached partially to central carbon atom. S_N2 mechanism involves the inversion of configuration due to back side attack of nucleophile known as *Walden Inversion of configuration*.

For example: *Hydrolysis of 2-bromobutane in presence of base to give (S)-butan-2-ol.*

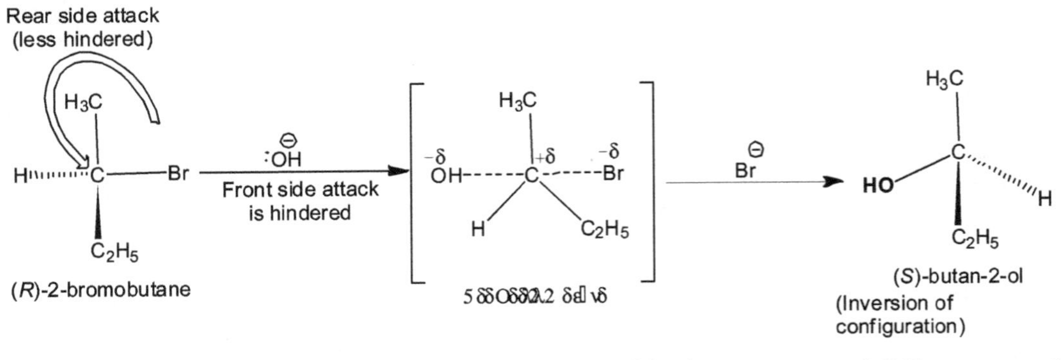

Rear side attack
(less hindered)

Front side attack
is hindered

(*R*)-2-bromobutane

(*S*)-butan-2-ol
(Inversion of configuration)

   The rate of reaction depends on concentration of both reactants and follows second order kinetics. The order of reactivity of haloalkanes is depend on stability of transition state.

a) A primary haloalkane undergoes S_N2 reaction more easily due to less steric hindrance in transition state and formation of more stable carbocation with lower energy.

b) A number of alkyl groups attached to carbon atoms makes the carbocation formed as more unstable due to steric hindrance produced by bulky groups, high energy transition state and slows down the rate of reaction. Order of reactivity of haloalkanes is as follows: $CH_3 > 1° > 2° > 3°$.

The effect of substituents on stability of transition state is given below:

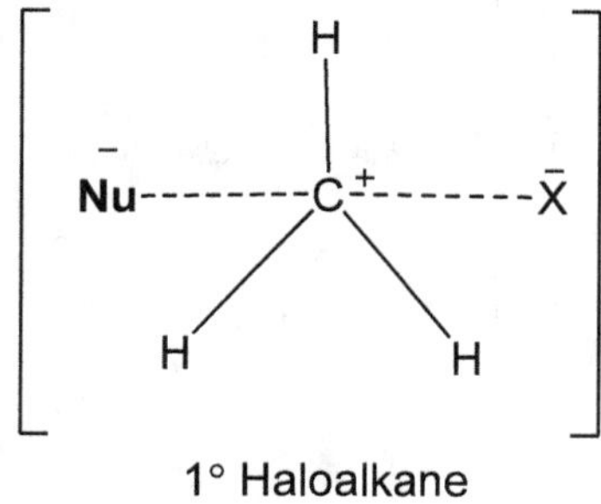

1° Haloalkane

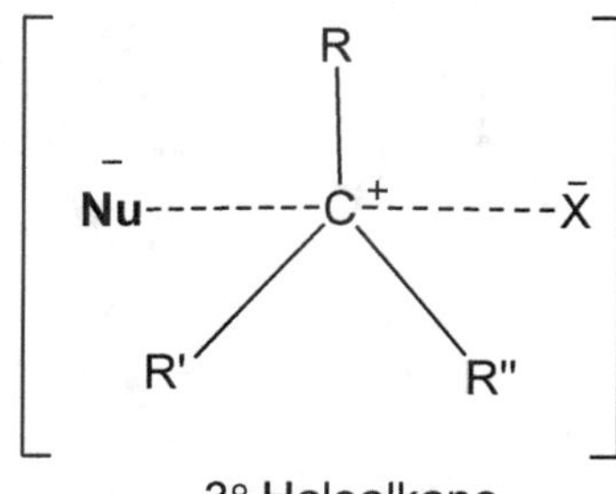

3° Haloalkane

138. Oppenauer Oxidation

Principle

The oxidation of secondary alcohols into ketones using another carbonyl compounds as an oxidant and catalyzed by aluminum tri-isopropanoxide (Al(O-i-Pr)$_3$) which is then reduced into corresponding alcohol is known as Oppenauer oxidation. Oppenauer in 1937 reported this reversible reaction. As it is the reversal of Meerwein-Ponndorf-Verley reduction, the interchange between alcohols and ketones in presence of an aluminum catalyst is known as the MPVO (Meerwein-Ponndorf-Verley-Oppenauer) reaction. When compared with other reported methods of oxidation, the Oppenauer oxidation is much safer for environment and is well tolerated by much other functionality includes unsaturated hydrocarbons, amines, carbonyls, halogens, and sulfur-containing groups. Reaction is highly selective and avoids oxidation of primary alcohol into carboxylic acid. It is common method for preparation of varieties of ketones.

However, Oppenauer oxidation also has problems in oxidation of primary alcohols and α-amino alcohols. The problems arise mainly due to Aldol condensation between formed aldehyde and ketone, deactivation of aluminum alkoxide by water, and the accompanying Tishchenko reaction. Hence, a benzaldehyde is often used as an oxidant in Oppenauer oxidation. The equilibrium can be controlled by amount of acetone, an excess of acetone favors the oxidation of alcohol. The failure for oxidation of α-amino alcohol is probably due to formation of stable cyclic complex between α-amino alcohol and aluminum alkoxide, which does not undergo further reaction. Hence, α-amino ketones are used as oxidants; for example: 1-methyl-4-piperidone.

General Reaction

a) Oppenauer oxidation of secondary alcohol into ketone and alcohol.

b) Oppenauer oxidation of secondary alcohol along with cyclohexanone in presence of aluminium tert-butoxide to give ketone and cyclohexanol.

Mechanism

Step 1: The isopropyl alcohol and aluminium tert-butoxide reacts to form Aluminium derivative of 2° alcohol.

$$3\ HO-\underset{\underset{H}{|}}{\overset{\overset{CH_3}{|}}{C}}-CH_3 \ + \ Al(OCCH_3)_3 \rightleftharpoons [(CH_3)_2CHO]_3Al \ + \ 3\ HO-\underset{\underset{CH_3}{|}}{\overset{\overset{CH_3}{|}}{C}}-CH_3$$

Isopropyl alcohol   Aluminium tert-butoxide   Aluminium derivative   tert-butyl alcohol

Step 2: The Aluminium derivative of 2° alcohol combines with acetone resulting in cyclic transition state which undergoes internal hydride ion transfer, resulting in the oxidation of alcohol to afford .ketone. As the reaction is reversal of MPV reduction, it involves the migration of a hydride ion within an acylic complex. The hydrogen transfer occurs through a six-membered transition state to give a new ketone and aluminum alkoxide.

$[(CH_3)_2CHO]_3Al$ + $H_3C-\overset{\overset{O}{||}}{C}-CH_3$ $\longrightarrow$

Aluminium derivative   Acetone
of 2° alcohol

Cyclic transition state

Transfer of hydride ion
Oxidation

$H_3C-\overset{\overset{O}{||}}{C}-CH_3$ + $(CH_3)_2CHO-Al[(OCHCH_3)_2]_2$

Acetone

Applications

The reaction has been useful in the preparation of ketones.

a) *Oppenauer oxidation is highly specific and selective for alcohols. The reaction is specifically used to synthesize ketones from 2° alcohols containing other oxidiazable functional groups (olefinic double bonds and phenolic groups).*

(3*E*,5*E*)-6-methylocta-3,5,7-trien-2-ol $\xrightarrow[\text{Acetone:benzene}]{Al(OCCH_3)_3}$ (3*E*,5*E*)-6-methylocta-3,5,7-trien-2-one

b) *Synthesis of α-ionone from α-ionol.*

$$\alpha\text{-ionol} \xrightarrow[\text{Acetone:benzene}]{\text{Al(OCCH}_3)_3} \alpha\text{-ionone}$$

c) *Aromatic alcohols are oxidized to aromatic ketones.*

$$\text{1-phenylethanol} \xrightarrow[\text{Acetone:benzene}]{\text{Al(OCCH}_3)_3} \text{Acetophenone}$$

139. Orton Rearrangement

Principle

The rearrangement of *N*-chloro acetanilide in presence of hydrochloric acid to yield mixture of *N*-acyl *o*-chloroanilide and *N*-acyl *p*-chloroanilide is known as Orton rearrangement. The reaction was broadly explored by Orton during 1899.

The rearrangement also takes place in presence of Lewis acid, for example: Silver tetrafluoroborate ($AgBF_4$); or in presence of light, for example: Irradiation from a mercury-vapor lamp in solid state. It was pointed that, when viscosity of solvent has been increased; both yield and *ortho/para* ratio of isomers has increased. It is interesting that in the presence of HNO_3, H_2SO_4, HCl, or $HClO_4$; *N*-nitro-2,4-dichloroaniline is converted into 2-nitro-4,6-dichloroaniline, whereas 2,4-dichlorobenzenediazonium bromide is obtained quantitatively in the presence of HBr. This rearrangement is an example of aromatic rearrangement.

Aromatic rearrangement: The rearrangement in which a group migrates from one position to another in an aromatic system or from a side chain on to the aromatic nucleus is known as Aromatic rearrangement.

General Reaction

N-chloroacetanilide   Dil. HCl   →   *o*-chloroacetanilide   +   *p*-chloroacetanilide

Mechanism

Intermolecular aromatic rearrangement

The basic principle involved in Orton rearrangement; a conjugate acid of aniline eliminates as electrophilic species which then reacts to yield *ortho* and *para* substituted product.

The reaction is assumed to proceed *via* a multistep processes, including the liberation of chlorine and subsequent electrophilic substitution. The former step is reversible, whereas the electrophilic substitution is irreversible. It is known that the Orton rearrangement is promoted simultaneously by proton and chloride. The migrating group, chlorine atom becomes completely detached from the molecule and then combines with chloride ion (Cl⁻) from HCl to from chlorine molecule.

Step 1:

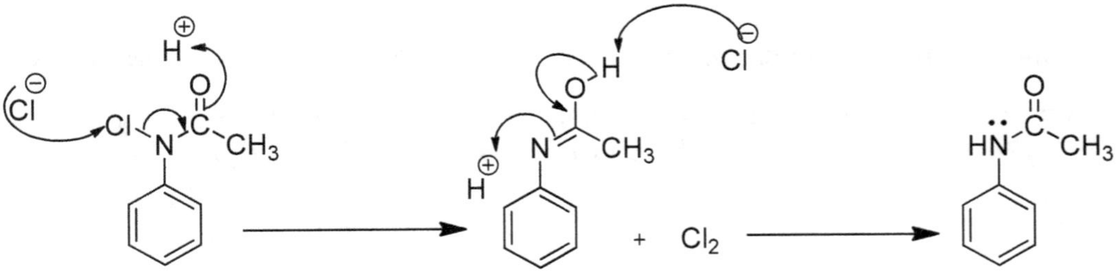

Acetanilide

Step 2: Aromatic chlorination. As only one molecule of chlorine is produced for every molecule of *N*-chloroacetanilide, the aromatic halogenations stops at monochlorination stage to afford *ortho* and *para* substituted product.

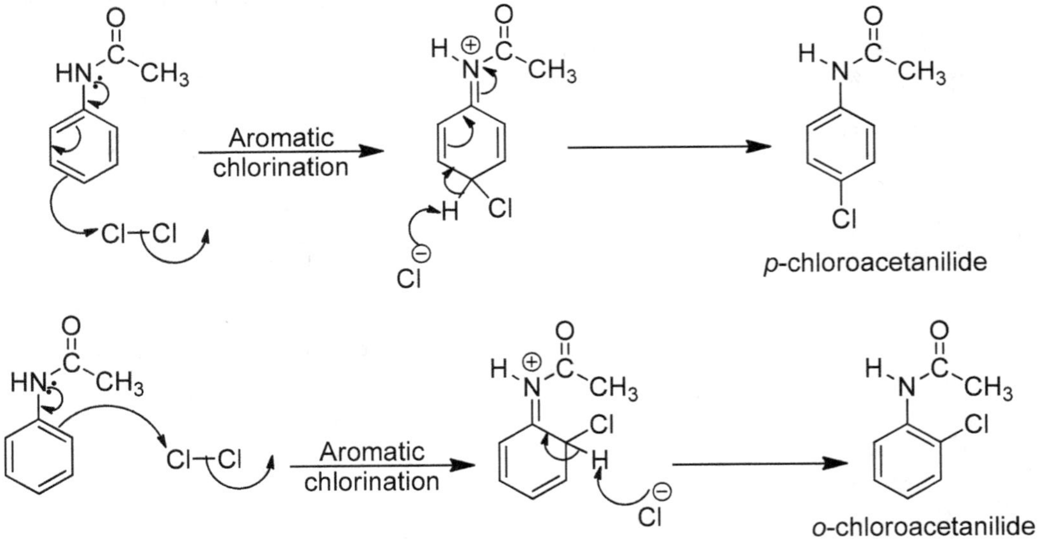

Aromatic chlorination

p-chloroacetanilide

Aromatic chlorination

o-chloroacetanilide

Applications

*a) The reaction is widely useful for the preparation of para-halo anilides. For example:
Synthesis of 4-methoxy-N-methyl-2-nitroaniline from N-(4-methoxyphenyl)-N-methylnitramide.*

N-(4-methoxyphenyl)-*N*-methylnitramide

HCl, CH$_3$OH / 85%

4-methoxy-*N*-methyl-2-nitroaniline

b) Rearrangement of (E)-N-(azulen-1-ylmethylene)aniline into (E)-N-((3-chloroazulen-1-yl)methylene)aniline and (E)-4-chloro-N-((3-chloroazulen-1-yl)methylene)aniline.

(*E*)-*N*-(azulen-1-ylmethylene)aniline

CuCl$_2$ / Dry CH$_3$CN

(*E*)-*N*-((3-chloroazulen-1-yl)methylene)aniline

+

(*E*)-4-chloro-*N*-((3-chloroazulen-1-yl)methylene)aniline

Principle

Overman in 1974 reported and explored a thermal or Hg(II) or Pd(II) promoted rearrangement of an allyl trichloroacetimidate into an allyl trichloroacetamide is considered as Overman rearrangement. It is also known to be *aza-oxa*-Cope rearrangement or Overman imidate rearrangement. Allyl trichloroacetimidate is readily available by reaction between allyl alcohol and trichloroacetonitrile in presence of catalytic amount of potassium hydride. The resulting allylic trichloroacetamides can be hydrolyzed or degraded into

a wide variety of nitrogen containing products, including amino sugars, nucleosides, peptides, amino acids, and many nitrogen heterocycles. The Overman rearrangement is generally carried out in dry and non-nucleophilic solvent to minimize the hydrolysis of allyl trichloroacetimidate. This reaction is useful for preparation of sterically hindered allylic amines that are difficult to obtain from conventional methods.

General Reaction

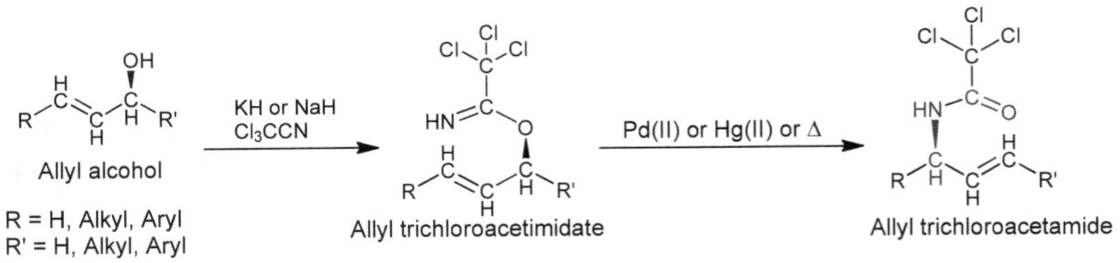

Allyl alcohol

R = H, Alkyl, Aryl
R' = H, Alkyl, Aryl

$\xrightarrow{\text{KH or NaH} \atop \text{Cl}_3\text{CCN}}$ Allyl trichloroacetimidate $\xrightarrow{\text{Pd(II) or Hg(II) or } \Delta}$ Allyl trichloroacetamide

Mechanism

It is a [3,3]-sigmatropic rearrangement with an excellent regiochemical and geometrical control as well as complete retention of stereochemistry.

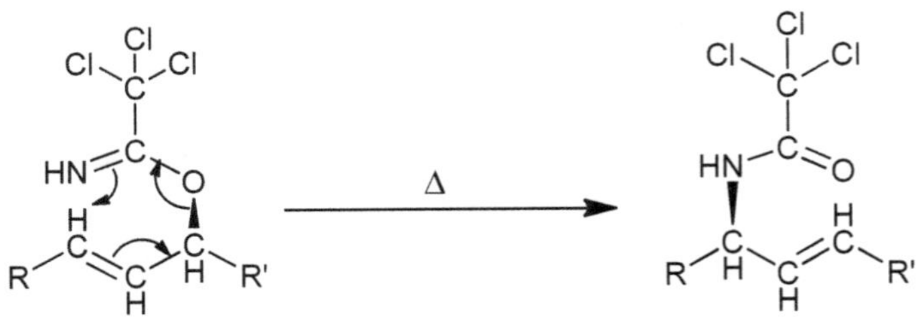

Applications

a) The rearrangement has broad range of application in organic chemistry for the synthesis of wide varieties of nitrogen-containing compounds.

N,N,1,3-tetramethyl-1,3,2-diazaphospholidin-2-amine

N-(1,3-dimethyl-2-oxido-1,3,2-diazaphospholidin-2-yl)-
N-(hex-1-en-3-yl)-4-methylbenzenesulfonamide

141. Paal-Knorr Furan Synthesis

Principle

An acid-catalyzed conversion of 1,4-dicarbonyl compounds into furan analogues was reported by Paal and Knorr in 1885; hence this reaction is known as Paal-Knorr Furan synthesis. The acidic catalysts used in this reaction are P_2O_5 in ethanol, *p*-TsOH or $ZnCl_2$ in acetic anhydride, and polyphosphoric acid. It is reported that *meso*-diastereomers with substituents between two carbonyl groups cyclize more easily when the substituents are electron donating groups (For example: alkyl); whereas the *dl*-isomers cyclize rapidly with bulky substituents. The reaction is believed to involve the protonation of carbonyl group that tautomerizes to enol, which attacks second protonated carbonyl group.

General Reaction

1,4-dicarbonyl compounds   Dialkyl derivative of Furan

Mechanism

Step 1: Enolization of dicarbonyl compounds.

Dicarbonyl compounds   Enol

Step 2: Intramolecular nucleophilic attack (ring closure).

Enol

Step 3: Protonation of cyclic product followed by removal of water molecule to give intermediate.

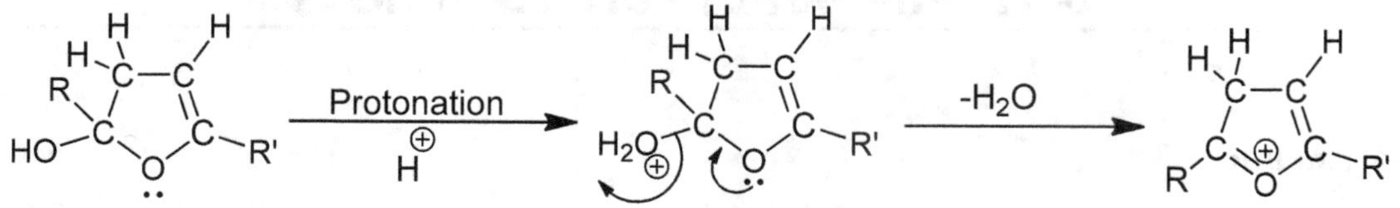

Step 4: Hydrolysis to form furan analogues.

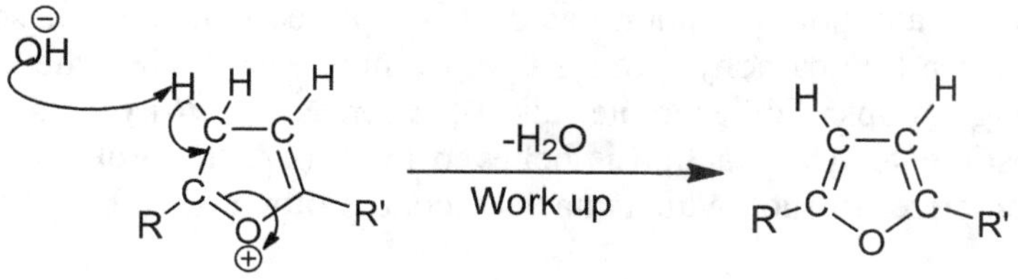

Applications

The reaction is useful in synthesis of furan derivatives.

a) Synthesis of 2,5-diphenylfuran.

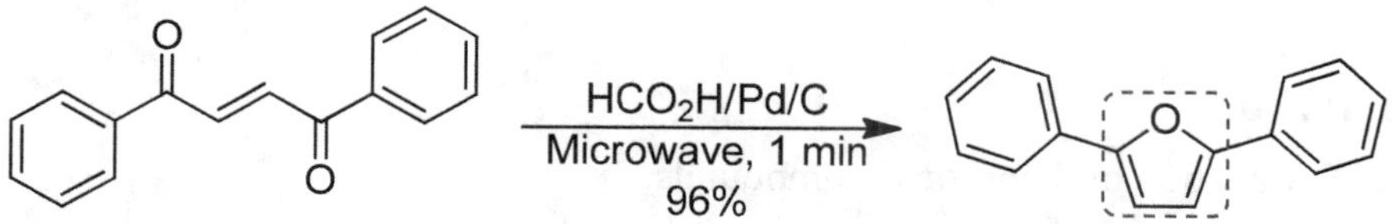

$$\text{(E)-1,4-diphenylbut-2-ene-1,4-dione} \xrightarrow[\substack{\text{Microwave, 1 min} \\ 96\%}]{\text{HCO}_2\text{H/Pd/C}} \text{2,5-diphenylfuran}$$

(E)-1,4-diphenylbut-2-ene-1,4-dione   2,5-diphenylfuran

b) Synthesis of 2,5-dimethylfuran.

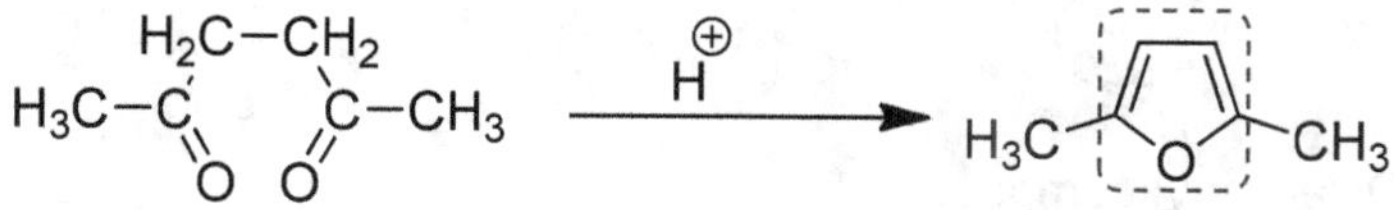

Hexane-2,5-dione   2,5-dimethylfuran

c) Synthesis of 2-ethyl-5-methylfuran

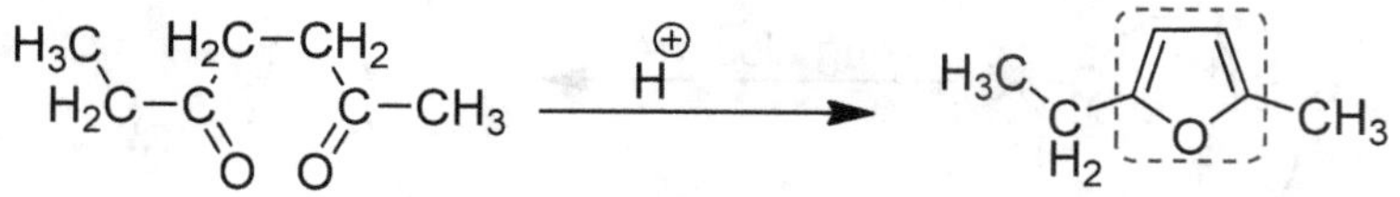

Heptane-2,5-dione   2-ethyl-5-methylfuran

142. Paal-Knorr Pyrrole Synthesis

Principle

An acid-catalyzed transformation of 1,4-dicarbonyl compounds and ammonia or primary amines into pyrrole derivatives is known as Paal-Knorr pyrrole synthesis. The reaction was reported by Paal and Knorr in 1884. The reaction is catalyzed in either a Brønsted acid or Lewis acid, by which the latter is especially useful for molecules with acid-sensitive/epimerizable functional groups. In this reaction, the 1,4-dicarbonyl compounds provide the four carbons of the pyrrole ring through double condensation with amines that contribute the nitrogen to the ring. However, the primary amines with an alkyl substituent at the α-position do not undergo this reaction. For example, no pyrrole was obtained by reaction with cyclohexyl amine.

General Reaction

1,4-dicarbonyl compounds $\xrightarrow{R''-NH_2}$ Pyrrole

Mechanism

Step 1: Attack of nucleophile (amine) on one of the carbonyl groups of 1,4-dicarbonyl compounds results in the formation of imine intermediate.

Dicarbonyl compounds $\xrightarrow{\text{Imine formation}}$ Imine

Step 2: Imine intermediate undergoes intramolecular ring closure reaction followed by removal of water molecule to give substituted pyrrole derivatives.

Imine $\xrightarrow{\text{Ring closure}}$ $\xrightarrow{-H^{\oplus}}$ $\xrightarrow{-H_2O}$

Applications

The reaction is applied in the synthesis of pyrrole analogues.

a) Synthesis of 2,5-dimethyl-1H-pyrrole.

Hexane-2,5-dione → $(NH_4)_2CO_3$ → 2,5-dimethyl-1*H*-pyrrole

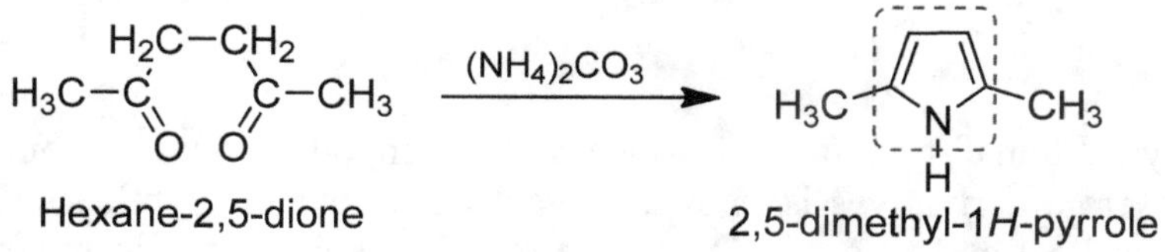

b) Synthesis of 2,5-dimethyl-1H-pyrrole.

Heptane-2,5-dione → CH_3-NH_2 → 2-ethyl-1,5-dimethyl-1*H*-pyrrole

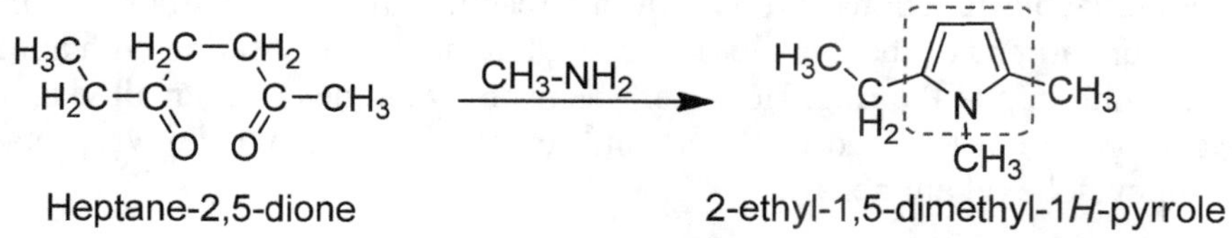

c) Synthesis of 1-(4-fluorophenyl)-2-methyl-5-(4-(methylsulfonyl)phenyl)-1H-pyrrole.

1-(4-(methylsulfonyl)phenyl)pentane-1,4-dione + 4-fluoroaniline → Toluene, Δ; TsOH, 79% →

1-(4-fluorophenyl)-2-methyl-5-(4-(methylsulfonyl)phenyl)-1*H*-pyrrole

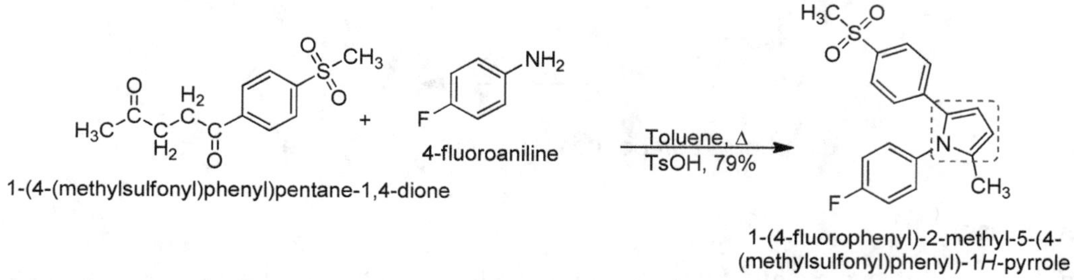

Principle

Pechmann and Duisberg in 1883 carried out synthesis of coumarins in acidic condition from phenols and β-ketoesters or β-keto carboxylic acids as condensation reaction is known as the Pechmann condensation or Pechmann cyclization. The outcome of reaction depends on nature of phenols, β-keto esters, and condensation reagents. For example, only phenols with electron-donating groups at *meta*-position of hydroxy group undergo Pechman condensation, whereas phenols as electron-withdrawing substitutents at *meta*-positions or with electron-donating groups at other positions fail for this reaction. The α-substituents and γ-substituents on β-keto esters retard this reaction. More often, the reaction is carried out in H_2SO_4 or 70% H_2SO_4, other acids, such as lewis acids, are also effective in promoting this reaction, $POCl_3$, P_2O_5, HF, sulfamic acid, TFA, triflic acid, polyphosphoric acid, $KHSO_4$, dipyridine copper chloride, $ZnCl_2$, and $AlCl_3$.

General Reaction

Mechanism

Pechmann coumarin condensation under acidic conditions, involves an esterification or trans-esterification, followed by a cyclization and dehydration to produce coumarin analogues. The detailed mechanism of reaction is illustrated here:

Step 1:

Step 2:

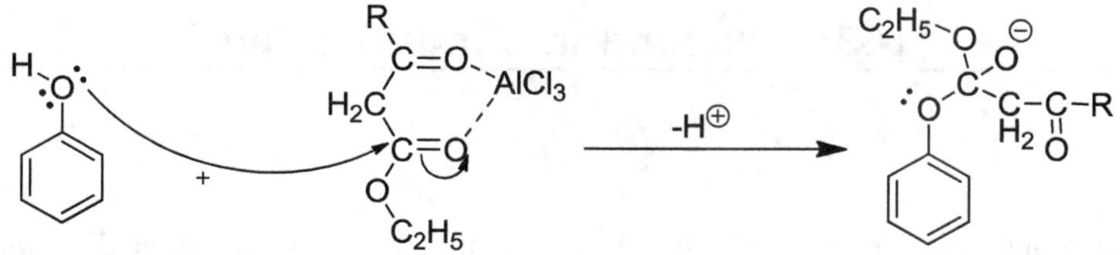

Step 3:

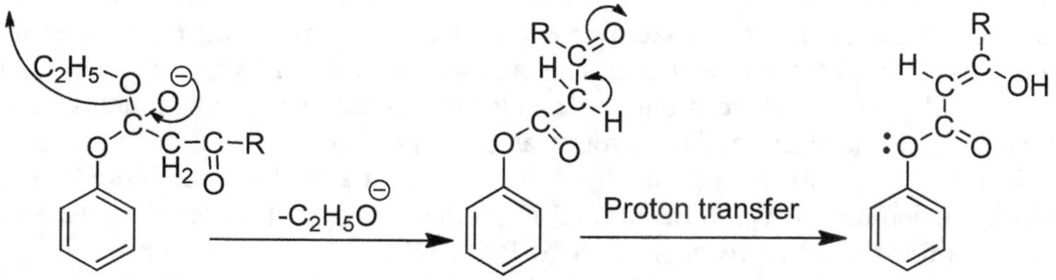

Step 4:

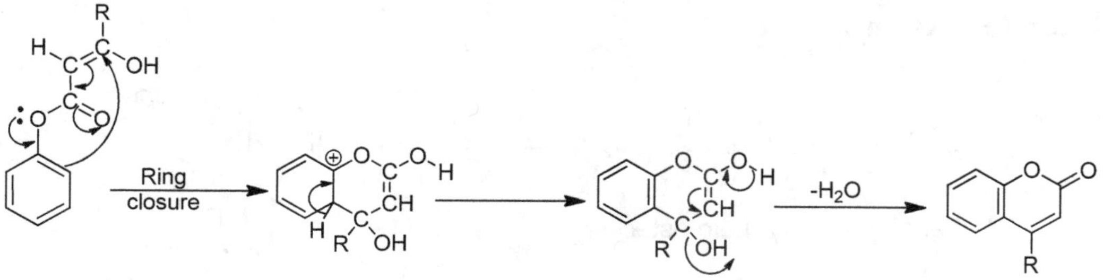

Applications

The reaction is useful in the synthesis of coumarin analogues.

a) Synthesis of 4-methyl coumarin.

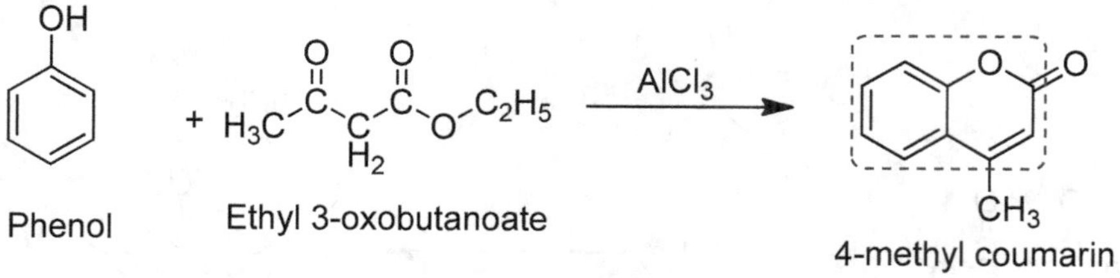

Phenol   Ethyl 3-oxobutanoate

4-methyl coumarin

b) Synthesis of coumarin analogues.

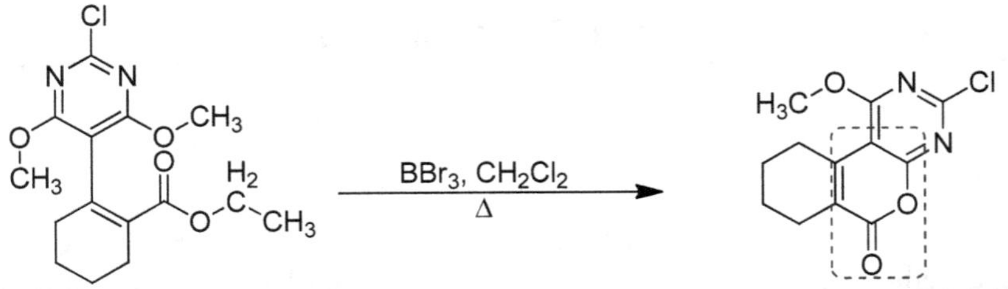

BBr$_3$, CH$_2$Cl$_2$, Δ

Ethyl 2-(2-chloro-4,6-dimethoxy
pyrimidin-5-yl)cyclohex-1-enecarboxylate

3-chloro-1-methoxy-7,8,9,10-tetrahydro-
6*H*-isochromeno[3,4-*d*]pyrimidin-6-one

144. Pechmann Pyrazole Synthesis

Principle

Pechmann in 1895 reported synthesis of pyrazole analgoues by 1,3-dipolar cycloaddition between a diazomethane (or other diazonium salt) and a molecule with carbon-carbon double bonds to pyrazoline and subsequent oxidation is known as Pechmann pyrazole synthesis. A carbonyl, ester, or cyano group that conjugates to olefinic double bond assists the formation of pyrazoline; whereas phenyl group decreases the reactivity. The mixing of diazomethane with α,β-unsaturated esters or acid, a σ-pyrazoline in which the nitrogen atom is linked to α-carbon of a carbonyl group is formed, which rearranges in presence of hydrochloric acid to give 2-pyrazoline containing carbon-nitrogen double bond in conjugation with carbonyl group. Upon oxidation in presence of with bromine, pyrazole derivatives are formed, although the dehydrogenation occurs without an oxidant present. In addition, a direct preparation of pyrazole from 1,3-Dipolar cycloaddition between diazomethane and acetylene was also developed by Pechmann.

General Reaction

$$H-C\equiv C-H \quad + \quad H_2C=N=N \longrightarrow \text{1}H\text{-pyrazole}$$

Acetylene or Ethyne   Diazomethane

Mechanism

1,3-dipolar cycloaddition reaction between diazomethane and acetylene (ethyne).

$$\xrightarrow[\text{cycloaddition}]{[3+2]}$$

1H-pyrazole

Applications

a) The reaction is used in the synthesis of pyrazoles followed by cyclopropane analogues upon decomposition of pyrazolines. For example: Transformation of (E)-1,4-diphenylbut-2-ene-1,4-dione into (1H-pyrazole-3,4-diyl)bis(phenylmethanone).

(*E*)-1,4-diphenylbut-2-ene-1,4-dione

+ $H_2C=N^+$:N^-
diazomethane

1. $CHCl_3$
2. Δ

(4,5-dihydro-1*H*-pyrazole-3,4-diyl)bis(phenylmethanone)

Br_2, $CHCl_3$

(1*H*-pyrazole-3,4-diyl)bis(phenylmethanone)

Principle

Condensation of aromatic aldehydes with anhydride of aliphatic carboxylic acid containing at least two α-hydrogens, in presence of same aliphatic acid results in the formation of α, β-unsaturated acids is known as Perkins condensation reaction. The reaction is given by only aromatic aldehydes.

General Reaction

Reaction of benzaldehyde with acetic anhydride in presence of sodium salt of acetic acid gives cinnamic acid.

Benzaldehyde + Acetic anhydride $\xrightarrow{CH_3COONa}$ Cinnamic acid (α, β-unsaturated acid)

Mechanism

Step 1: Formation of carbanion by abstraction of α-hydrogen from acid anhydride by base. The anion of sodium salt of acetic acid (acetate ion) acts as a base and easily removes α-hydrogen from the anhydride.

Step 2: Attack of carbanion formed in step-1 on carbonyl carbon of benzaldehyde to from reactive intermediate.

Step 3: Intramolecular acetyl shift. The migration of acetyl group occurs through formation of cyclic intermediate to alkoxy oxygen.

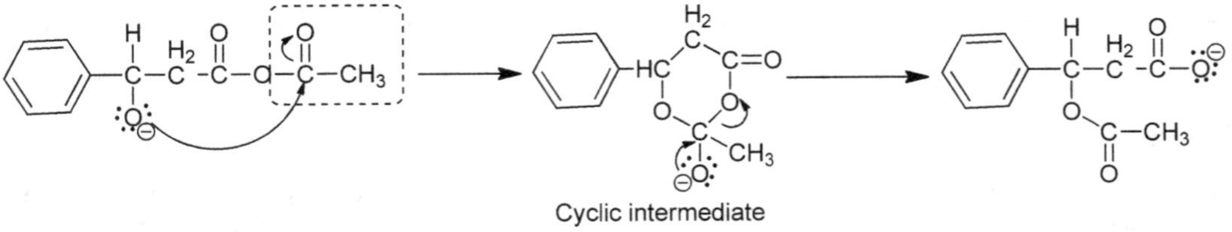

Cyclic intermediate

Step 4: Loss of acetate ion from β-carbon. In the presence of a base, the α-hydrogen is abstracted which in turn causes the removal of a good leaving group to form α,β-unsaturated carboxylic acid.

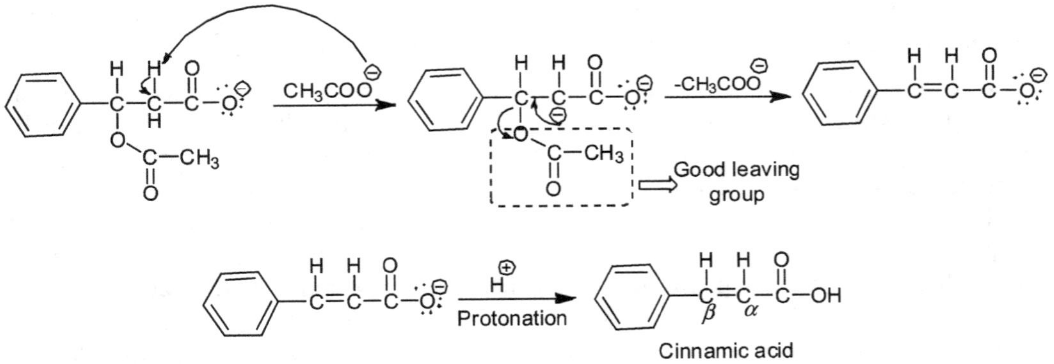

Good leaving group

Cinnamic acid

Applications

a) *Furfural may undergo Perkin reaction to form furylacrylic acid in 65-70% yield.*

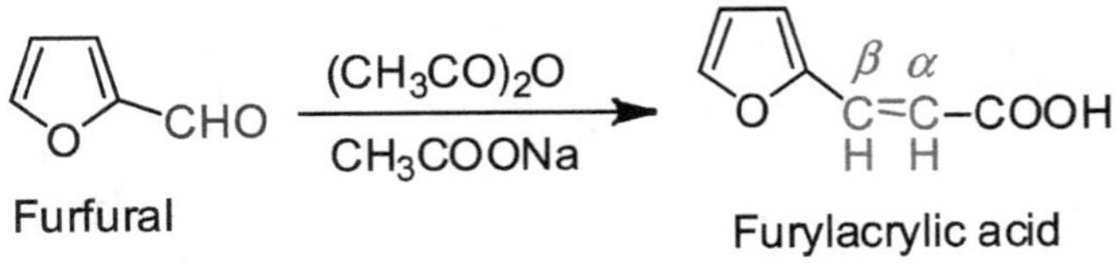

Furfural

Furylacrylic acid

b) *Synthesis of coumarin from 2-hydroxybenzaldehyde.*

2-hydroxybenzaldehyde   3-(2-hydroxyphenyl)acrylic acid   Coumarin

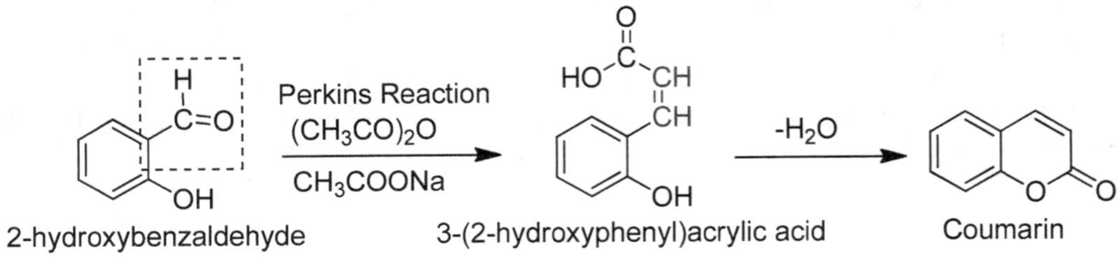

146. Pinacol-Pinacolone Rearrangement

Principle

Fittig in 1860 reported acid-promoted *1,2*-rearrangement of vicinal diols to corresponding aldehydes or ketones. *1,2*-diols with complete substitution like 2,3-dimethyl-2,3-butane diol are known as pinacols. These compounds (pinacols) undergo dehydration and rearrangement in acidic condition to form ketones, the reaction is known as Pinacol-Pinacolone rearrangement. The reaction is often carried out in H_2SO_4, including the application of Brønsted and Lewis acids. For example: Brønsted acids HCOOH, $HClO_4$, CH_3COOH with a catalytic amount of H_2SO_4; Lewis acids includes BF_3 in acetic acid, $ZnCl_2$ in acetic anhydride, $MgBr_2$, and $AlCl_3$ in the absence of solvent.

It is very important to know which hydroxyl group will be dehydrated and which group will migrate in both symmetric and asymmetric pinacols to increase utility of reaction in organic synthesis. The regioselectivity is determined by formation of primarily formed carbocations, the acid used, and the migratory aptitudes of the substituents on the vicinal diols. It is possible that in an asymmetric pinacol, the decrease of the basicity of one hydroxyl group caused by its neighboring substituents might facilitate the protonation on the other hydroxyl group. In addition, it has been reported that for certain pinacol rearrangements, the treatment by Brønsted and Lewis acids leads to different pinacolones. The migratory aptitudes of substituents on diols are determined by their electron density and steric hindrance. The groups with high electron density are easier to migrate than are the groups with electron deficiency, because such migration groups will rearrange to the new-formed carbocation. Hence cyclopropyl, vinyl, 2-thienyl, and 2-furyl groups are good migratory groups, whereas 2 and 3-pyridyl groups have smaller migratory aptitudes than a phenyl group. In addition, a good migratory group might not move in the pinacol rearrangement unless it can reach the adjacent carbon atom by a single movement from the opposite face of the leaving hydroxyl group, involving the Walden Inversion. Solvent and temperature will affect the outcome of pinacol rearrangement. Pinacol rearrangement can occur regularly in primary or secondary alcohols but not in tertiary alcohols because tertiary alcohols will be protonated and hydrated. The rearrangement of aziridines under similar conditions is known as the *aza*-pinacol rearrangement, whereas α-hydroxy epoxides undergo semi-pinacol rearrangements to give aldols. Moreover, a similar rearrangement involving a double bond is referred to as the vinylogous pinacol rearrangement.

General Reaction

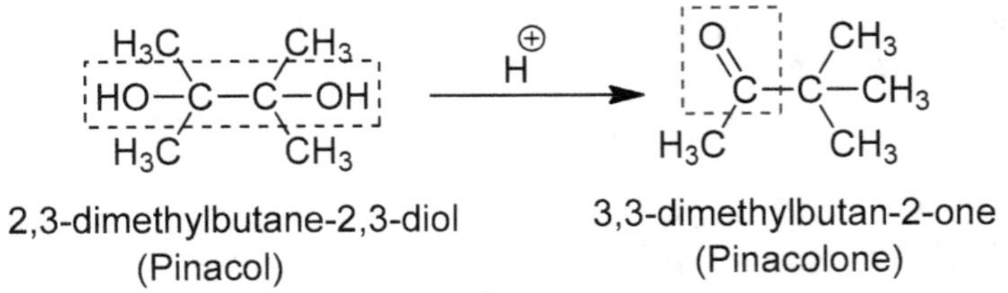

2,3-dimethylbutane-2,3-diol
(Pinacol)

3,3-dimethylbutan-2-one
(Pinacolone)

Mechanism

The reaction believed to proceed by following mechanism:

Step 1: Protonation of pinacol to form protonated pinacol.

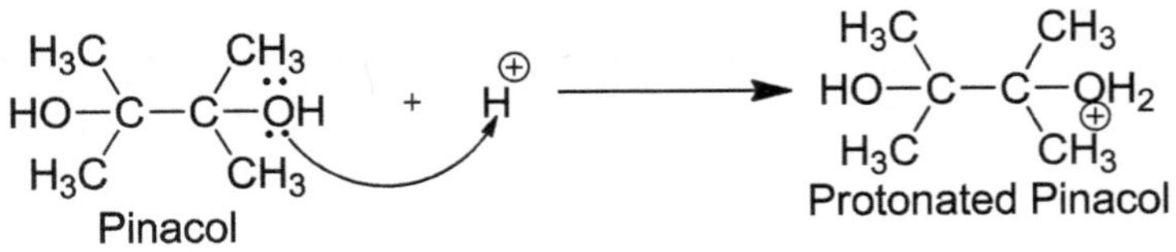

Pinacol

Protonated Pinacol

Step 2: Formation of carbocation by loss of water molecule from the protonated pinacol.

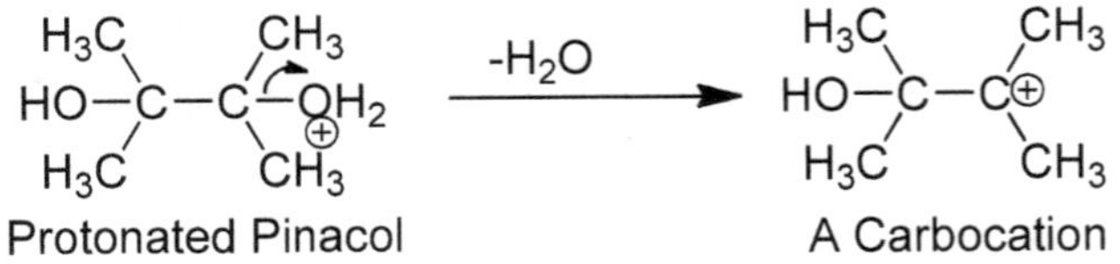

$-H_2O$

Protonated Pinacol

A Carbocation

Step 3: Rearrangement of carbocation by 1,2-shift to give protonated ketone.

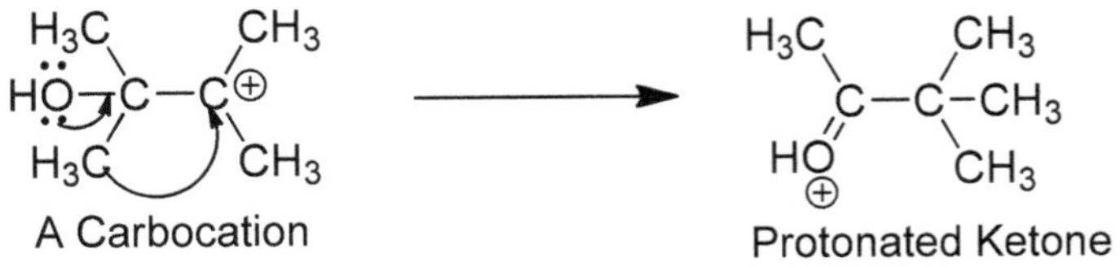

A Carbocation

Protonated Ketone

Step 4: Formation of pinacolone by loss of proton from protonated ketone.

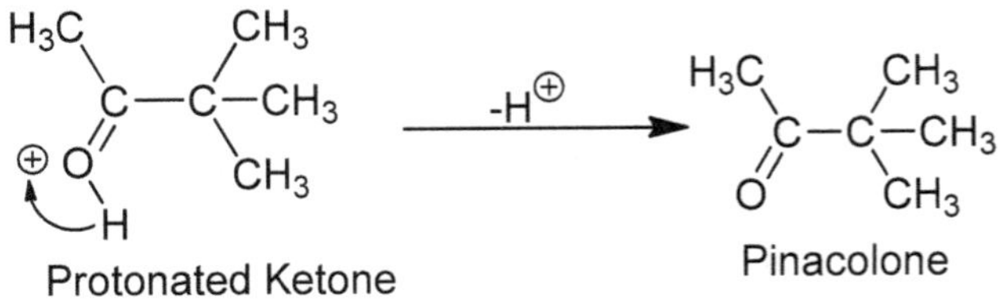

$-H^{\oplus}$

Protonated Ketone

Pinacolone

Applications

a) Pinacol-Pinacolone rearrangement has been effectively applied to cyclic 1,2-diols form conversion into ketones.

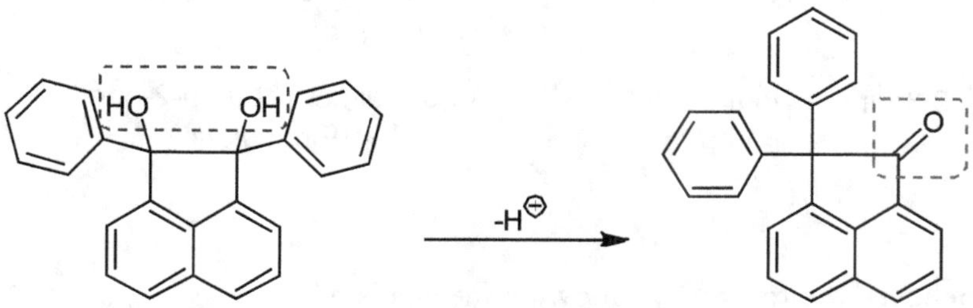

1,2-diphenyl-1,2-dihydroacenaphthylene-1,2-diol
(Cyclic 1,2-diols)

2,2-diphenylacenaphthylen-1(2*H*)-one
(Cyclic ketones)

b) Cyclic ketones can be converted into pinacols which on rearrangement gives spiro-ketones.

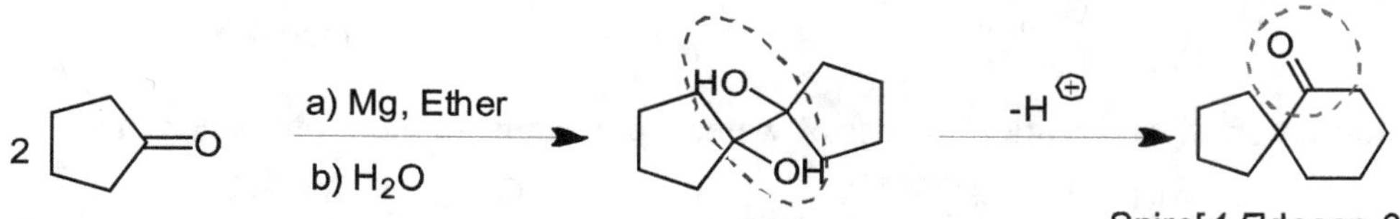

Cyclopentanone

[1,1'-bi(cyclopentane)]-1,1'-diol
(Pinacol)

Spiro[4.5]decan-6-one

147. Prins Reaction

Principle

Prins in 1917 reported acid-catalyzed condensation between aldehydes and olefins to produce 1,3-dioxanes, 1,3-diols, allyl alcohols, or 1,3-diesters, depending on type of alkenes and condition of reaction is known as Prins reaction. The reaction is performed in an aqueous or anhydrous medium, and the reaction is possible with aliphatic, aromatic, α,β-unsaturated aldehydes or cyclic acetals. The acid can be either Brønsted acid or Lewis acid, by which commonly used Bronsted acid is 10-65% sulfuric acid, and Lewis acid can be $AlCl_3$, $ZnCl_2$, BF_3, etc. In the reaction, the commonly used aldehyde is formaldehyde.

General Reaction

For example: Prins reaction between cyclohexene and formaldehyde to give 2-(hydroxymethyl)cyclohexanol.

Mechanism

Step 1: Protonation of aldehyde.

Step 2: Olefinic addition into protonated aldehyde to give 1,3-diols.

Step 3: Nucleophilic addition of formaldehyde to intermediate gives 1,3-diesters.

Intermediate

Applications

a) The reaction has wide application in synthesis of 1,3-dioxanes, 1,3-diols, allyl alcohols, or 1,3-diesters. For example: Synthesis of 4-(4-methoxyphenyl)-1,3-dioxane.

1-methoxy-4-vinylbenzene   Formaldehyde

4-(4-methoxyphenyl)-1,3-dioxane

148. Pummerer Rearrangement

Principle

Pummerer in 1909 reported an acid or an anhydride promoted transformation of sulfoxide bearing an α-proton to α-substituted sulfide through a sulfenium (or thionium) intermediate is known as Pummerer rearrangement. Pummerer cyclization is helpful for the synthesis of heterocyclic compounds through the addition of an intramolecular nucleophile to sulfenium intermediate. Brønsted or Lewis acids are also helpful to activate the reaction, such as p-toluenesulfonic acid and diethylaminosulfur trifluoride. In this reaction, the formed sulfenium (or sulfonium) intermediate is very electrophilic, which can even react with some weak nucleophiles like xylene and anisole.

General Reaction

Sulfoxide
R = Alkyl, Aryl

α-substituted sulfide

R, R' = Alkyl, Aryl

Mechanism

(a) Rearrangement *via* carbonium intermediate involving a concerted cyclic elimination of acetic acid, elimination of acetic acid with an external base or elimination to form a sulfonium ylide;

(*b*) Internal transfer of the acetoxy group through either cyclic rearrangement, 1,2-shift of ylide, homolytic dissociation and recombination, or nucleophilic displacement of ylide.

Applications

a) *The reaction is used for synthesis of an aldehyde by hydrolysis of α-acyloxysulfide, the formation of vinyl sulfide by elimination of α-acyloxy group, and the synthesis of glyoxal from β-ketosulfoxide. The formed sulfide is used as intermediate for reactions such as Diels-Alder cycloaddition and Michael Addition.*

b) *Synthesis of (1R)-1-2-(p-tolylsulfanyl)phenylethan-1-ol.*

(*S*)-1-ethyl-2-(*p*-tolylsulfinyl)benzene   (1*R*)-1-2-(*p*-tolylsulfanyl)phenylethan-1-ol

149. Reformatsky Reaction

Principle: Reformatsky in 1887 reported this condensation reaction. The carbonyl compounds (aldehyde and ketones) react with α-halo ester (esters with halogen atom at α-carbon) in presence of zinc to yield β-hydroxy ester is known as Reformatsky reaction. Reformatsky's protocol is known to have several drawbacks, such as:

1) Problem in its initiation and regulation because of heterogeneity;

2) Requirement for fresh preparation of zinc reagent;

3) Low reproducibility and access only to the thermodynamic products;

4) Low yields because of enolization of Reformtasky reagent and many associated side reactions under the reaction condition, including the self condensation of carbonyl compounds and α-bromoester; elimination of α-bromoester; and retro-Reformatsky fragmentation.

General Reaction

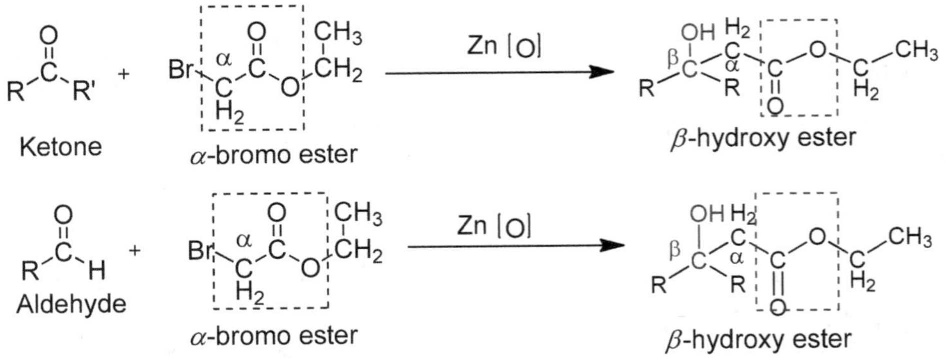

Ketone + α-bromo ester $\xrightarrow{\text{Zn [O]}}$ β-hydroxy ester

Aldehyde + α-bromo ester $\xrightarrow{\text{Zn [O]}}$ β-hydroxy ester

Mechanism

Preparation of organozinc reagent

Oxidative addition of zinc into α-haloester results in the formation of organo zinc intermediate. Zinc with its shaired pair of electrons added to the α-carbon atom. Carbon is not able to hold electron pair and it is shifted to produce Π-bond. The inserted complex between zinc and α-bromoester is generally referred to as the Reformatsky reagent.

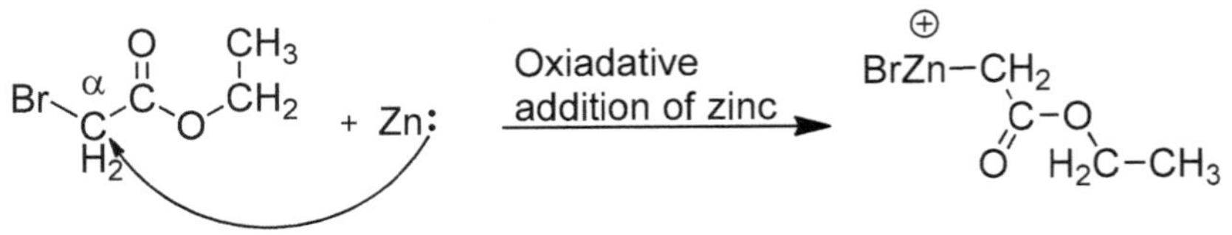

α-bromo ester $+$ Zn: $\xrightarrow{\text{Oxiadative addition of zinc}}$ Reformatsky reagent

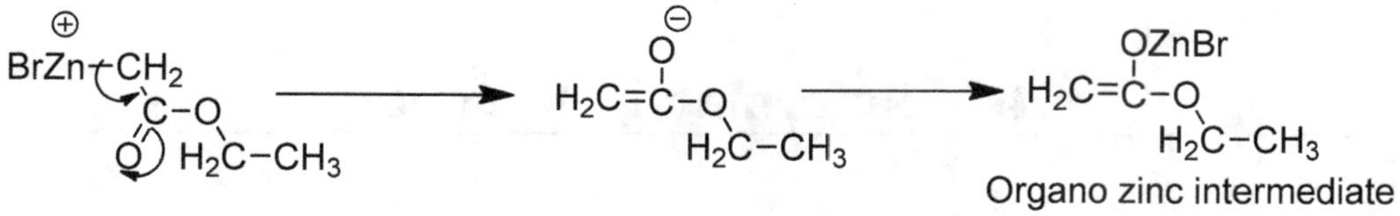

Organo zinc intermediate

Step 1: Nucleophilic addition of organo zinc intermediate into carbonyl carbon atom to form complex.

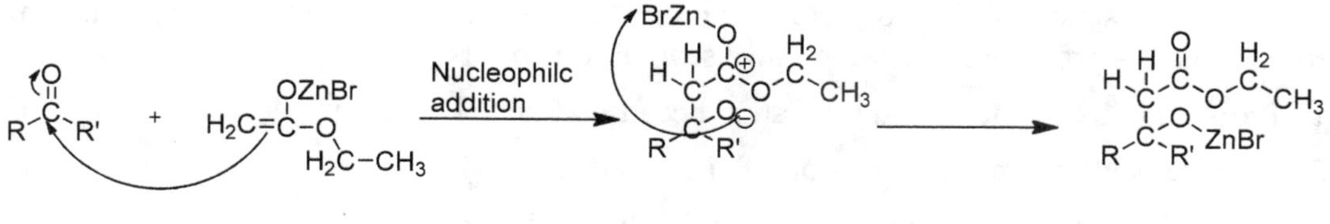

Nucleophilc addition

Step 2: Acidic hydrolysis to afford β-hydroxy ester.

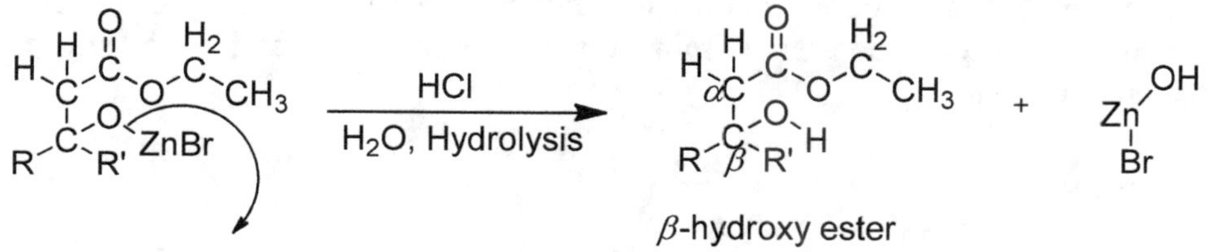

HCl

H_2O, Hydrolysis

β-hydroxy ester

Applications

a) The reaction is used in the synthesis of spiro-α-methylene-γ-butyrolactone from keto steroid.

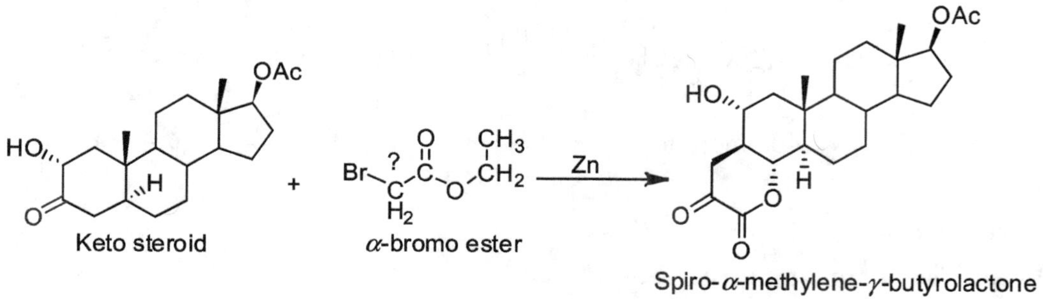

Keto steroid

α-bromo ester

Spiro-α-methylene-γ-butyrolactone

b) *Synthesis of natural products: The reaction is useful in synthesizing natural products, such as geranic esters. Citral synthesis is carried out by reacting 6-methylhept-5-en-2-one and α-iodo ester.*

6-methylhept-5-en-2-one

α-iodo ester

Zn

β-hydroxy ester

Acetic anhydride

$-H_2O$

a) Hydrolysis,
b) Distillation with calcium salt with HCOONa

Citral

150. Reimer-Tiemann Reaction

Principle: The reaction was initially reported by Reimer and subsequently extended by Reimer and Tieman in 1876. Sodium phenoxide, on reaction with chloroform in alkaline medium, results in the formation of *o*-hydroxybenzaldehyde (salicylaldehyde) and *p*-hydroxybenzaldehyde is commonly known as Reimer-Tiemann reaction. It is an electrophilic substitution reaction where the electrophile involved is dichlorocarbene. Dichlorocarbene is generated by the reaction of chloroform and sodium hydroxide. The attack of dichlorocarbene at strongly activated ortho and para positions in phemoxide ion results in the formation of ortho and para products. Salicylaldehyde is formed as major product due to intramolecular hydrogen bonding.

General Reaction

Phenol + CHCl$_3$ + 3NaOH $\xrightarrow{60°C}$ *o*-hydroxybenzaldehyde + *p*-hydroxybenzaldehyde + 3NaCl + 2H$_2$O

a) Synthesis of 4-hydroxy-3-methoxybenzaldehyde.

2-methoxyphenol $\xrightarrow[60°C]{CHCl_3, NaOH}$ 4-hydroxy-3-methoxybenzaldehyde

b) Synthesis of 3,4-dihydroxybenzaldehyde.

Pyrocatechol $\xrightarrow{CHCl_3, NaOH}$ 3,4-dihydroxybenzaldehyde

The reaction proceeds *via* following steps:

Step 1: Generation of electrophile (Formation of dichlorocarbene-*1,1*-elimination).

Dichlorocarbene
(Electrophile)

Step 2: Reaction of electrophile with phenoxide ion.

Phenoxide ion

o-dichloromethylphenoxide

Step 3: Hydrolysis in alkaline medium to give salicylaldehyde which is stabilized by intramolecular hydrogen bonding.

salicylaldehyde

Analogously, the electrophile generated in this reaction is reacts at para position of phenoxide ion to give *p*-hydroxybenzaldehyde.

Applications

The reaction is very useful in the synthesis of different substituted aromatic aldehydes.

a) Synthesis of piperonal from catechol.

Catechol

3,4-dihydroxybenzaldehyde

Piperonal

b) Synthesis of 1H-indole-3-carbaldehyde from 1H-indole.

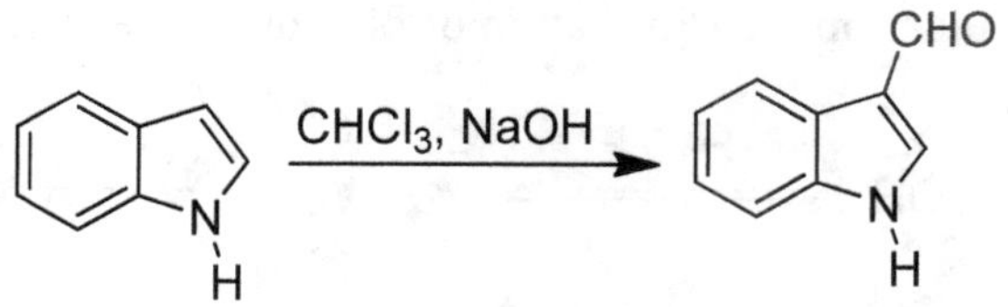
CHCl3, NaOH
CHO
N
H
N
H

1*H*-indole 1*H*-indole-3-carbaldehyde

c) Synthesis of 2-hydroxy-1-naphthaldehyde from napthalen-2-ol

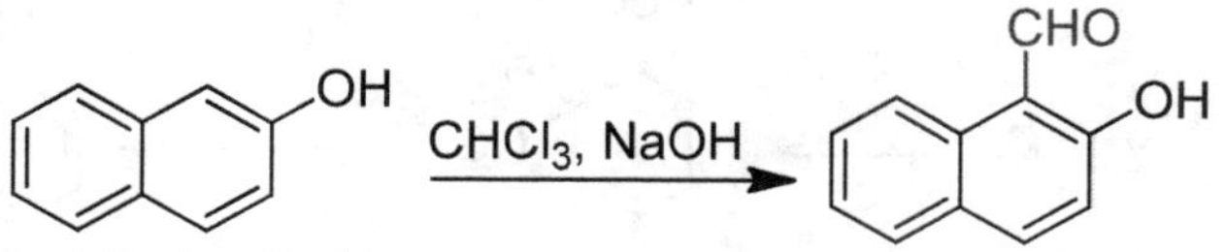
OH
CHCl3, NaOH
CHO
OH

Naphthalen-2-ol 2-hydroxy-1-naphthaldehyde

d) Synthesis of Salicylic acid from phenol.

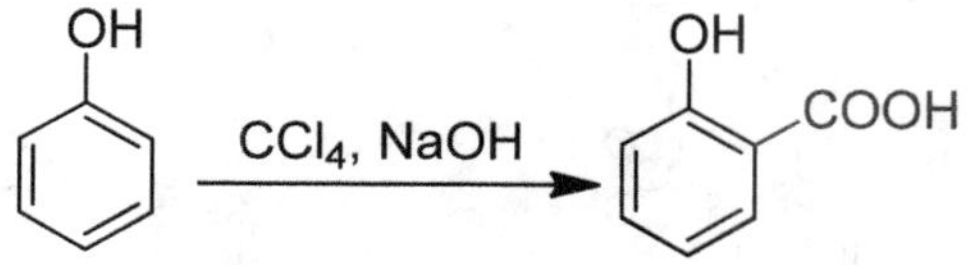
OH
CCl4, NaOH
OH
COOH

Phenol Salicylic acid

151. Reissert Indole Synthesis

Principle

Reissert in 1896 invented and explained multistep synthesis of indoles from *o*-nitrotoluene through condensation with oxalic ester into *o*-nitrophenylpyruvic ester, reduction of nitro group to an amino group followed by cyclization to indole-2-carboxylic acid and final decarboxylation is known as Reissert indole synthesis. In a reaction, *o*-nitrotoluene was deprotonated with sodium ethoxide or potassium ethoxide, where as in this reaction potassium ethoxide offers better results than sodium ethoxide. The resulting *o*-nitrophenylpyruvic ester was reduced by zinc in acetic acid, stannous chloride in methanol, or alkaline ferrous hydroxide, or may be hydrogenated by Pd/C in ethanol.

General Reaction

Mechanism:

Step 1:

Step 2:

Step 3:

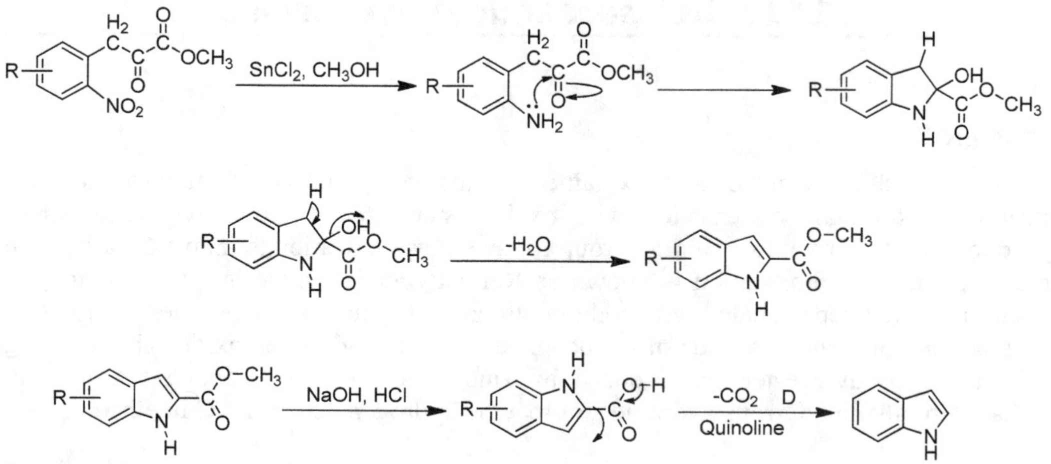

Applications

a) The reaction has broad application in synthesis of various indole analogues. For example: Synthesis of methyl 6-chloro-4-iodo-1H-indole-2-carboxylate.

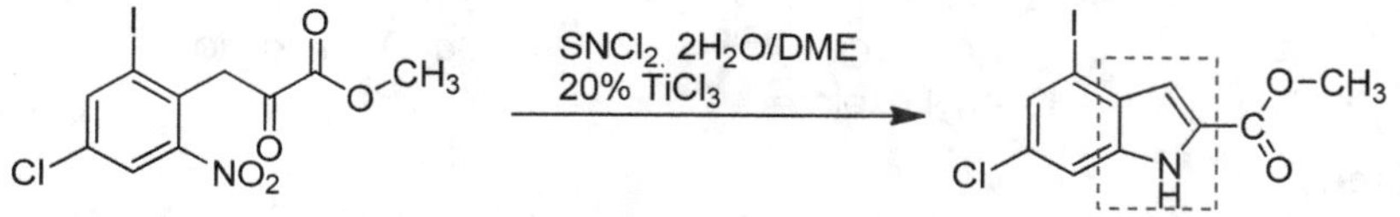

Methyl 3-(4-chloro-2-iodo-6-nitrophenyl)-2-oxopropanoate

Methyl 6-chloro-4-iodo-1*H*-indole-2-carboxylate

152. Retro-Diels-Alder Reaction

Principle

Diels and Thiele reported a thermally [π4s+π2s] cycloreversion of an organic compound with a double bond in a six-membered ring, leading to the formation of a diene and a dienophile, this is the reversed process of *Diels-Alder cycloaddition* reaction; hence known as *retro*-Diels-Alder fragmentation, or *retro*-Diels-Alder reaction. Although *Diels-Alder cycloaddition* is thermodynamically reversible, but *Diels-Alder* cycloadducts are fairly stable and *retro* Diels-Alder reaction does not occur regularly, as demonstrated by examples of the resistance of norbornenes and norbornadienes toward thermolysis. This is because two stronger σ-bonds must break for the formation of two new 2π-bonds. However, as the unimolecular *retro*-Diels-Alder reaction is dominated by enthalpy of activation with small contribution from entropy of activation, most *retro*-Diels-Alder reactions take place at very high temperature, if the resulting dienes and dienophiles are more stable.

General Reaction

Thermolysis of cyclohexene into buta-1,3-diene and ethene

Mechanism

Applications

a) Synthesis of 2-butyl-1H-imidazole.

153. Retro-Ene Reaction

Principle

Perkins and Cruz in 1927 synthesized undecylenic acid from ricinoleic acid, the major component of castor oil. This reaction is an intramolecular thermolysis of organic compounds with an unsaturated functional group involving the transfer of a γ-hydrogen to unsaturated center through six-membered transition state to yield both ene and enophile; hence known to be *retro*-ene reaction or *retro*-ene elimination. It is an example of pericyclic reactions involving [1,5]-H sigmatropic migration. The unsaturated groups involves alkenyl, alkynyl, allenyl, cyclopropyl, carbonyl, diazo, and iminyl, 1,6-dienes, allyltrimethylsilanes, acetylenic ethers, allylic and propargylic diazenes, allylic sulfinic acid, β-hydroxyethyl pyrazine, cyclic α-hydroxynitrosamines, metastable immonium salts, allenic esters, allenic amides, etc. The reaction involves concerted mechanism with the stereochemical integrity preserved.

General Reaction

Unsaturated compound

R = H, Alkyl, Aryl; R' = H, Alkyl, Aryl
Y = C, N; X = C, N, O, Si

a) Intramolecular thermolysis of 4-isopropoxypent-2-yne into propan-2-one and 2-methylpenta-2,3-diene.

4-isopropoxypent-2-yne   Propan-2-one   2-methylpenta-2,3-diene

Mechanism

The concerted mechanism of retro-ene reaction is illustrated here. The reaction involves concerted six-member transition state.

Applications

a) The reaction has widely applicable in synthetic chemistry. For example: Synthesis of (E)-N-(2-methylpropylidene)-1,1-diphenylmethanamine.

(*E*)-3-(benzhydrylimino)-2,2-dimethyl-1-phenylpropan-1-one

methyl benzoate

(*E*)-N-(2-methylpropylidene)-1,1-diphenylmethanamine

Principle

Rapson and Robinson in 1935 reproted Michael addition of cyclohexanones into methyl vinyl ketone followed by intramolecular aldol condensation to yield six-membered α,β-unsaturated ketones is known as Robinson annulations. The formation of α,β-unsaturated ketones resulting from the attachment of four carbon units to a ketone's carbonyl and α-carbon atom. Similarly, the reaction involving a nitrogen atom is called *aza*-Robinson annulation.

General Reaction

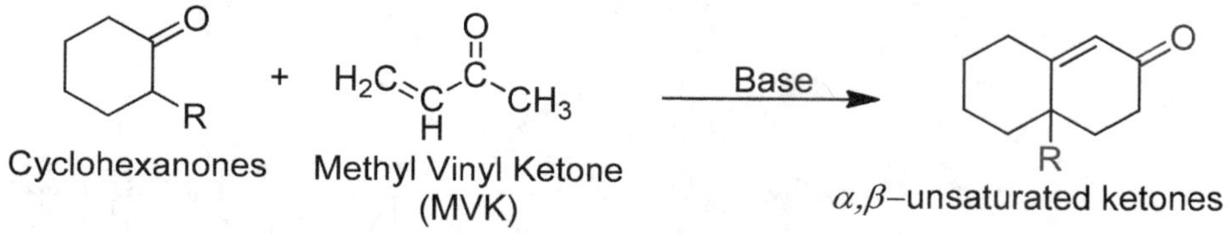

Cyclohexanones   Methyl Vinyl Ketone (MVK)   Base α,β-unsaturated ketones

Mechanism

The reaction involves the following steps:

Step 1: The generation of an enolate from the ketone.

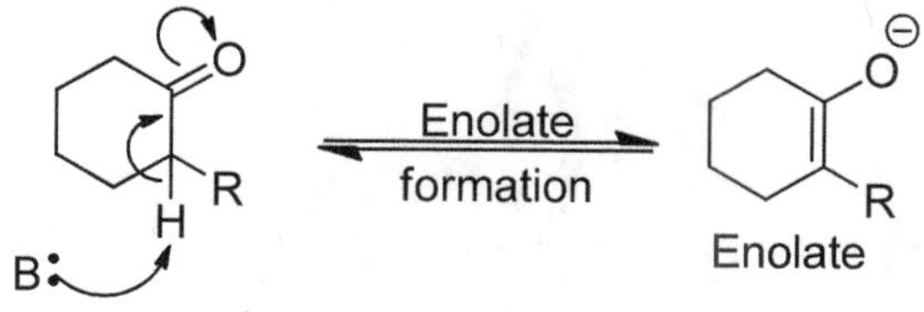

Enolate formation   Enolate

Step 2: *Michael addition* of the enolate to the α,β-unsaturated ketone followed by isomerization to give stable intermediate.

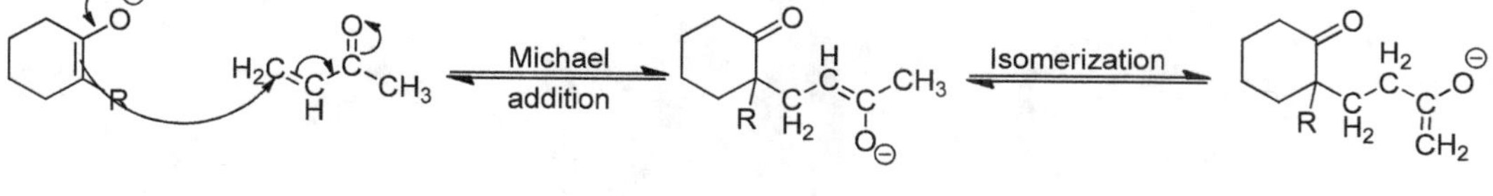

Michael addition   Isomerization

Step 3: *Aldol condensation* and dehydration to yield α,β-unsaturated ketone.

Applications

The reaction has been widely used for the preparation of α,β-unsaturated ketone, especially useful in the preparation of steroids.

a) Formation of α,β-unsaturated ketone.

Ethyl 2-oxocyclohexane carboxylate

Methyl Vinyl Ketone (MVK)

α,β-unsaturated ketone

b) Synthesis of (1R,4R)-3,4,5,6,7,8-hexahydro-1,4-ethanonaphthalen-2(1H)-one.

Cyclohexanone

Cyclohex-2-enone

CH_2Cl_2, 40° C, 8 h, 57%

3 TfOH, P_4O_{10}, Microwave

(1R,4R)-3,4,5,6,7,8-hexahydro-1,4-ethanonaphthalen-2(1H)-one

155.   Rosenmund Reduction

Principle: Rosenmund in 1918 reported formation of aldehyde through reduction of acid halides by reaction with hydrogen in presence of barium sulphate poisoned with palladium catalyst is known as Rosenmund reduction. Without poison, the resulting aldehyde may be further reduced to alcohol. Hence the reaction is carried out in varieties of modifiers such as sodium acetate, nitrogenous bases (For example: quinoline with sulfur, *N,N*-dimethylaniline, ethyldiisopropylamine, 2,6-dimethylpyridine), and sulfur-containing molecules (For example: thiourea, thiophene, and dibenzothiophene) for resistance of the aldehyde toward further reduction. Because presence of modifiers is believed to block sites with the most catalytic activity and prevent undesired reactions. Because of its intrinsic problems of reproducibility, the Rosenmund reduction has been widely replaced by the more reliable Brown hydroboration, using lithium tributoxyborohydride as the reducing reagent.

General Reaction

$$R-COCl + H_2 \xrightarrow{Pd/BaSO_4} R-CHO + HCl$$

Acid chloride → Aldehyde

a) Benzoyl chloride on catalytic reduction in presence of Pd/BaSO$_4$ affords benzaldehyde.

Benzoyl chloride $+ H_2 \xrightarrow{Pd/BaSO_4}$ Benzaldehyde $+$ HCl

b) Acetyl chloride on catalytic reduction in presence of Pd/BaSO$_4$ results acetaldehyde.

$$H_3C-COCl + H_2 \xrightarrow{Pd/BaSO_4} H_3C-CHO + HCl$$

Acetyl chloride → Acetaldehyde

Mechanism

Step 1: Poisoned palladium catalyst.

Step 2: Oxidative addition of palladium catalyst and ligand exchange.

Step 3: Reductive elimination to produce aldehyde and regeneration of catalyst.

Applications

a) Preparation of 2,2-dichloroacetyl chloride from 2,2,2-trichloroacetyl chloride.

2,2,2-trichloroacetyl chloride

2,2-dichloroacetyl chloride

b) Preparation of 2-([1,1'-biphenyl]-2-yl)acetaldehyde from 2-([1,1'-biphenyl]-2-yl)acetyl chloride.

2-([1,1'-biphenyl]-2-yl)acetyl chloride

2-([1,1'-biphenyl]-2-yl)acetaldehyde

156. Ruff Degradation

Principle

Ruff in 1892 described transformation of an aldose to its analog of one less carbon atom (pentoses or *D*-arabinose) by oxidation of the original aldose into aldonic acid followed by the decarboxylation of the corresponding calcium salt of aldonic acid with hydrogen peroxide in the presence of ferric acetate is known as Ruff degradation. For example, *D*-arabinose is easily prepared from calcium *4*-gluconate, and *D*-lyxose is degraded from calcium *D*-galactonate. In addition, this reaction can be used for the general conversion of α-hydroxy acids into aldehydes. Ruff degradation has been widely used for the structural determination and the preparation of new sugars, but it suffers from difficulty in separating the sugar from the gross quantities of organic and inorganic materials presenting in reaction mixture.

General Reaction

α-*D*-glucose (aldose) $\xrightarrow{Br_2/H_2O}$ aldonic acid $\xrightarrow{\text{a) } CaCO_3 \text{ b) } H_2O_2/Fe^{3+}}$ *D*-arabinose

Mechanism

$\xrightarrow{\underset{HOBr}{Br_2/H_2O}}$ $\xrightarrow{-HBr}$

Applications

The ruff degradation is helpful in the structural determination of carbohydrate.

157. Sabatier-Senderens Reduction

Principle

In 1899 Sabatier and Senderens reported nickel catalyzed hydrogenation of unsaturated organic compounds (For example: ketones, aldehydes, alkenes and aromatics) into corresponding saturated compounds such as alcohols, hydrocarbons by passing the vapor of organics and hydrogen over hot, finely divided nickel is commonly known as Sabatier-Senderens reduction. Sabatier in 1912 won the Nobel Prize for this one of the most practically useful hydrogenation reactions.

General Reaction

a) Carbon monoxide is reduced to methane in presence of nickel.

$$CO \xrightarrow[D]{H_2/Ni} CH_4$$

Carbon monoxide → Methane

b) Ketones have been reduced to alcohols.

Ketones $\xrightarrow[D]{H_2/Ni}$ Alcohols

c) Unsaturated organic molecules reduced to saturated compounds.

Unsaturated organic molecules $\xrightarrow[D]{H_2/Ni}$ Saturated compounds

R = Alkyl; R' = H, Alkyl

Mechanism

The detailed mechanism of reduction by using nickel catalyst is outlined below:

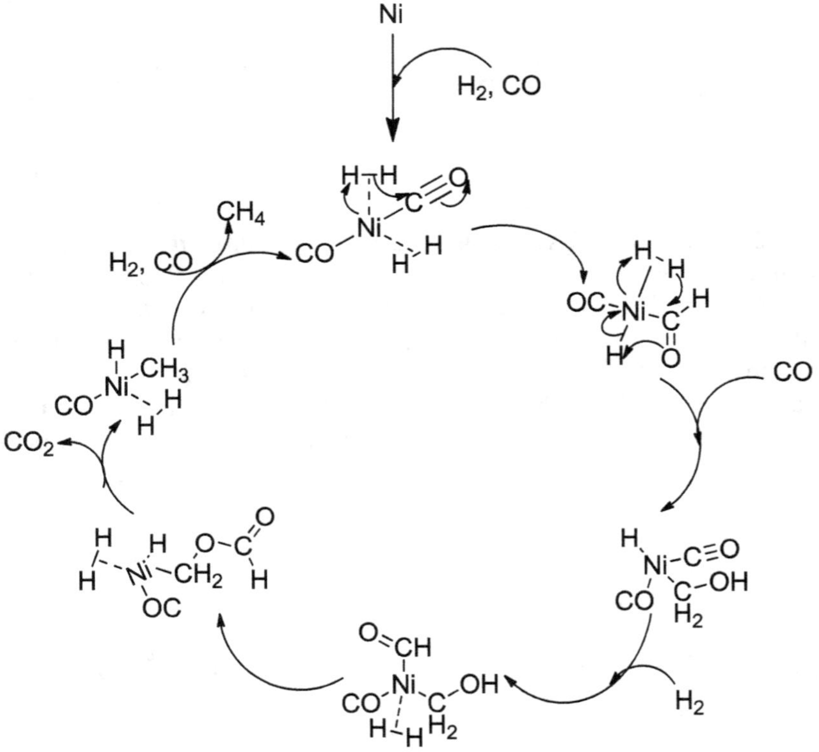

Applications

a) *The reaction is useful for transformation of carbon monoxide or carbon dioxide into organic compounds. The reaction has wide application in hydrogenation of organic molecules. For example: Hydrogenation of cotton seed oil.*

$$Cotten\ seed\ oil \xrightarrow{\ H_2/Ni\ on\ silica\ black,\ 180°C\ } Partial\ saturated\ oil$$

158. Sandmeyer Isatin Synthesis

Principle

Sandmeyer in 1919 performed synthesis of isatins by multicomponent condensation of trichloroacetaldehyde (chloral), hydroxylamine, and a primary aromatic amine to afford α-isonitrosoacetanilide followed by electrophilic cyclization in presence of concentrated sulfuric acid is known as Sandmeyer isatin synthesis. During this reaction, the oximino intermediate can be isolated by extraction with warm 2N sodium hydroxide followed by the precipitation with a 2N hydrochloric acid. The reaction fails to convert oximino intermediates arising from nitroanilines, 2-fluoroanilines, and 2,4-difluoroanilines into corresponding isatin analogues.

General Reaction

Mechanism

Step 1: Formation of oximino intermediate by reaction of chloral with hydroxyl amine.

Step 2: Formation of α-isonitrosoacetanilide and electrophilic cyclization in presence of concentrated sulfuric acid to form isatin.

Applications

a) Isatins are broadly used as starting material for synthesis of different organic molecules such as quinolines, acridines, and indophenazines. For example: Synthesis of methyl 2,3-dioxoindoline-4-carboxylate.

Methyl 3-aminobenzoate   2,2,2-trichloroacetaldehyde

Methyl 2,3-dioxoindoline-4-carboxylate

b) Synthesis of 6-methoxy-5-methyl-7-(methylperoxy)indoline-2,3-dione.

2,3-dimethoxy-4-methylaniline   2,2,2-trichloroacetaldehyde

6-methoxy-5-methyl-7-
(methylperoxy)indoline-2,3-dione

159. Sandmeyer Reaction

Principle

Two step synthesis of aryl halides or cyanides from primary aromatic amines through formation of diazonium salts of corresponding amines with nitrous acid and transformation of diazo compounds into aryl halides or cyanides with cuprous halides or cyanide is referred as Sandmeyer reaction. Sandmeyer in 1884 also reported the synthesis of phenylacetylene from benzenediazonium chloride and cuprous acetylide. The synthesis of aryl nitrile by this reaction is known as Sandmeyer nitrile synthesis. The haloalkane in combination with cuprous halide is known as Sandmeyer reagent. A halogen present into aromatic ring originates from cuprous halide instead of diazonium halide, as supported by formation of *p*-dichlorobenzene from *p*-chlorobenzenediazonium bromide and cuprous chloride as well as production of *p*-chloro-bromobenzene from *p*-chlorobenzenediazonium chloride and cuprous bromide. It should be remember that iodonation of aromatic diazonium salt does not need presence of cuprous iodide. In this reaction quantitative yield of aryl halide is obtained due to formation of some side products, including biaryl, azobenzene, and phenol.

General Reaction

a) Anthranilic acid on diazotization followed by Sandmeyer reaction gives o-chlorobenzoic acid.

Mechanism

Step 1: Formation of aryl diazonium salt.

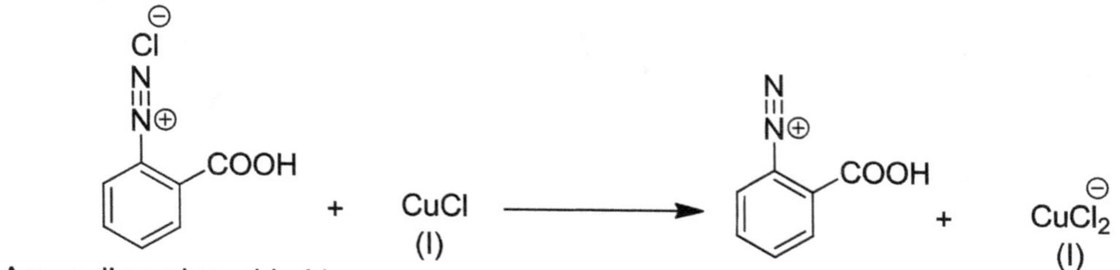

Arene diazonium chloride + CuCl (I) ⟶ + CuCl₂ (I)

Step 2: Reaction of aryl diazonium salt with cuprous chloride to give chloro-benzoic acid.

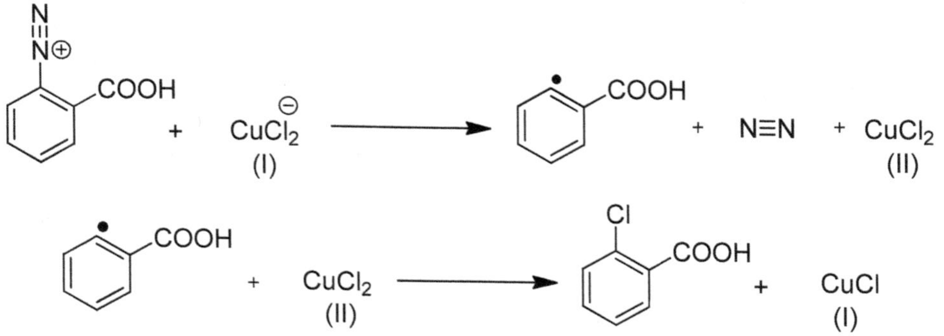

+ CuCl₂ (I) ⟶ + N≡N + CuCl₂ (II)

+ CuCl₂ (II) ⟶ + CuCl (I)

Applications

The reaction has broad application in preparation of aryl halides and nitriles.

a) Synthesis of o-chloro toluene from o-toluidine.

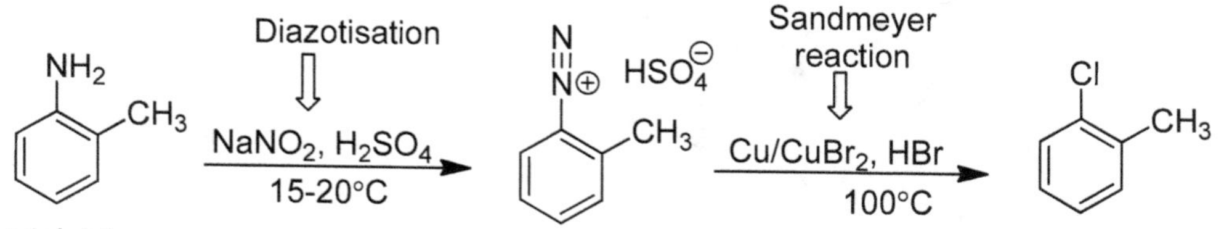

o-toluidine $\xrightarrow[\text{15-20°C}]{\text{NaNO}_2,\ \text{H}_2\text{SO}_4}$ Arene diazonium salt $\xrightarrow[\text{100°C}]{\text{Cu/CuBr}_2,\ \text{HBr}}$ o-chloro toluene

Diazotisation — Sandmeyer reaction

b) Synthesis of m-chlorobenzene from m-phenylenediamine.

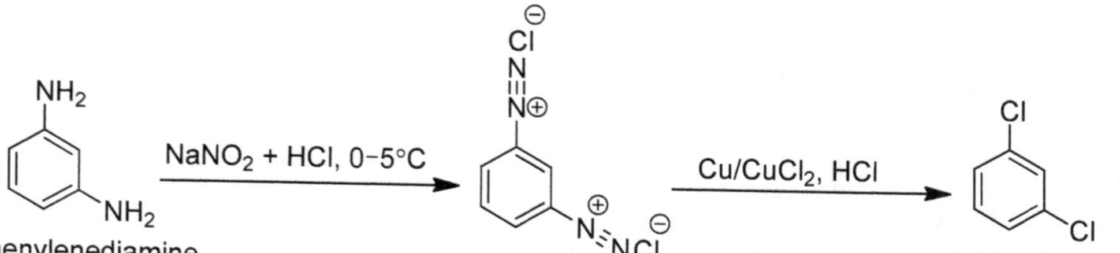

m-phenylenediamine $\xrightarrow{\text{NaNO}_2 + \text{HCl},\ 0\text{-}5°C}$ $\xrightarrow{\text{Cu/CuCl}_2,\ \text{HCl}}$ m-dichlorobenzene

c) *Synthesis of 2-chloronapthalene from β-naphthol.*

Principle

A rule for predicting the regio-selectivity of alkene formation by elimination of secondary or tertiary haloalkanes, in which hydrogen is eliminated preferentially from the carbon atom bearing the smallest number of hydrogens is known as Saytzeff rule (or Zaitsev rule). The rule was first reported by Saytzeff (Zaitsev) in 1875. The regioselectivity to give a more stable alkene is referred as Saytzeff regiospecificity, and corresponding alkene is known as Saytzeff product. The Saytzeff rule does not apply to the pyrolysis of esters in gas phase (> 400°C) and the elimination of onium salts, such as the quaternary ammonium salts and substituted sulfonium salts. Similarly, the size of the base applied in the elimination also affects the regioselectivity between the Saytzeff and Hofmann rules. The steric effect may also result in Saytzeff product from elimination of an ammonium salt.

General Reaction

Dehydrohalogenation of haloalkanes (Saytzeff elimination)

Haloalkanes undergo elimination reaction (dehydrohalogenation) on heating with potassium hydroxide in presence of ethanol (alcoholic KOH) to form alkenes.

In haloalkanes, β-elimination occurs, that is, hydrogen is removed from β-carbon (carbon adjacent to halogen bearing carbon) during dehydrohalogenation.

$$\text{Haloalkane} \xrightarrow{\text{KOH/C}_2\text{H}_5\text{OH}} \text{RCH}=\text{CH}_2 + \text{HX}$$

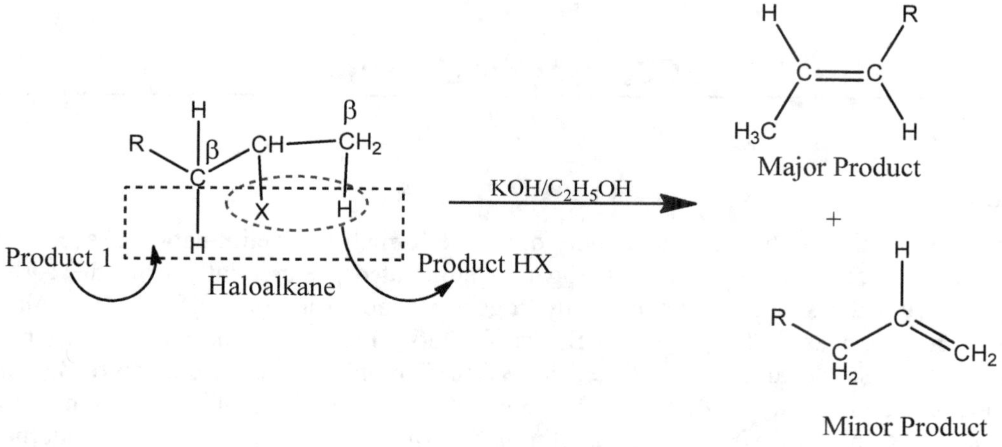
R
H
β
CH
CH2
C
X
H
H
Product 1
Haloalkane
Product HX
KOH/C2H5OH
H
R
C=C
H3C
H
Major Product
+
H
R
C
C
H2
CH2
Minor Product

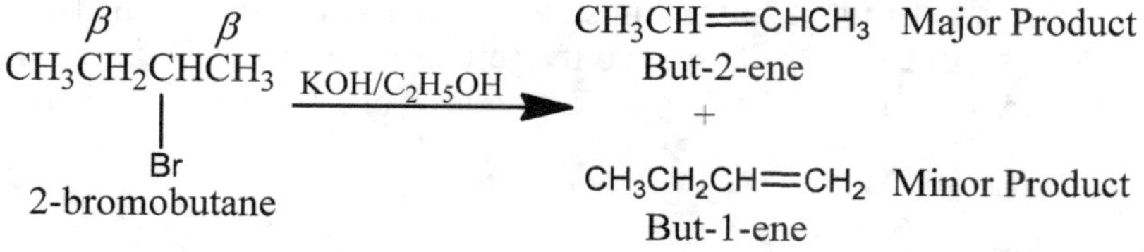
β β
CH3CH2CHCH3
Br
2-bromobutane
KOH/C2H5OH
CH3CH=CHCH3 Major Product
But-2-ene
+
CH3CH2CH=CH2 Minor Product
But-1-ene

Statement

[1] This rule states that whenever there is possibility of formation of two alkenes by elimination, the formation of more substituted alkene is favored. In other words, the hydrogen is removed from β-carbon having less number of hydrogens.

[2] In general, more the number of alkyl substitutents attached to olefinic carbons, higher is the stability of that alkene.

[3] A tetra-substituted alkene is more stable than tri-substituted alkene; tri-substituted is more stable di-substituted alkene.

Applications

a) *A rule has been widely used for the determination of regioselectivity of elimination. For example: Transformation of 2-bromo-2-methylbutane into 2-methylbut-2-ene and 2-methylbut-1-ene.*

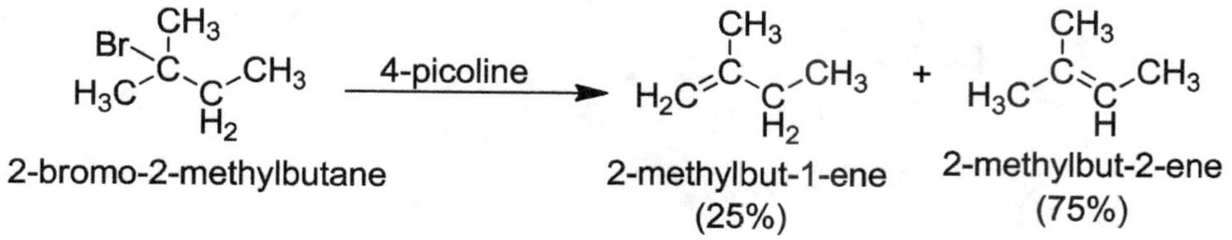
Br
CH3
H3C
C
C
CH3
H2
2-bromo-2-methylbutane
4-picoline
CH3
H2C=C
C
CH3
H2
2-methylbut-1-ene
(25%)
+
CH3
H3C
C=C
CH3
H
2-methylbut-2-ene
(75%)

b) *Conversion of 1-(4-methoxyphenyl)-3-phenylpropyl acetate into (E)-1-methoxy-4-(3-phenylprop-1-en-1-yl)benzene.*

Δ, 70°C

1-(4-methoxyphenyl)-3-phenylpropyl acetate

(*E*)-1-methoxy-4-(3-phenylprop-1-en-1-yl)benzene

161. Schmidt Reaction

Principle

Schmidt in 1923 reported a reaction between hydroazoic acid (azo-imide) and carboxylic acid or a carbonyl compound through *1,2*-migration to yield primary amine or amide; hence known as Schmidt reaction. More specifically, the reaction of hydroazoic acid and carboxylic acid to produce primary amine with one less carbon atom *via* decarboxylation is known as Schmidt rearrangement. In particular, the reaction between hydroazoic acid and carbonyl derivatives, gives two possible amides from a ketone or formamide from an aldehyde is also referred as Schmidt rearrangement or Schmidt transformation. Schmidt reaction has been extensively explored for hydroazoic acid or alkyl azide and other molecules that can generate carbocation intermediates under acidic conditions, such as olefins, tertiary alcohols, and some secondary alcohols such as benzhydrols. The reaction between hydroazoic acid and ketone is much more complicated, yielding two possible amides by migration of different alkyl or aryl groups, and ratio of two amides depends on strength of acid and migratory aptitude of the two groups. Aryl group is preferred over alkyl group for migration from aminodiazonium ion, and a migrating tendency correlates with Hammett substituent constant, in which an electron-donating group on aryl facilitates the migration while an electron-withdrawing group decreases a migrating ability.

General Reaction

$$R-COOH \ + \ HN_3 \ \xrightarrow{H_2SO_4} \ R-NH_2 \ + \ CO_2 \ + \ N_2$$

Acid   Hydrazoic acid   Amine

$$R-CHO \ + \ HN_3 \ \xrightarrow{H_2SO_4} \ R-CN \ + \ R\text{-}N(H)\text{-}CHO$$

Aldehyde   Hydrazoic acid   Cyanide   N-formyl amine

Ketone   Hydrazoic acid   Amide   + N_2

Alcohol   Hydrazoic acid   Imine

Mechanism

For the degradation of carboxylic acid, the reaction involves an acyl azide intermediate, in a manner analogous to Curtius rearrangement. Like, acid-catalyzed esterification of alcohol, the carbonyl oxygen is protonated rather than the hydroxyl oxygen.

In Schmidt reaction of ketones, the reaction proceeds by two different mechanisms, both involve the α-hydroxyhydrazidium ion arising from the nucleophilic addition of hydroazoic acid to protonated carbonyl group. The resulting α-hydroxyhydrazidium ion may rearrange to amides along with the evolution of nitrogen, and ratio of amides produced depends on inherent migratory aptitudes of both groups and abilities of group acting as stationary group to stabilize positive charge. Alternatively, the α-hydroxyhydrazidium ion undergoes dehydration to form iminodiazonium ion, and the stereospecific migration of the group *anti-periplanar* to the azido group produces an acylium ion, which upon hydration forms the corresponding amide.

In this mechanism, the ratio of amides depends on the population of *syn-* and *anti-*iminodiazonium ions arising from different steric repulsions in the transition state of dehydration, which cannot isomerizes but may interchange into each other only through the hydration and dehydration process. If the hydration and dehydration process is fast enough, then two iminodiazonium ions can quickly reach equilibrium, this situation is controlled by the migratory tendency of substituents in hydroxyhydrazinium ion.

Schmidt degradation of a carboxylic acid

Schmidt reaction with carbonyl compounds

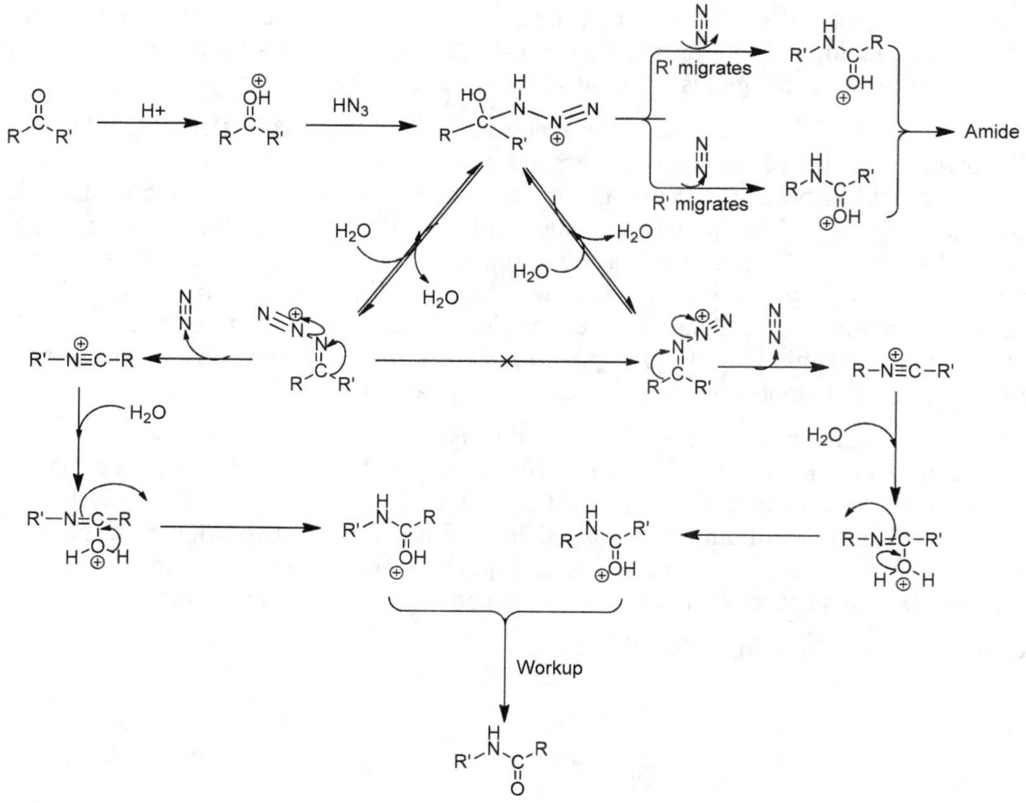

Applications

a) Schmidt reaction is versatile and has advantages of simplicity, readily available reactants, mild reaction conditions, and a certain level of functional group tolerance. The degradation of carboxylic acid has been used for the preparation of amines, amino acids, and structural elucidation through ^{14}C labeling. The reaction with carbonyl compounds has an even wider application in organic synthesis. *For example: Synthesis of methanamine from ethanoic acid.*

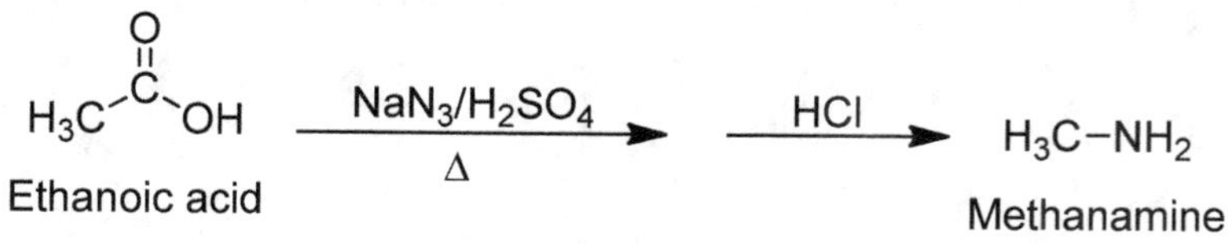

b) *Synthesis of 4-hexyl-4-azahomoadamantane.*

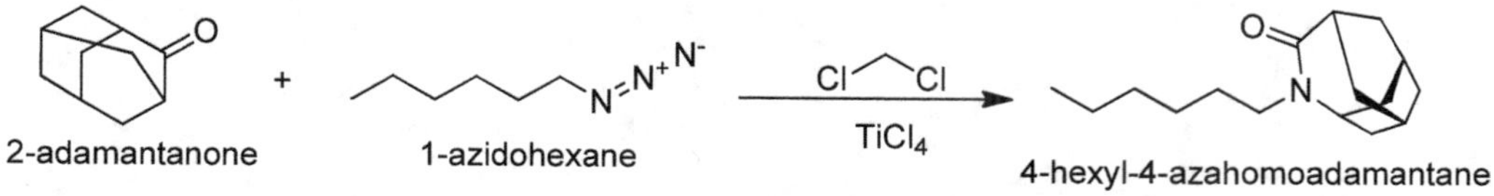

Principle

Schotten in 1884 reported acylation of alcohols and amines from acyl halide or anhydride in an aqueous alkaline solution; the reaction is further explored by Baumann in 1886; hence referred as Schotten-Baumann reaction or acylation. The synthesis of benzoyl ester under these conditions is known as Schotten-Baumann benzoylation.

General Reaction

$$Ar-NH_2 \quad + \quad Cl-\underset{O}{\overset{}{C}}-Ar \quad \xrightarrow{\text{NaOH}} \quad Ar-\underset{H}{N}-\underset{O}{\overset{}{C}}-Ar$$

Aromatic amine Aryloxy chloride Amide analogues

a) Phenols react with benzoyl chloride in presence of bases to form phenyl esters (Phenyl benzoate).

Phenol + Benzoyl chloride $\xrightarrow{\text{NaOH}}$ Phenyl benzoate

Mechanism

Step 1: Attack of aniline on carbonyl carbon of benzoyl chloride.

Step 2: Loss of proton (Rapid transfer of proton towards electron rich nitrogen atom).

Rapid proton transfer

Step 3: Removal of HCl (Formation of benzamide).

-HCl

Applications

a) *The nucleophilic substitution of aromatic amines with acid chlorides results in the formation of N-substituted amides. The reaction of aniline with benzoyl chloride is carried out in alkaline medium.*

Aniline   +   Benzoyl chloride   NaOH   *N*-phenylbenzamide

Principle

Skraup in 1880 reported and explained synthesis of quinolines through reaction of primary aromatic amines with glycerol in presence of an oxidizing agent in concentrated sulfuric acid is commonly referred as Skraup quinoline synthesis. In particular, the *ortho*- or *para*-substituted primary aromatic amines give expected quinolines; on the other hand, the *meta*-substituted aromatic amines lead to the formation of a mixture of products. When an unstable aromatic amine has been used; the higher yield of quinoline can be obtained from an azo compound of the corresponding aromatic amine. The reaction has been modified by the addition of various chemicals (For example: acetic acid, boric acid, $FeSO_4$, thorium, vanadium, or iron oxides) to moderate the reaction, because the initial Skraup reaction condition is exothermic with uncontrollable violence. Now a days, Skraup reaction can be carried out in protic acid (For example: H_2SO_4, H_3PO_4, TsOH, and $HClO_4$) or in the presence of a Lewis acid, such as $Sc(OTf)_3$, $SnCl_4$, $Yb(OTf)_3$, and $ZnCl_2$.

General Reaction

Aniline + Glycerol $\xrightarrow[\text{—NO}_2]{H_2SO_4}$ Quinoline

Mechanism

Step 1: Formation of Acrolein.

Step 2: Conjugate addition of aniline to acrolein to give 3-(phenylamino)propanal.

3-(phenylamino)propanal

Step 3: Intramolecular electrophilic addition followed by dehydration to give 1,2-dihydroquinoline, which on oxidation gives quinoline.

1,2,3,4-tetrahydroquinolin-4-ol

1,2-dihydroquinoline

Quinoline

Applications

The reaction is widely applicable in synthesis of quinoline analogues, phenanthrolines and biquinolines.

a) Synthesis of 3-bromo-6-nitroquinolin-8-ol.

2-bromoacrolein

2,2,3-tribromopropanal

3-bromo-6-nitroquinolin-8-ol

b) Synthesis of 10-hydroxy-2,2,4-trimethyl-1H-chromeno[3,4-f]quinolin-5(2H)-one.

nitro-coumarin

8-amino-6H-benzo[c]chromen-6-one

10-hydroxy-2,2,4-trimethyl-1H-chromeno[3,4-f]quinolin-5(2H)-one

164. Smiles Rearrangement

Principle

Henriques in 1894 reported this intramolecular nucleophilic substitution which was later explored by Smiles in 1930. An intramolecular nucleophilic substitution attacking on an aromatic system bearing an activating electron-withdrawing group at *o* or *p*-position to the reaction center connected to heteroatom and is accompanied by the migration of an aromatic ring from the heteroatom binding to the reaction center to more nucleophilic heteroatom is known as Smiles rearrangement. Similarly, rearrangement initiated by photo irradiation discovered in 1970 is photo-Smiles rearrangement or photochemical Smiles rearrangement. In Smiles rearrangement, the two carbon atoms connecting to two heteroatoms before and after the rearrangement are usually a part of the aromatic system, but may also be a part of an aliphatic component. The leaving heteroatom or group can be -O-, -S-, -SO-, -SO$_2$; and the attacking nucleophile can be the conjugate base of -OH, -NH$_2$, -CONH-, -SO$_2$NH-, while usual activating electron-withdrawing group is nitro or sulfonyl.

General Reaction

$$X = S, SO, SO_2, O, CO_2$$
$$YH = OH, NHR, SH, CH_2R, CONHR$$
$$Z = NO_2, SO_2R$$

Conversion of 2-((2-nitrophenyl)sulfonyl)phenol into 2-(2-nitrophenoxy)benzenesulfinate in presence of base.

2-((2-nitrophenyl)sulfonyl)phenol   2-(2-nitrophenoxy)benzenesulfinate

Mechanism

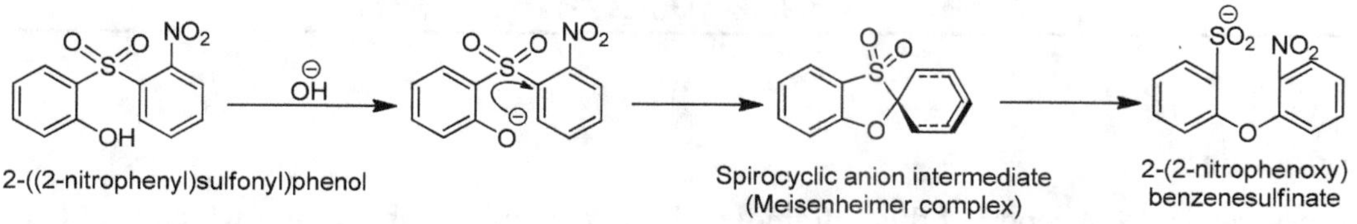

2-((2-nitrophenyl)sulfonyl)phenol

Spirocyclic anion intermediate
(Meisenheimer complex)

2-(2-nitrophenoxy)
benzenesulfinate

Applications

The reaction is useful for synthesis of various organic intermediates.

a) Synthesis of 1,2-bis(3,5-dimethoxyphenyl)ethane.

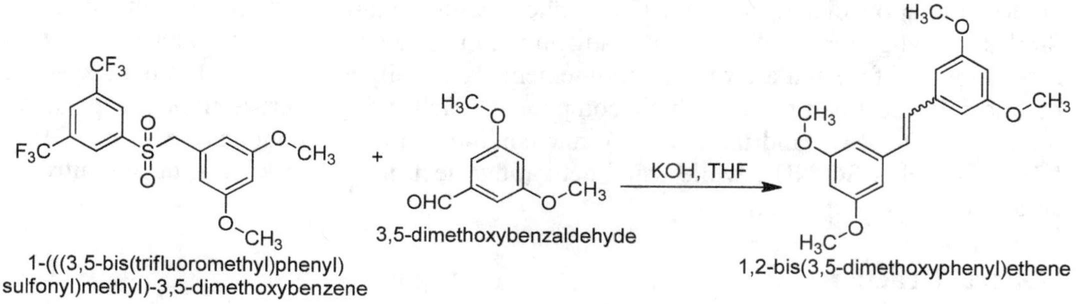

1-(((3,5-bis(trifluoromethyl)phenyl)
sulfonyl)methyl)-3,5-dimethoxybenzene

3,5-dimethoxybenzaldehyde

KOH, THF

1,2-bis(3,5-dimethoxyphenyl)ethene

b) Synthesis of 3-((methylsulfonyl)methoxy)-4-phenyl-1,2,5-oxadiazole.

2-((4-phenyl-1,2,5-oxadiazol-3-
yl)sulfonyl)ethanol

NaOH,
Acetone/H2O

CH3OH/CH3I

3-((methylsulfonyl)methoxy)-4-phenyl-1,2,5-oxadiazole

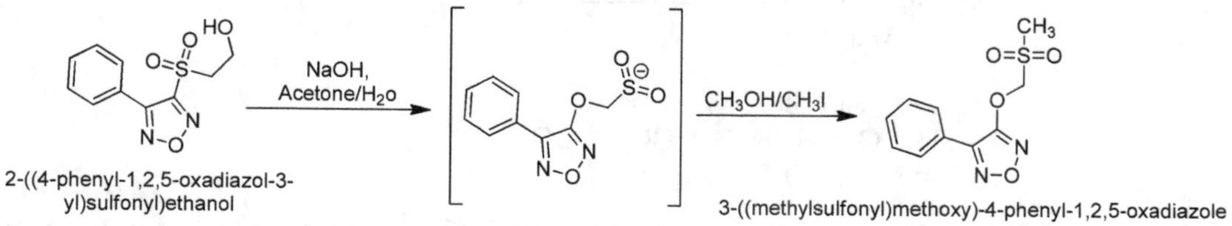

165. Sommelet Reaction

Principle

The conversion of α-halomethyl aromatic compound (benzyl halide) into the corresponding aldehyde *via* treatment with hexamethylenetetramine (HMTA) followed by mild acidic hydrolysis is first reported by Sommelet in 1913. Hence the reaction is commonly referred as Sommelet reaction. The reaction is applicable to α-halomethyl group containing aromatics, which often gives corresponding aldehydes with 60-70% of yields, and reaction occurs in presence of stoichiometric or small amount of water. In addition, Sommelet reaction is also applicable for primary or secondary benzyl amines and converts them to aldehydes; when compared secondary amines are less reactive than primary amines. The aromatic compounds with two chloromethyl groups *ortho* to each other generally give a nitrogenous compound rather than the expected dialdehyde under the Sommelet reaction condition.

General Reaction

a) Benzyl chloride and hexamethylenetetramine in presence of aqueous acid gives quaternary ammonium salt which decomposes on acidic workup to afford benzaldehyde.

b) Formation of 1-naphthaldehyde from 1-(chloromethyl) naphthalene and hexamethylenetetramine.

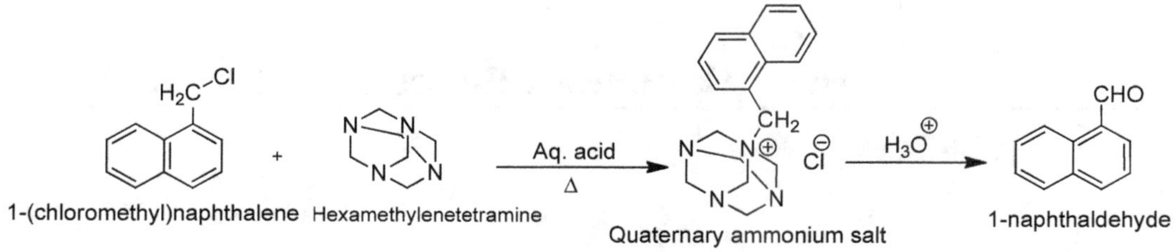

1-(chloromethyl)naphthalene   Hexamethylenetetramine

Aq. acid
Δ

Quaternary ammonium salt

$H_3O^{\oplus}$

1-naphthaldehyde

Mechanism

Step 1: Nucleophilic substitution to give quaternary ammonium salt followed by hydrolysis to give aromatic primary amine.

S_N2 reaction

H_2O

Aromatic primary amine

Quaternary ammonium salt

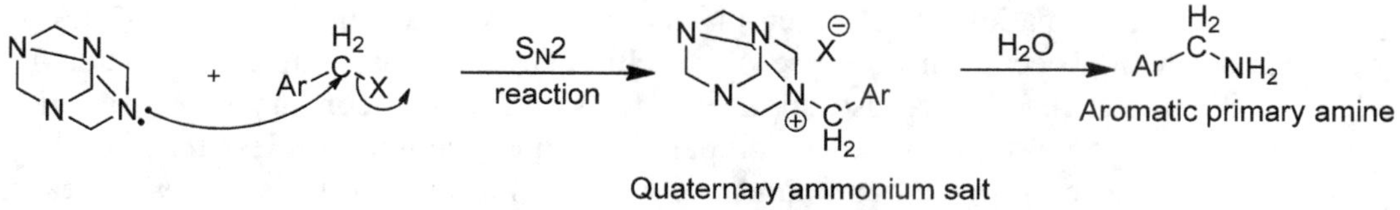

Step 2: Hexamethylenetetramine on acidic hydrolysis gives methaniminium.

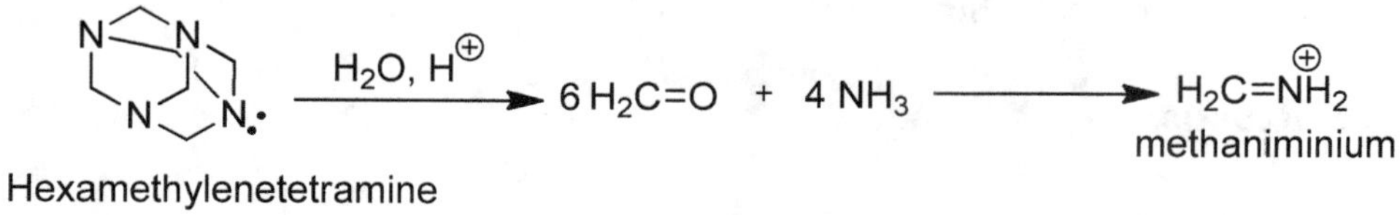

$$H_2O, H^{\oplus} \longrightarrow 6\,H_2C=O \;+\; 4\,NH_3 \longrightarrow H_2C=\overset{\oplus}{N}H_2$$

methaniminium

Hexamethylenetetramine

Step 3: Reaction between methaniminium and aromatic primary amine.

$$Ar-\overset{H}{\underset{H}{C}}-NH_2 \;+\; H_2\overset{\oplus}{C}=NH_2 \xrightarrow{\text{Hydride transfer}} Ar-\overset{\oplus}{C}=NH_2 \;+\; CH_3-NH_2$$

Imine

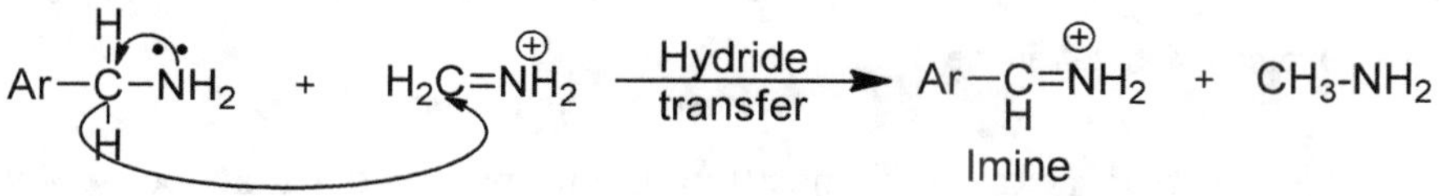

Step 4: Conversion of imine into aromatic aldehyde.

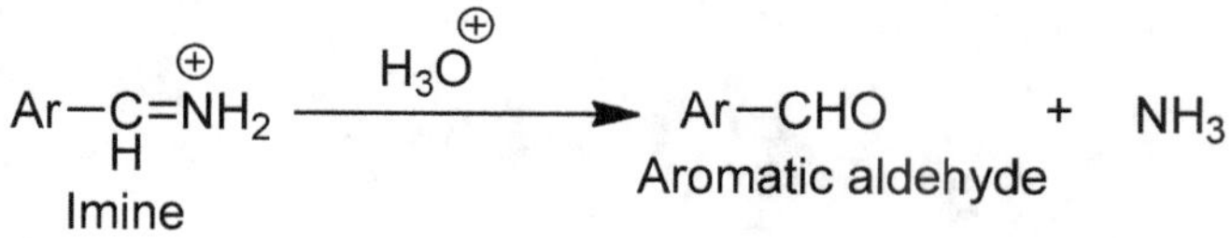

$$Ar-\overset{\oplus}{\underset{H}{C}}=NH_2 \xrightarrow{H_3O^{\oplus}} Ar-CHO \;+\; NH_3$$

Imine   Aromatic aldehyde

Applications

The reaction is useful in the synthesis of aromatic aldehydes.

a) Synthesis of thiophene-2-carbaldehyde.

b) Synthesis of nicotinaldehyde.

Principle

An intramolecular [2,3]-sigmatropic rearrangement of benzyl quaternary ammonium salt to *ortho*-substituted benzyl dialkylamines by means of the treatment with a strong base such as alkali metal amide is first reported by Sommelet in 1937 and further explored by Hauser in 1951 is known as Sommelet-Hauser rearrangement. The α-carbon of the alkyl group connecting to the quaternary nitrogen atom will be deprotonated by a strong base (NaNH$_2$, NaOMe, and KO-*t*-Bu), and the resulting carbanion attacks the *ortho*-position of the aromatic ring to form the rearranged product through o-dialkylamino methylene cyclohexadiene intermediate. If nitrogen atom is a member of a cyclic ring, then a ring-enlarged product may be resolved. It should be pointed out that it is also possible for the migratory group to move to *p*-position. It is interesting that the rearrangement intermediate undergoing the re-aromatiz

ation to give the final product also gives competitive cyclopropanation product.

General Reaction

Mechanism

Step 1: Proton is first removed from more acidic benzyl position to give nitrogen ylide (I).

Step 2: The nitrogen ylide is in equilibrium with a second nitrogen ylide (II) formed in small amount due to its stability. The ylide (II) is highly reactive species.

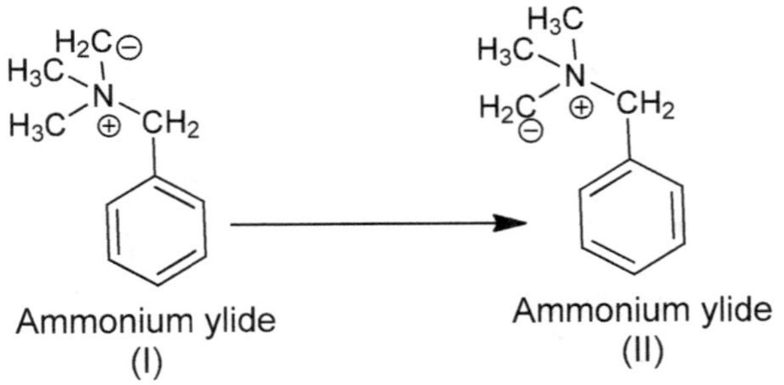

Ammonium ylide
(I)

Ammonium ylide
(II)

Step 3: Ylide (II) undergoes 2,3-sigmatropic rearrangement to generate exomethylene derivative.

Ammonium ylide (II)

[2,3]-sigmatropic
rearrangement

Exomethylene derivative

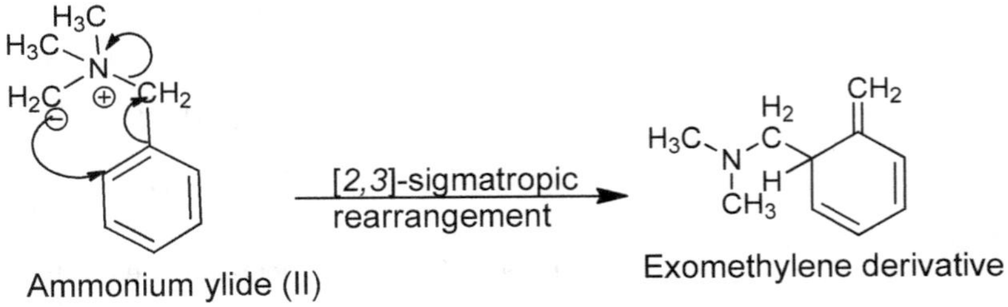

Step 4: Exomethylene derivative gives product after loss of proton.

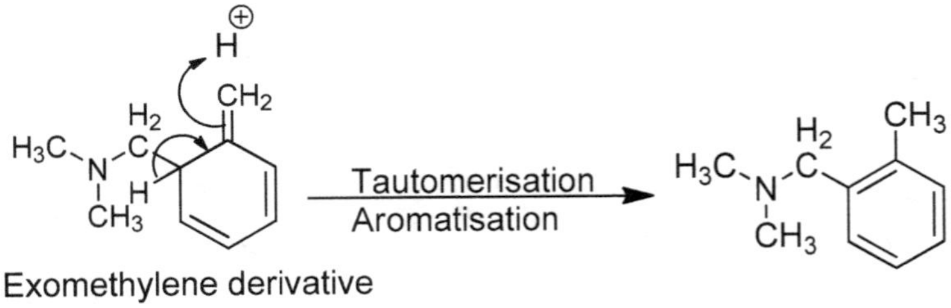

Exomethylene derivative

Tautomerisation
Aromatisation

Applications

a) Synthesis of N,N-dimethyl-1-(2,3,4,5,6-pentamethylphenyl)methanamine.

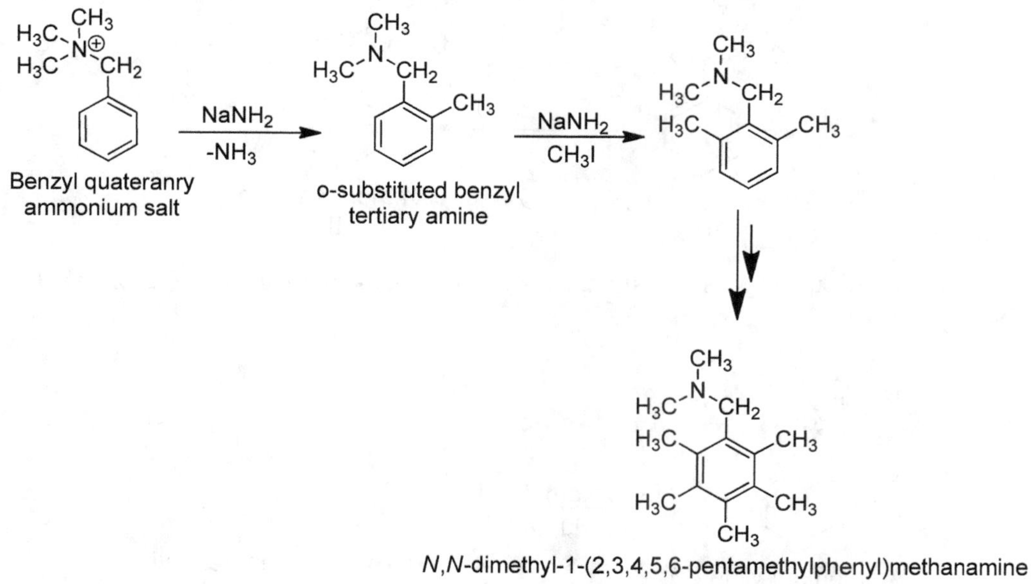

NaNH2
-NH3
Benzyl quateranry ammonium salt
o-substituted benzyl tertiary amine
NaNH2
CH3I
N,N-dimethyl-1-(2,3,4,5,6-pentamethylphenyl)methanamine

167. Staudinger [2+2] Cycloaddition Reaction

Principle

Staudinger in 1907 reported synthesis of 2-azetidinones (*β*-lactams) by the [2+2] cycloaddition of ketenes and imines is known as Staudinger [2+2] cycloaddition reaction. The convenience of the Staudinger reaction arises from not only the simple and mild reaction conditions but also the availability of imines and ketenes, and the ketenes can be simply synthesized by the reaction of acyl chlorides with triethylamine. The reaction is very fast and smoothly takes place at a low temperature without an initiator or a catalyst present, and the reaction rate is dependent on the nucleophilicity of the imines. In addition, the *cis*-substituted imines are much more reactive than the *trans*-imines in cycloaddition reaction. Staudinger [2+2] cycloaddition is not pericyclic reaction but it is a stepwise reaction involving a nucleophilic attack of an imine on ketene to form a zwitterionic intermediate, which cyclizes conrotatorily to form the *β*-lactam. *β*-lactams obtained from monosubstituted ketenes by cycloaddition are observed in *cis*-configurations. The stereo-specificity is attributed to a conrotatory transition structure that incorporates donor in 3-out ward and vinylic moiety in 4-in ward position, where the direct ring closure, favored in accordance to the least motion principle, results in the formation of the *cis*-*β*-lactam; and isomerization of imine geometry before closure of ring leads to formation of corresponding *trans*-*β*-lactam.

General Reaction

Ketenes + Imines → 2-azetidinones (β-lactams)

Mechanism

In general, the Staudinger [2+2] cycloaddition, initiated by the nucleophilic attacking of an imine on the *sp*-hybridized carbon atom of ketene followed by the conrotatory cyclization of the azadiene intermediate. The mechanism of the Staudinger [2+2] cycloaddition between methyl ketene and *N*-methyl imine is illustrated here:

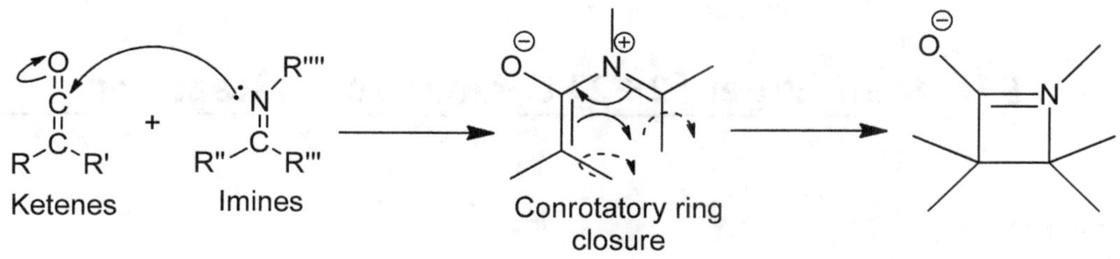

Applications

The formed β-lactams through this reaction are the structural components of the widely used penicillins, cephalosporins, thienamycine, nocardicins, aztreonam, and carumonam.

a) Synthesis of β-lactams.

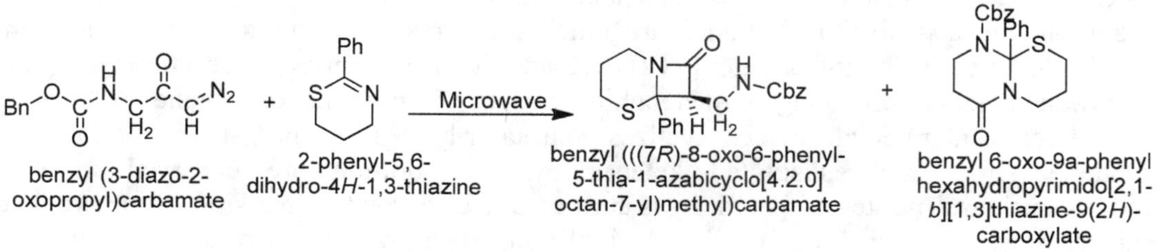

Principle

Staudinger in 1919 reported and explained the reaction between an azide and triphenylphosphine to obtain an iminophosphorane (*aza*-ylide) which on hydrolysis gives a primary amine called as Staudinger reaction. In Staudinger reduction, the reduction of azides to primary amines takes place under mild conditions. In addition, the coupling between intermediate *aza*-ylide and an ester to give an amide or peptide is known to be the Staudinger ligation, where joining of two molecules take place. The Staudinger reaction is a simple, high yield reaction where iminophosphorane intermediate hydrolyzes in presence of water. The thermal stabilities of iminophosphoranes from different phosphines vary, due to the repulsive interactions between lone pair electrons on phosphorus and nitrogen.

General Reaction

$$R-\overset{\ominus}{N}-\overset{\oplus}{N}\equiv N \ + \ \text{Triphenylphosphine} \ \xrightarrow[\text{H}_2\text{O}]{\text{Reduction}} \ R-NH_2 \ + \ \text{Triphenylphosphine oxide}$$

Azides + Triphenylphosphine → Primary amines + Triphenylphosphine oxide

Reduction of azidomethane into methanamine

$$H_3C-\overset{\ominus}{N}-\overset{\oplus}{N}\equiv N \ + \ \text{Triphenylphosphine} \ \xrightarrow[\text{H}_2\text{O}]{\text{Reduction}} \ H_3C-NH_2 \ + \ \text{Triphenylphosphine oxide}$$

Azidomethane + Triphenylphosphine → Methanamine + Triphenylphosphine oxide

Mechanism

Step 1: Nucleophilic substitution reaction between azide and triphenylphosphine.

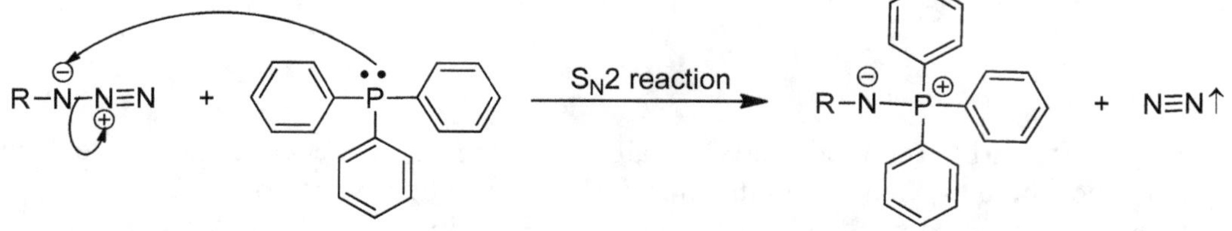

Step 2: An iminophosphorane is converted to primary amines.

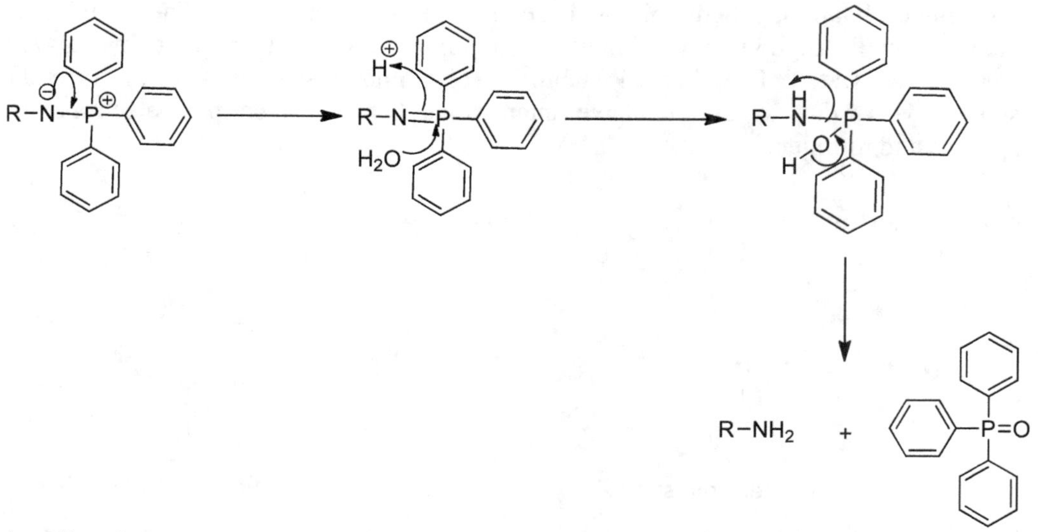

$$R{-}NH_2 \quad + \quad Ph_3P{=}O$$

Applications

a) The reaction is useful in the preparation of peptide bonds in a living system. For example: preparation of aniline from azidobenzene and triphenylphosphine.

Azidobenzene

Triphenylphosphine

Aniline

Triphenylphosphine oxide

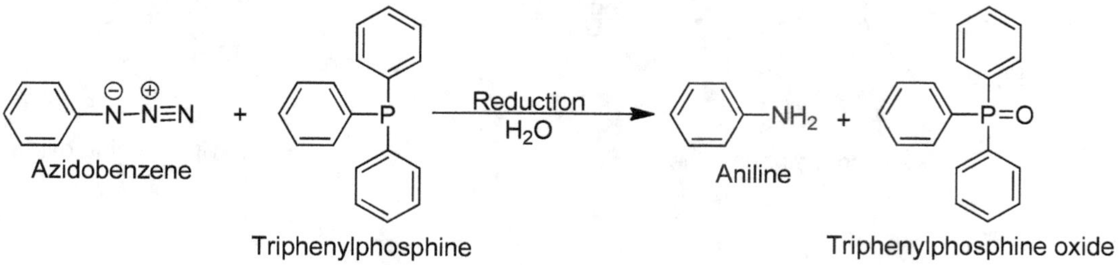

b) Synthesis (E)-ethyl 2-(3-hydroxy-3,4-dimethylcyclopentylidene)acetate.

PPh₃, THF, Δ

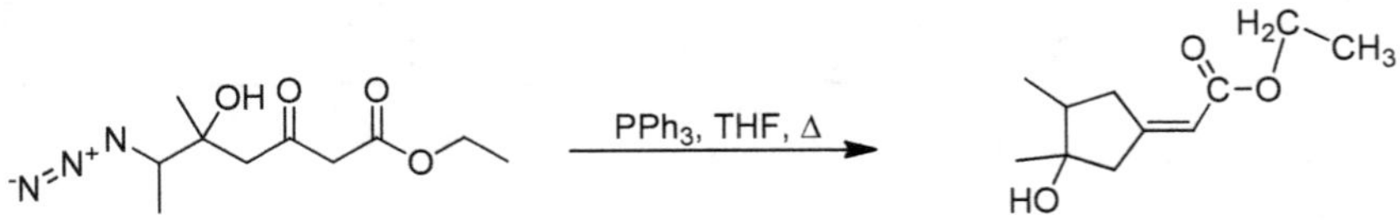

Ethyl 6-azido-5-hydroxy-5-methyl-3-oxoheptanoate   (*E*)-ethyl 2-(3-hydroxy-3,4-dimethylcyclopentylidene)acetate

169. Stephen Reaction

Principle

The reaction of nitriles with anhydrous stannous chloride (saturated with dry HCl gas) to obtain crystalline "aldimine stannichloride" followed by hydrolysis to afford aldehyde is known to be Stephen reaction reported by Stephen in 1925. The success of reaction based on precipitation of the ether insoluble crystalline "aldimine stannichloride", isolated from the reaction mixture, washed with ether, and then hydrolyzed to produce aldehyde. The reaction is suitable for aromatic nitriles compared to aliphatic nitriles.

General Reaction

$$Ar-C\equiv N \xrightarrow[\text{ii) } H_2O/H^{\oplus}]{\text{i) } SnCl_2/HCl} Ar-CHO \quad + \quad NH_4Cl$$

Aromatic nitriles   Aromatic aldehyde

a) Synthesis of benzaldehyde from benzonitrile.

$$C_6H_5-C\equiv N \xrightarrow[\text{ii) } H_2O/H^{\oplus}]{\text{i) } SnCl_2/HCl} C_6H_5-CHO \quad + \quad NH_4Cl$$

Benzonitrile   Benzaldehyde

b) Formation of acetaldehyde from acetonitrile.

$$H_3C-C\equiv N: \; + \; HCl \longrightarrow H_3C-\underset{NH}{\overset{Cl}{C}} \xrightarrow{2H} H_3C-\underset{NH \cdot HCl}{\overset{H}{C}}$$

Acetonitrile   Acetimidoyl chloride   Ethanimine hydrochloride

$$H_3C-\underset{NH}{\overset{H}{C}} \; + \; H_2O \longrightarrow H_3C-\underset{O}{\overset{H}{C}} \; + \; NH_3$$

Ethanimine   Acetaldehyde

Mechanism

The detailed mechanism of Stephen reaction is illustrated here:

$$\text{Benzonitrile} \quad C\equiv N + SnCl_2 + 2HCl \longrightarrow \text{Phenylmethanimine} \quad + \left[SnCl_4 \xrightarrow{HCl} H_2SnCl_6 \right]$$

Benzonitrile · Phenylmethanimine · Chlorostannic acid

$$2 \text{ Phenylmethanimine} + H_2SnCl_6 \longrightarrow \text{Araldiminium hexachlorostannate (complex)}$$

Phenylmethanimine · Araldiminium hexachlorostannate (complex)

$$\text{Araldiminium hexchlorostannate (complex)} \xrightarrow{2H_2O} 2 \text{ } CHO + (NH_4)_2SnCl_6$$

Araldiminium hexchlorostannate (complex) · Benzaldehyde

Applications

The reaction is widely applicable in synthesis of different aldehydes.

a) Synthesis of 2-phenylacetaldehyde.

$$\text{2-phenylacetonitrile} \xrightarrow[\text{2. } H_2O, NaHCO_3, \Delta]{\text{1. } SnCl_2/HCl/Et_2O} \text{2-phenylacetaldehyde}$$

2-phenylacetonitrile · 2-phenylacetaldehyde

b) Synthesis of dodecanal.

$$\text{Dodecanenitrile} \xrightarrow[\text{2. } H_2O]{\text{1. } SnCl_2/HCl/Et_2O} \text{Dodecanal}$$

Dodecanenitrile · Dodecanal

170. Stevens Rearrangement

Principle

An intramolecular thermal [1,2]-electrophilic migration of an alkyl group from the heteroatom of an ylide to the adjacent carbanion center upon treatment of strong base, resulting in the formation of tertiary amine is reported and explained by Stevens 1928; hence known as Stevens rearrangement. Stevens rearrangement is commonly observed on ammonium and sulfonium salts, heteroatom of ylides such as nitrogen, sulfur, oxygen (oxonium ylide), phosphorus, arsenic or antimony. Phosphorus ylide undergoes Stevens rearrangement in harsh conditions is known as phospha-Stevens rearrangement. Generally, Stevens rearrangement occurs for ylides bearing acidic α-proton devoid of β-hydrogens, otherwise Hofmann elimination may take place.

General Reaction

N-benzyl-N,N-dimethyl-2-oxo-2-phenylethanaminium bromide

2-(dimethylamino)-1,3-diphenylpropan-1-one
(α–dimethylamino-β-phenylpropiophenone)

Mechanism

Step 1: Formation of ylide.

N-benzyl-N,N-dimethyl-2-oxo-2-phenylethanaminium

phenyllithium

Ylide

Step 2: Generation of α-dimethylamino-β-phenylpropiophenone.

Ylide

[1,2-shift]

α–dimethylamino-β-phenylpropiophenone

Applications

a) The reaction has been widely used in organic synthesis.

2-methoxy-*N*,*N*-dimethyl-2-oxo-1-phenyl-*N*-
((trimethylsilyl)methyl)ethanaminium

methyl 3-(dimethylamino)-
2-phenylpropanoate

methyl 3-(dimethylamino)-
3-phenylpropanoate

171. Stobbe Condensation

Principle

The condensation reaction between carbonyl compounds and dialkyl ester of diacid in presence of strong base such as sodium ethoxide in an alcoholic solution to produce half ester is reported by Stobbe in 1893; hence popularly known as Stobbe condensation. Ketones are found to react equally well with succinates in Stobbe condensation in comparison to aldehydes. In addition, the hydroxyl moiety of resulting carboxyl group in substituted itaconate (Stobbe product or Stobbe acid ester) actually originates from oxygen atom in corresponding carbonyls. Stobbe condensation involving ketone usually gives higher yield and purer substituted itaconate if potassium t-butoxide or sodium hydride is used in short reaction period rather than sodium ethoxide. The driving force for Stobbe condensation is the generation of γ-lactone.

General Reaction

Carbonyl compounds + Diethyl succinate $\xrightarrow{C_2H_5ONa}$ Sodium salt of alkylidenesuccinic acid + C_2H_5OH

Sodium salt of alkylidenesuccinic acid $\xrightarrow[ii)\ HCl]{i)\ NaOH}$ Alkylidene succinic acid

Propan-2-one + Diethyl succinate $\xrightarrow{C_2H_5ONa}$ Sodium salt of alkylidenesuccinic acid + C_2H_5OH

Sodium salt of alkylidenesuccinic acid $\xrightarrow[ii)\ HCl]{i)\ NaOH}$ 2-(propan-2-ylidene)succinic acid

Mechanism

Step 1: Formation of carbanion.

Step 2: Addition of carbanion into carbonyl carbon atom.

Step 3: Lactonization.

Step 4: Ring opening followed by acidic workup to give final product.

Applications

a) *The reaction has been used for synthesis of cyclopentanone as well as the carbonyl compounds with propanionic acid as the side chain. The reaction is also useful for the synthesis of estrone derivatives. Synthesis of (Z)-3-benzylidene-4-(diphenylmethylene) dihydrofuran-2,5-dione.*

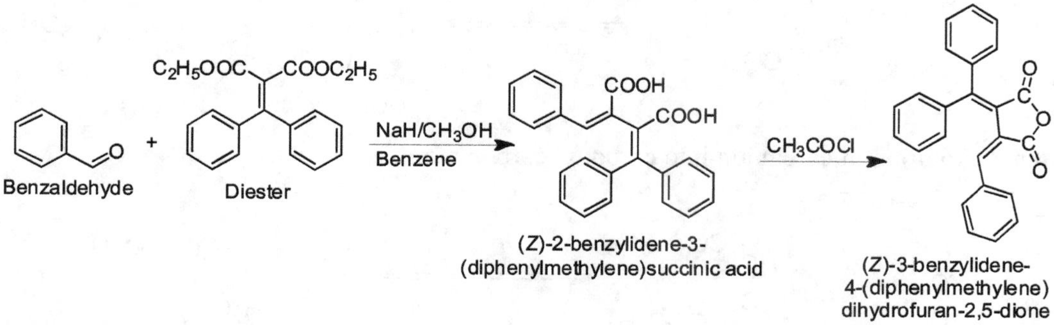

172. Stork Enamine Reaction

Principle

Stork reported and explained enamine promoted alkylation or acylation at the α-carbon of the carbonyl compounds in 1954. Therefore, enamine-mediated alkylation is referred as Stork Enamine alkylation, and acylation is as Stork acylation. In particular, enamine based addition is reffered as Michael-Stork addition. The rate of alkylation in this reaction is concerned with the basicity and steric hindrance of secondary amine, where the six-membered cyclic enamines are slightly stronger bases than saturated tertiary amines, whereas the five and seven-membered cyclic enamines are much stronger bases than the saturated ones.

General Reaction

2-methylcyclohexanone   Pyrrolidine   1-(6-methylcyclohex-1-en-1-yl)pyrrolidine   8-methyl-4,4a,5,6,7,8-hexahydronaphthalen-2(3H)-one

Mechanism

Step 1: Conjugate addition.

1-(6-methylcyclohex-1-en-1-yl)pyrrolidine   but-3-en-2-one   Conjugate addition

Step 2: Isomerization.

Isomerization

Step 3: Ring closure and formation of product.

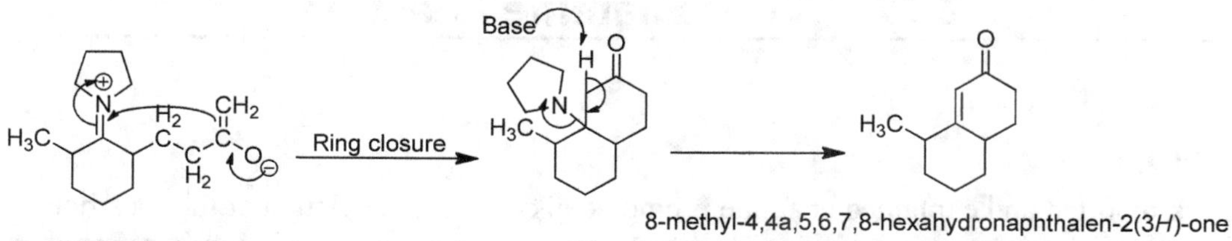

8-methyl-4,4a,5,6,7,8-hexahydronaphthalen-2(3*H*)-one

Applications

a) The reaction has wide applicability in organic synthesis. For example: Synthesis of 4-(cyclopent-1-en-1-yl)morpholine.

Cyclopentanone   Morpholine   *N,O*-bis(trimethylsilyl)acetamide   4-(cyclopent-1-en-1-yl)morpholine

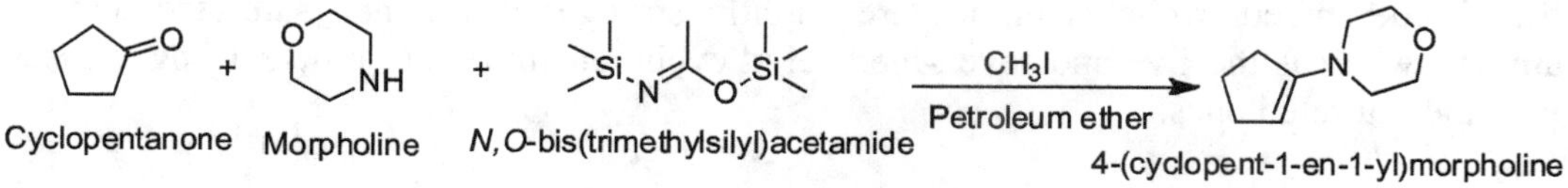

173. Strecker Amino Acid Synthesis

Principle

Strecker in 1850 detailed the first multicomponent reaction in organic chemistry. He explained simple and efficient method for synthesis of α-amino acids involving the formation of imines from aldehydes and ammonia (or amines), cyanation of imines from hydrogen cyanide or equivalents, followed by hydrolysis to α-amino acid. Therefore, the reaction is known as Strecker amino acid synthesis. He also discovered *de novo* synthesis of α-amino acids in presence of a catalyst or a chiral additive is known as asymmetric Strecker reaction.

Knoevenagel and Bucherer explored Strecker's method for synthesis of α-amino acids from amine, potassium cyanide, and aldehyde (or ketone) bisulfites; extended further by Zelinsky and Stadnikoff using potassium cyanide and ammonium chloride is known as *Zelinsky-Stadnikoff Reaction*. In addition, the Strecker reaction in presence of a chiral cyclohexadiamine catalyst is termed Jacobsen-Strecker reaction.

General Reaction

Mechanism

The reaction involves the formation of an imine (or Schiff base, if an aromatic amine is used), and the nucleophilic addition of cyanide to imine intermediate forms α-amino nitrile. The hydrolysis of the α-amino nitrile gives α-amino acid. Mechanism is illustrated for reaction between aldehyde, ammonia, and hydrogen cyanide.

Step 1: Generation of ammonia.

$$NH_4Cl \;+\; NaCN \;\rightleftharpoons\; :NH_3 \;+\; HCN \;+\; NaCl$$

Step 2: Formation of iminium ion.

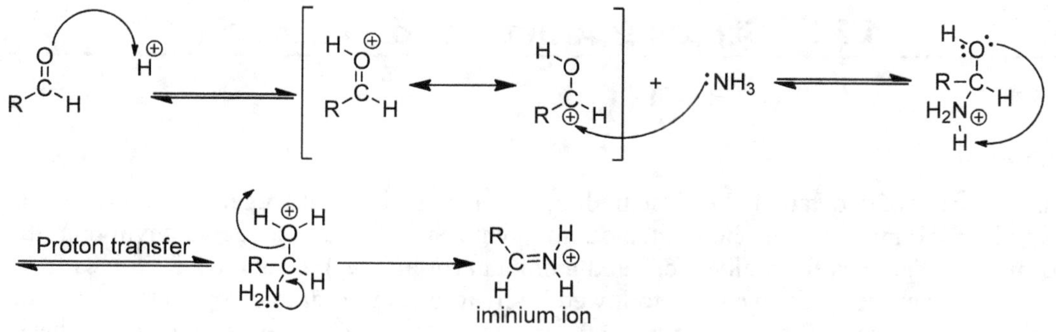

Proton transfer

iminium ion

Step 3: Nucleophilic addition of cyanide ion to imine intermediate forms α-amino nitrile which on hydrolysis gives α-amino acids.

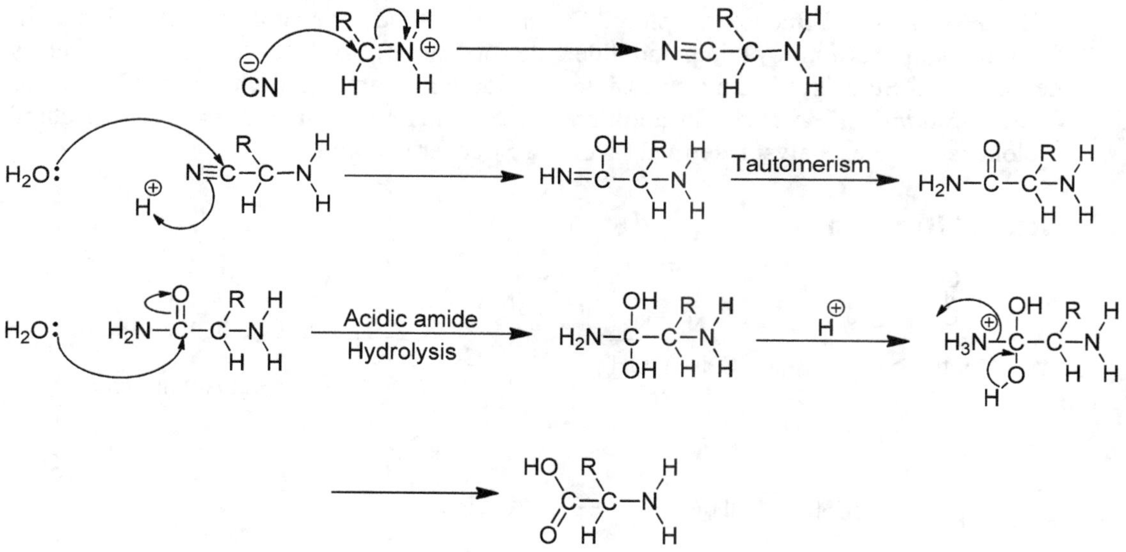

Tautomerism

Acidic amide Hydrolysis

Applications

a) The Strecker synthesis has application in preparation of α-amino acids as well as synthesis of peptides.

1,1,1-trifluoropropan-2-one   2-methylpropane-   (Z)-2-methyl-*N*-(1,1,1-trifluoropropan-
                                 2-sulfinamide   2-ylidene)propane-2-sulfinamide
                                                       (sulfinimine)

N-((*S*)-2-cyano-1,1,1-trifluoropropan-2-yl)-2-methylpropane-   (*S*)-2-amino-3,3,3-trifluoro-2-methylpropanoic acid
2-sulfinamide

174. Suzuki Coupling

Principle

Suzuki and Miyaura in 1979 discovered palladium catalyzed cross-coupling between an aryl or vinyl halide and an aryl or vinyl borane, boronic acid or boronic ester is known as the Suzuki-Miyaura cross coupling or Suzuki reaction. In addition, the use of the Suzuki coupling for the synthesis of polyconjugated polymers from aryl dihalide and aryl diboronic acid is called the Suzuki polycondensation. The palladium catalysts comprises of (a) palladium catalyst such as $Pd(PPh_3)_4$, $Pd(PPh_3)_2Cl_2$, $Pd(PCy_3)_2Cl_2$, $Pd(dppf)Cl_2$, $Pd(PPh_3)_2Cl_2$, $DPEPhosPdCl_2$, $Pd(dppb)Cl_2$, $Pd(dppe)Cl_2$, $Pd(t\text{-}Bu_3P)_2$, $[Pd(i\text{-}Bu)(t\text{-}Bu_3P)]_2$, $[Pd(\eta3\text{-}C_3H_5)Cl]_2$, $Pd(N,N\text{-dimethyl }\beta\text{-alaninate})_2$, $Pd(Ph_2PCH_2CO_2)_2$, and $Pd_2(dba)_3$ (dba = dibenzylidene acetone); (b) an inorganic palladium compound in combination with various ligands, such as $Pd(OAc)_2$ or $PdCl_2$ in association with DtBPF, electron-rich biaryl dialkylphosphines, triethylenediamine (Dabco), pyrazole-tethered phosphine ligand, diazabutadiene, N,N-dicyclohexyl-1,4-diazabutadiene, carbenes, $R_2PN:P(i\text{-}BuNCH_2CH_2)_3N$, tetran-butylammonium bromide (TBAB)/polyethylene glycol (PEG-400), thiourea, PEG-300, iminophosphorane-phosphane ($Ph_2PC_6H_4OC_6H_4PPh_2=NP(O)(OPh)_2$, isonitrile, and 2-pyridinealdoxime, (c) cyclic palladium catalysts, such as palladacycles; and (d) solid-supported palladium catalysts, such as Pd/C, MCM-41-supported sulfur ligand, silica templated palladium hollow spheres, and palladium associated with polymer bound phosphorus ligands. Many of these catalytic systems are very effective for the Suzuki coupling, such as the combination of $Pd(OAc)_2$ and Dabco, which is a stable, inexpensive, and highly efficient catalytic system.

General Reaction

$$R{-}X \quad + \quad R{-}\underset{R''}{\overset{\overset{\displaystyle R'}{|}}{B}} \quad \xrightarrow[\text{NaOR'''}]{L_2Pd(O)} \quad R{-}R'$$

Organic halides   Organoboranes   Alkanes

Mechanism

Mechanism of Suzuki coupling for general alkyl halide and boron ligand is displayed here:

Step 1: Suzuki coupling involves an initial oxidative addition of electrophilic aryl halide to a Pd(0) complex.

$$R{-}X \quad + \quad L_2Pd(O) \quad \xrightarrow[\text{addition}]{\text{Oxidative}} \quad R{-}\underset{L}{\overset{X}{\underset{|}{\overset{|}{Pd}}}}{-}L$$

Step 2: Addition of base.

$$R' \quad + \quad NaOR''' \xrightarrow{\text{Addition of base}} R-B^{\ominus}-R'' \text{ (with } R' \text{ top, } OR''' \text{ bottom)}$$

Step 3: Transmetalation of other coupling moiety from boron atom to form a new Pd(II) complex.

$$\xrightarrow{\substack{\text{Transmetalation} \\ \text{isomerization}}}$$

Step 4: Reductive elimination from palladium (II).

$$\xrightarrow{\substack{\text{Reductive} \\ \text{elimination}}} R-R' \quad + \quad L_2Pd(O)$$

Applications

a) *Suzuki coupling is popular coupling reactions in organic synthesis because of advantages, including the general availability, stability (for both air and water), low toxicity of boron reagents, and the tolerance for different functional groups. In addition, the Suzuki coupling of aryl perfluoroalkylsulfonates has shown a high reactivity, good stability, and resistance toward hydrolysis.*

b) *The reaction has broad applications in organic synthesis.*

Dimethyl 2-((1-(ethoxycarbonyl)-4-iodo-1*H*-indol-3-yl)methyl)malonate

+

(*E*)-2-(3-methylbut-1-en-1-yl)benzo[*d*][1,3,2]dioxaborole

Pd(OAc)$_2$, Ph$_3$P, K$_2$CO$_3$
THF, CH$_3$OH
70°C, 15 h, 80%

(*E*)-dimethyl 2-((1-(ethoxycarbonyl)-4-(3-methylbut-1-en-1-yl)indolin-3-yl)methyl)malonate

175. Swern Oxidation

Principle

Oxidation of alcohols to the corresponding carbonyl compounds using carbonyl chloride, dimethyl sulfoxide, and quenching with triethylamine. The reaction was initially reported by Daniel Swern. The reaction is well known for its mild character and wide tolerance of functional groups. The primary and secondary alcohols are oxidized to an aldehyde or ketone. Oxidation of alcohols by activated dimethyl sulfoxide was reported by Swern *et al* in 1978.

General Reaction

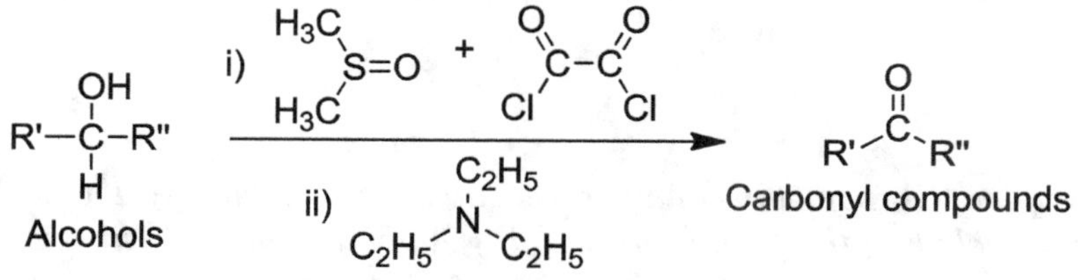

Transformation of propan-2-ol into propan-2-one

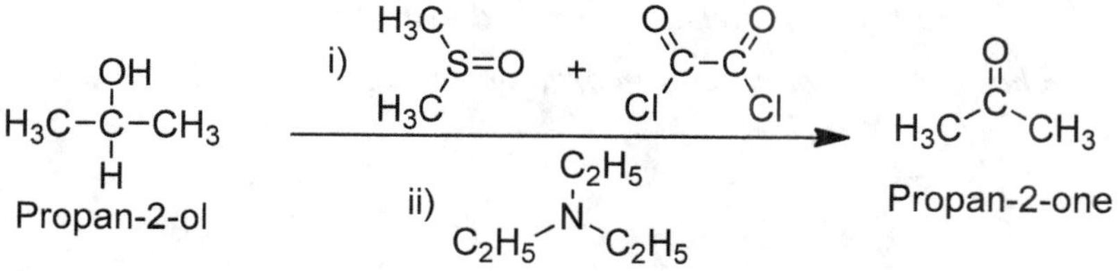

Mechanism:

Step 1:

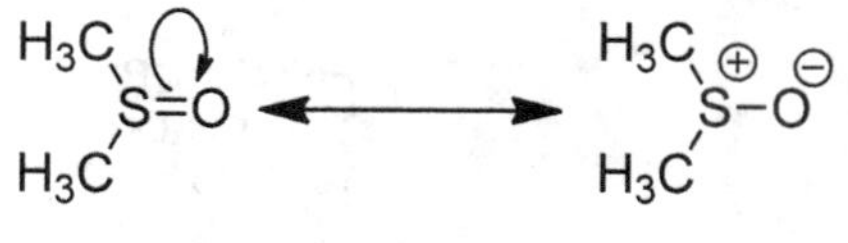

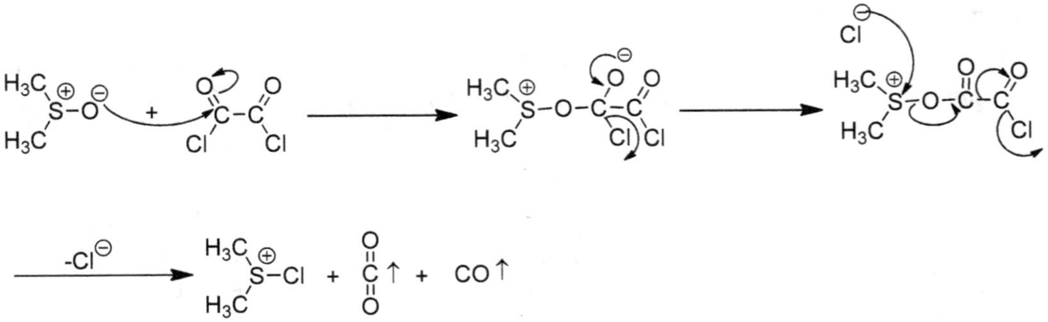

Step 2: Formation of alkoxysulphonium ion

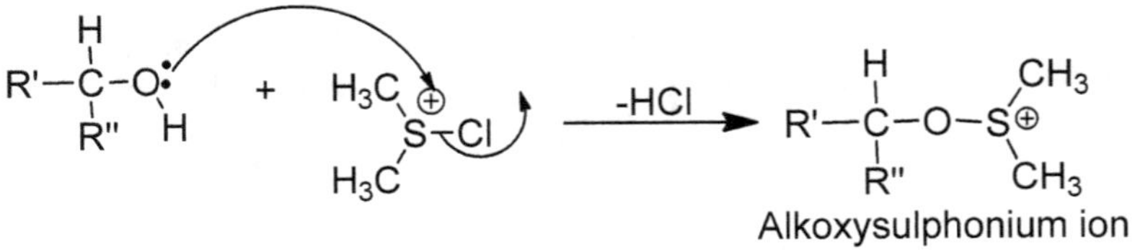

Alkoxysulphonium ion

Step 3: Deprotonation

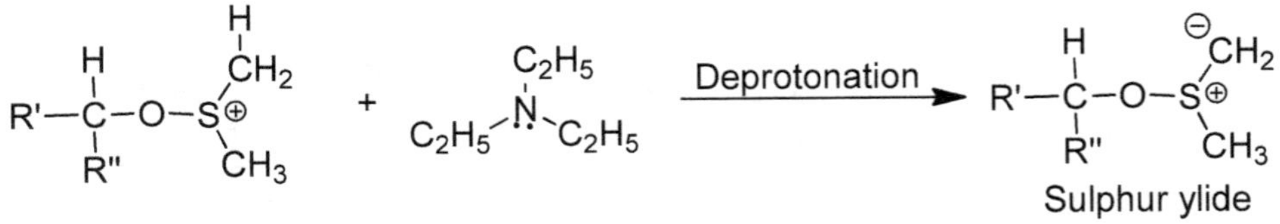

Sulphur ylide

Step 4: Intramolecular deprotonation to give carbonyl compounds.

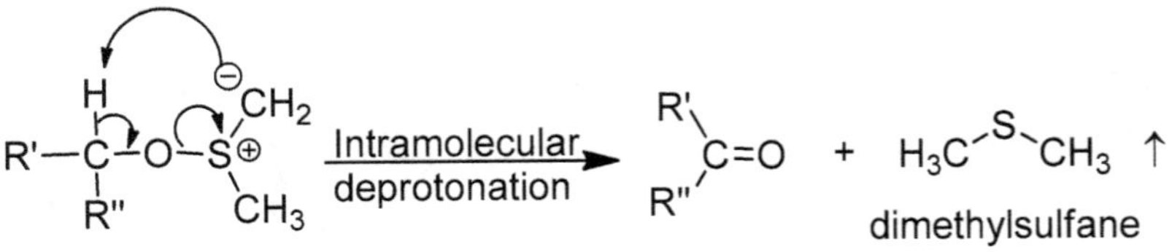

dimethylsulfane

Applications

a) *The reaction is very useful in synthesis of wide varieties of carbonyl compounds. For example: Synthesis of (2S,3S,5S)-2-hexyl-3-((2,3,3-trimethylbutan-2-yl)oxy)-5-undecylcyclopentanone.*

Reagents: $(COCl)_2$, DMSO, Et_3N, 89%

(1R,2R,3S,5S)-2-hexyl-3-((2,3,3-trimethylbutan-2-yl)oxy)-5-undecylcyclopentanol → (2S,3S,5S)-2-hexyl-3-((2,3,3-trimethylbutan-2-yl)oxy)-5-undecylcyclopentanone

176. Tiemann Cyanohydrin Amination

Principle

Tiemann in 1880 reported transformation of cyanohydrins into α-amino nitrile in presence of alcoholic ammonia in reversed process is known as Strecker-Tiemann reaction. The direct addition of hydrogen cyanide to aldimine to form aminonitrile was reported by Tiemann, and further extended for to the addition of hydrogen cyanide to ketimine, hydrazones, semicarbazones, and Schiff bases. The resulting amino nitriles are hydrolyzed to α-amino acids or converted into hydantoins.

General Reaction

Cyanohydrin
R = Alkyl, Aryl

α-amino nitrile

4-chloro-2-hydroxybutanenitrile

2-amino-4-chlorobutanenitrile

Mechanism

It has been assumed that in presence of alcoholic ammonia, the cyanohydrins decomposes back to carbonyl compounds (aldehyde or ketone) and hydrogen cyanide, which further reacts with ammonium ion followed by elimination of water molecule gives ammonium cyanide. The transiently generated carbonyl intermediate then reacts with ammonium ion followed by the addition of cyanide, as displayed here.

Applications

a) *The reaction is helpful for the synthesis of α-amino nitriles, which can be transformed into the corresponding α-amino acids and hydantoins.*

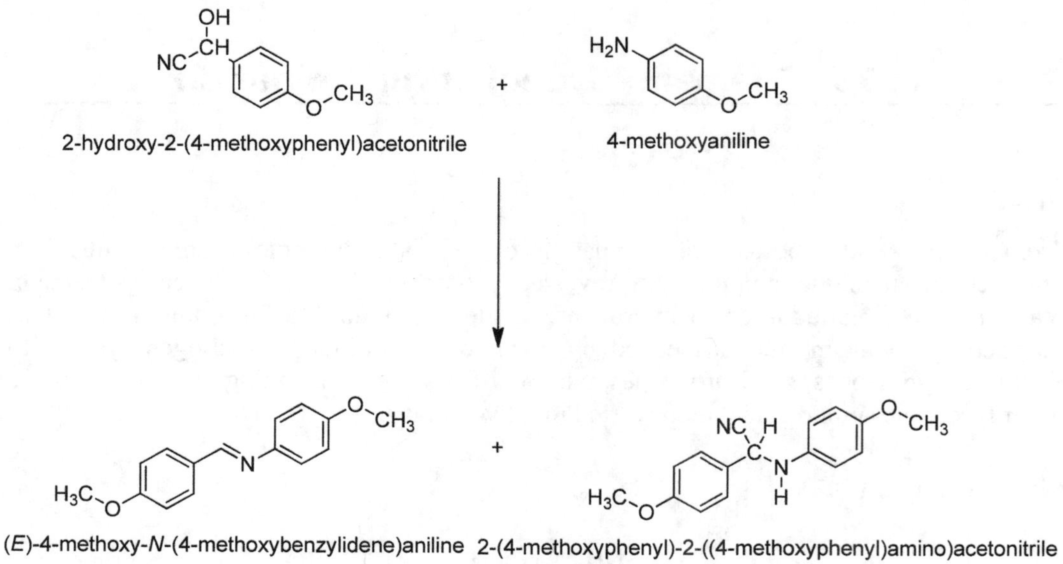

2-hydroxy-2-(4-methoxyphenyl)acetonitrile

4-methoxyaniline

(*E*)-4-methoxy-*N*-(4-methoxybenzylidene)aniline

2-(4-methoxyphenyl)-2-((4-methoxyphenyl)amino)acetonitrile

Principle

Tiemann and Pinnow in 1891 explained conversion of amidoximes into corresponding substituted ureas when treated with benzenesulfonyl chloride or acetic anhydride followed by hydrolysis is called as Tiemann amidoxime-urea rearrangement.

General Reaction

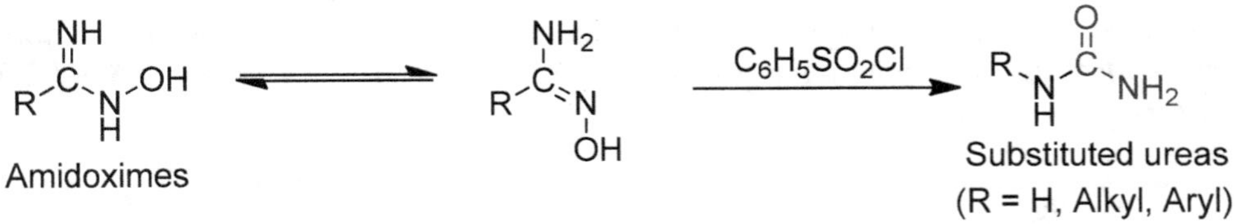

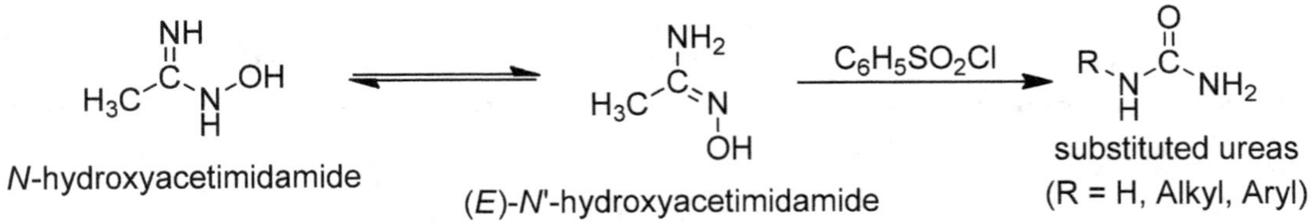

Amidoximes

Substituted ureas
(R = H, Alkyl, Aryl)

N-hydroxyacetimidamide

(E)-N'-hydroxyacetimidamide

substituted ureas
(R = H, Alkyl, Aryl)

Mechanism

Step 1:

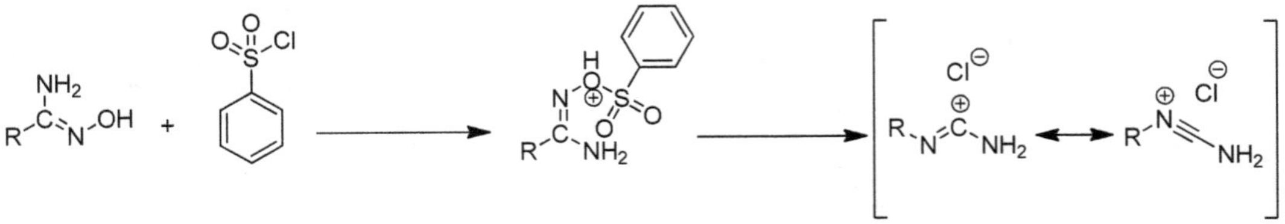

Step 2:

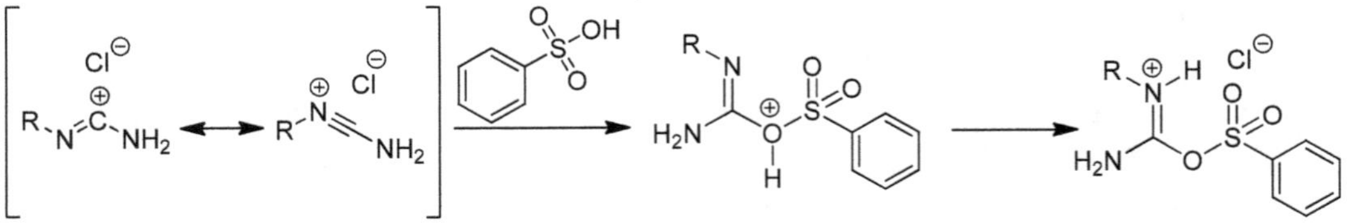

Step 3:

Applications

a) The reaction is useful in synthesis of substituted ureas, cyanamides, oxathiadiazoles, and carbodiimides. For example: N-phenyliminoforyl-N-phenylhydroxylamine has been converted into N,N'-diphenylurea when treated with acetic anhydride in nearly quantitative yield. In some cases, the amidoxime itself may undergo such rearrangement, as shown by the presence of urea in the aqueous solution of formamidoxime at high temperatures. Synthesis of 3,4-diphenyl-3H-1,2,3,5-oxathiadiazole-2-oxide.

(Z)-3-chloro-N'-hydroxy-N-phenylbenzimidamide   3,4-diphenyl-3H-1,2,3,5-oxathiadiazole 2-oxide

178. Tishchenko Reaction

Principle

Tishchenko in 1906 reported disproportionation of nonenolizable aldehydes to produce esters in the presence of aluminum alkoxide or sodium alkoxide is referred as Tishchenko dimerization or Tishchenko reaction. The esters obtained through interaction of two equivalents of aldehyde are Tishchenko esters. Tishchenko reaction also possible between an aldehyde and another carbonyl compound (aldehyde or ketone), and the reduction of ketones by a nonenolizable aldehyde to produce esters is known as Tishchenko reduction. Sometimes, when the enolizable aldehydes are used, the Tishchenko reaction is competing with Aldol condensation, in presence of a basic catalyst; the resulting aldols can be reduced through Tishchenko reaction. This clubbing of reversible Aldol and irreversible Tishchenko reaction is referred as Aldol-Tishchenko reaction.

General Reaction

$$2 \text{ RCHO} + \text{Al}(\text{OC}_2\text{H}_5)_3 \rightleftharpoons \text{RCOOCH}_2\text{R}$$

Aldehyde   Aluminium ethoxide   Ester

a) Preparation of ethyl acetate

$$2 \text{ CH}_3\text{CHO} + \text{Al}(\text{OC}_2\text{H}_5)_3 \rightleftharpoons \text{CH}_3\text{COOCH}_2\text{CH}_3$$

Ethanal   Aluminium ethoxide   Ethyl acetate

b) Preparation of propyl propionate

$$2 \text{ CH}_3\text{CH}_2\text{CHO} + \text{Al}(\text{OC}_2\text{H}_5)_3 \rightleftharpoons \text{CH}_3\text{CH}_2\text{COOCH}_2\text{CH}_2\text{CH}_3$$

Propanal   Aluminium ethoxide   Propyl propionate

Mechanism

Step 1:

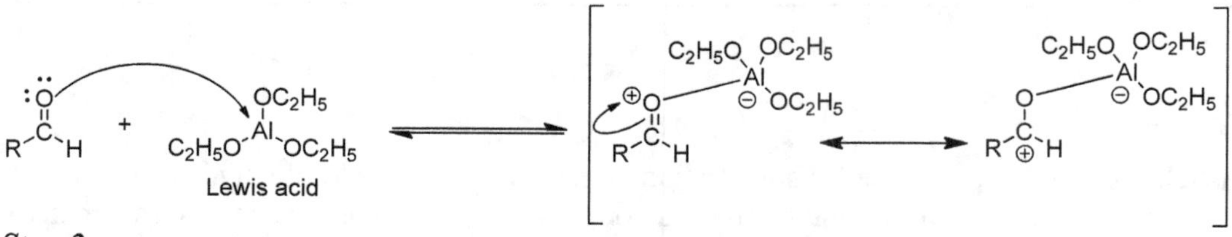

Step 2:

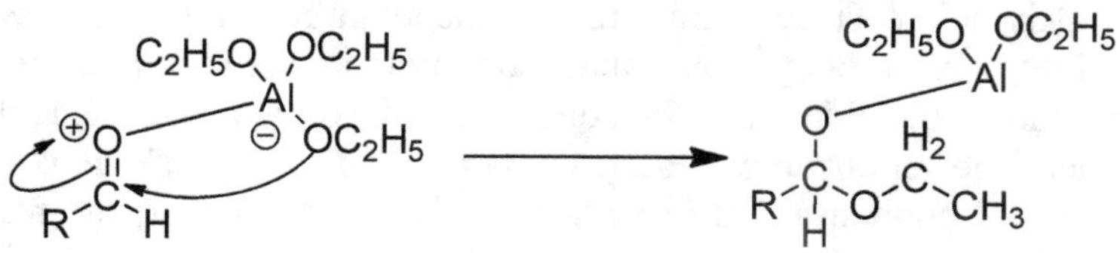

Step 3:

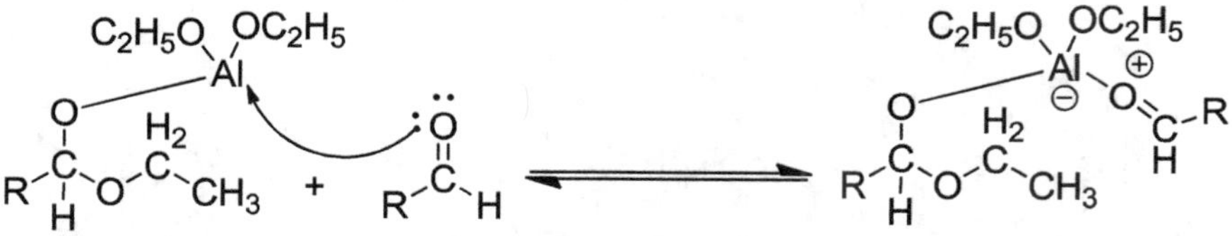

Step 4:

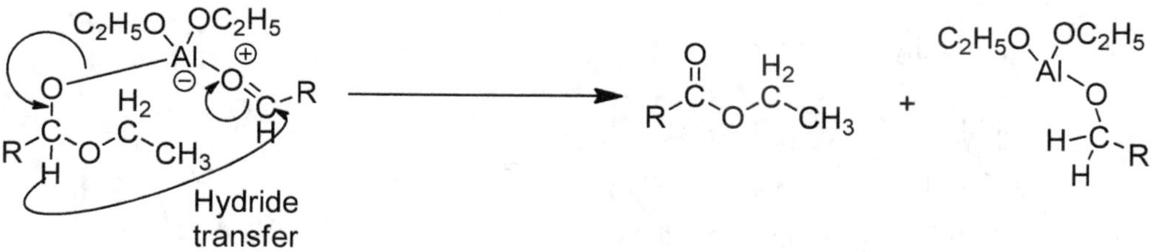

Hydride
transfer

Applications

a) Synthesis of (2R,3R,4S,5S)-3,5-dihydroxy-2,4-dimethyl-1,5-diphenylpentyl benzoate.

Benzaldehyde + Pentan-3-one — Aldol reaction / TiCl4 → 1-hydroxy-2-methyl-1-phenylpentan-3-one — Ti(OBu)₄ / Cinchonine →

(2R,3R,4S,5S)-3,5-dihydroxy-2,4-dimethyl-
1,5-diphenylpentyl benzoate

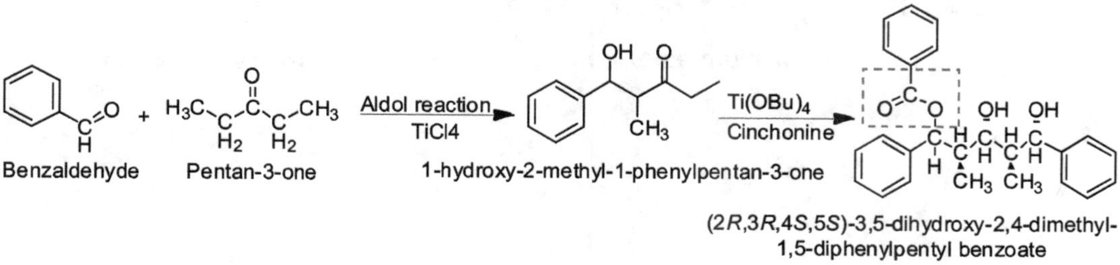

b) Synthesis of 2-(1-hydroxy-2-methylpropan-2-yl)-5,5-dimethyl-1,3-dioxan-4-ol.

Isobutyraldehyde + Formaldehyde $\xrightarrow{Et_3N}$ 3-hydroxy-2,2-dimethylpropanal $\xrightarrow{65?C}$ 2-(1-hydroxy-2-methylpropan-2-yl)-5,5-dimethyl-1,3-dioxan-4-ol

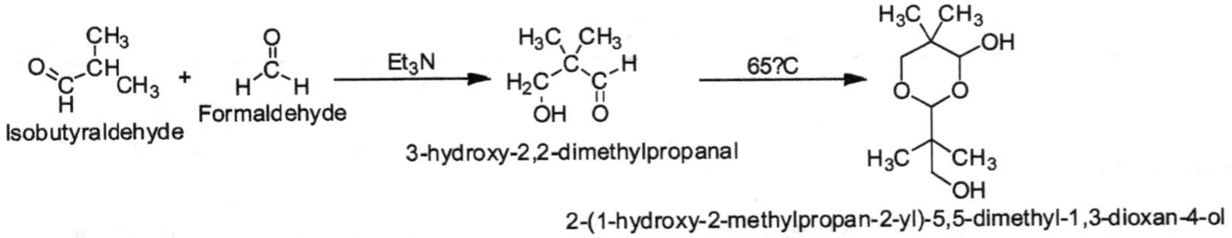

179. Ugi Reaction

Principle

Ugi in 1959 reported multicomponent reaction of an amine, a carbonyl compound, a carboxylic acid, and an isonitrile to give an α-acylamino amide with four different substituent's is known as Ugi reaction. This reaction is also referred as Ugi condensation, Ugi four-component condensation (4CC), Ugi four-component reaction (U-4CR), or Ugi multiple-component condensation. In addition, when a compound bearing two reactive elements is used in the Ugi reaction, the corresponding multicomponent reaction is known as the Ugi four-center three-component reaction or Ugi 4-center 3-component reaction (U-4C-3CR). Similarly, the reaction with five reactive centers involving three reactants is known as the Ugi five-center three-component reaction (U-5C-3CR). In these multicomponent reactions, the amines can be ammonia, mono and disubstituted amines, hydroxylamine, hydrazine and the carbonyl compounds (aldehydes or ketones). In addition, many other compounds have been successfully used for this reaction such as water, thiosulfates, hydrogen selenide, hydrazoic acid, hydrogen cyanate, thiocyanate, aminocyanic acid, alkoxycarboxylic acids, and thioacids.

General Reaction

Formation of peptides.

Mechanism

Step 1:

Imine

Step 2:

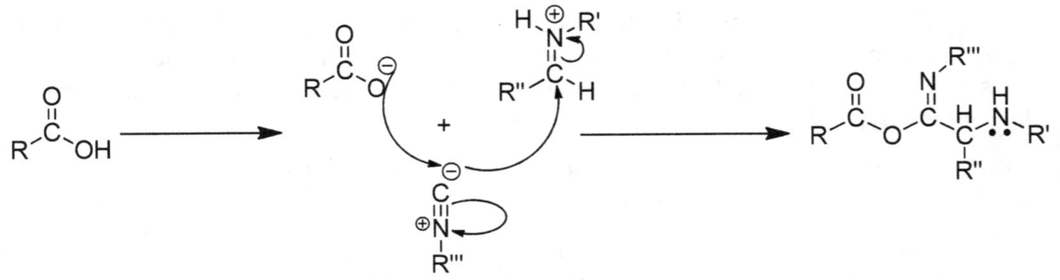

Step 3:

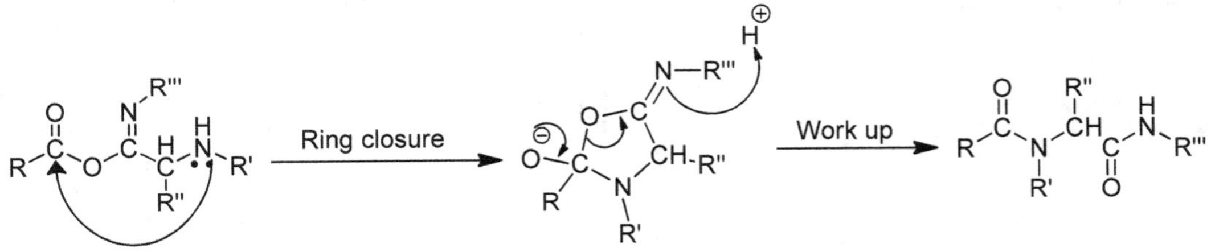

Applications

The variety of chemical libraries has been synthesized by Ugi reaction, as well as compounds such as simple α-amino acids and other heterocycles can also be synthesized such as β-lactam antibiotics, benzodiazepines, morpholines, tetrazoles, diketopiperazines, and α-aminobutyrolactones

a) *Synthesis of N-cyclohexyl-2-(4-nitrophenyl)-2-((1R,6S)-8-oxo-7-azabicyclo[4.2.0]octan-7-yl)acetamide.*

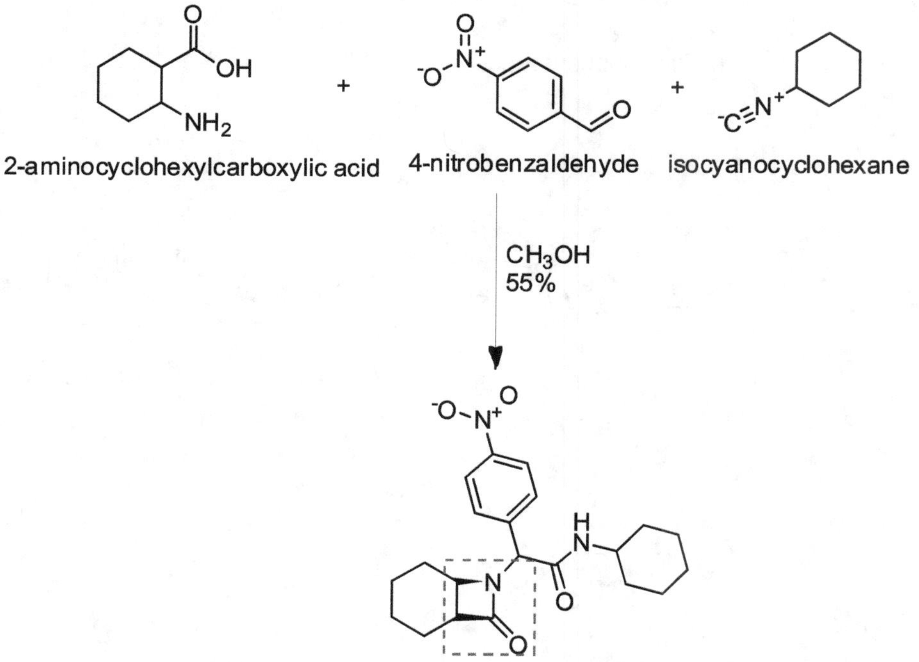

N-cyclohexyl-2-(4-nitrophenyl)-2-((1*R*,6*S*)-8-oxo-7-azabicyclo[4.2.0]octan-7-yl)acetamide

180. Ullmann Acridine Synthesis

Principle

Ullmann in 1902 reported and explained reaction of *N*-arylanthranilic acid with an acylating agent [polyphosphoric acid (PPA)] followed by reduction and dehydration to give acridine analogues is known as Ullmann acridine synthesis. In this reaction, different acylating reagents has been used such as $POCl_3$, H_2SO_4, $(CF_3CO)_2O$, HCl, TFA, or PPA. In addition, the acridines with a substituent at position 3 is not synthesized from 2-chlorobenzoic acid and 3-aminobenzoic acid because the cyclization of the resulting diphenylamine yields a mixture of 1- and 3-substituted products, which cannot be easily separated. The acridones can be converted into the corresponding acridines through the reduction by Na/BuOH or Al/Hg/NaOH and dehydration by $FeCl_3$/HCl. In addition, the acridones by reaction with Grignard reagent followed by dehydration gives 9-substituted acridines.

General Reaction

N-arylanthranilic acid derivatives
R', R" = H, Alkyl, Aryl

Acridine derivatives

Mechanism

The reaction is a simple aromatic acylation. However, regardless of the acylating reagents used, condition of ring closure has little effect on the product composition, indicating that the cyclization occurs through an acylcarbonium ion. Stepwise mechanism for the synthesis of acridines is illustrated below:

Step 1:

Step 2:

Step 3:

Applications

The reaction has broad applications in synthesis of acridines.

a) Synthesis of 7-ethyl-9-oxo-9,10-dihydroacridine-4-carboxylic acid.

2-((3-ethylphenyl)amino)isophthalic acid

7-ethyl-9-oxo-9,10-dihydroacridine-4-carboxylic acid

b) Synthesis of 7-ethylacridine-4-carboxylic acid.

7-ethyl-9-oxo-9,10-dihydroacridine-4-carboxylic acid

7-ethylacridine-4-carboxylic acid

181. Ullmann Reaction

Principle

Ullmann in 1901 reported the homocoupling of aryl halides in presence of copper/nickel/palladium to obtain biaryls. Hence the reaction is popularly known as Ullmann cross-coupling reaction. A reaction is similar to Wurtz-Fittig reaction in which biaryl or polyaryl derivatives are synthesized by condensing aromatic halides (other than fluorides) with themselves, with aniline or with sodium phenolates by heating with copper powder or copper-bronze at an elevated temperature (above 150 °C). The reaction is most successful with aryl iodides, however, aryl chlorides and bromides do not undergo reaction unless an electronegative group is substituted in the reactant. The electronegative group activates the halogen substituents. The reaction gives better yield in presence of DMF as solvent.

General Reaction

Mechanism

Step 1: Generation of free radical followed by reaction with copper iodide to give phenylcopper(II) iodide.

Step 2: Reaction of phenylcopper(II) iodide with aromatic halide (iodobenzene) to give 1,1'-biphenyls.

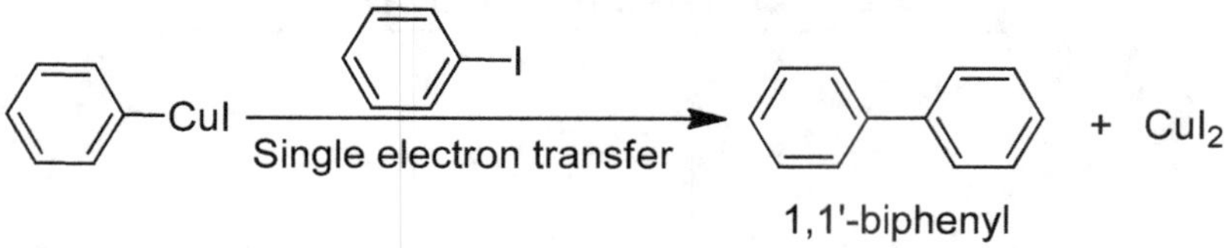

Single electron transfer

1,1'-biphenyl

Applications

a) Preparation of polycyclic hydrocarbons. Ullmann reaction has been used in synthesizing many polycyclic hydrocarbons For example: Synthesis of perylene from 1,8-diiodonaphthalene.

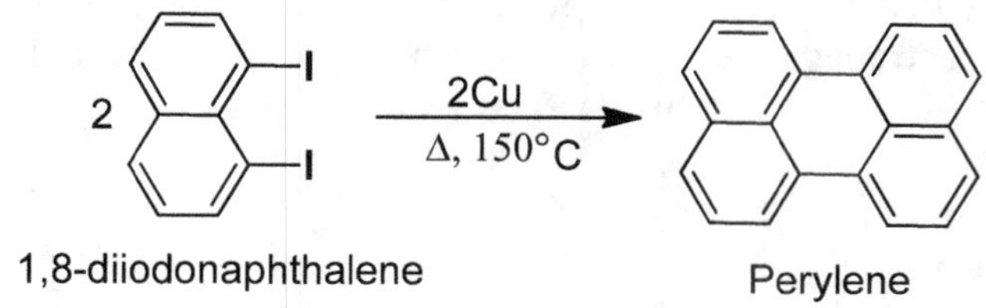

1,8-diiodonaphthalene   Perylene

b) Synthesis of 2,7-dimethoxy-9,10-dihydrophenanthrene.

1,2-bis(2-iodo-5-methoxyphenyl)ethane 2,7-dimethoxy-9,10-dihydrophenanthrene

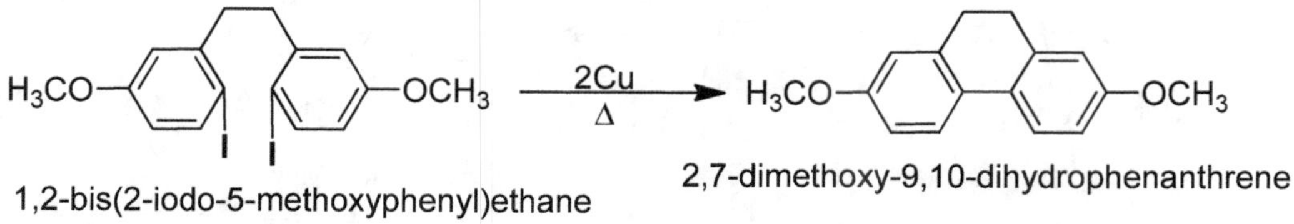

182. Victor Meyer Reaction

Principle

Victor Meyer in 1872 carried out the synthesis of nitroalkane from haloalkane and silver nitrite is known as Victor Meyer reaction. The reaction is performed in presence of petroleum ether or dialkyl ether in a temperature range of 80-110°C, to produce mixtures of nitroalkane and alkylnitrite, due to the bidenate nature of nitrite. The concentration of nitroalkane and alkylnitrite varies, depending on the nature of haloalkane and metal nitrite. For example: the yields of nitroalkanes fall progressively when silver nitrite reacts with primary, secondary, and tertiary haloalkanes; whereas the yields of corresponding alkyl nitrite increase in an inverse order. The change of product ratio is due to involvement of both steric and electronic effects. It is proposed that the formation of C-N bond in nitroalkane has higher steric hindrance than the formation of the C-O bond in alkyl nitrite; in parallel, the 3° haloalkane is less electrophilic than the 1° haloalkane, and the oxygen atom has more negative charge than the nitrogen atom in silver nitrite; therefore, alkyl nitrite is favored for 3° haloalkane over 1° haloalkane.

General Reaction

$$R_1\!-\!\underset{X}{\overset{R_3}{\underset{|}{\overset{|}{C}}}}\!-\!R_2 \ + \ AgNO_2 \ \xrightarrow{\text{Ether, } \Delta} \ R_1\!-\!\underset{NO_2}{\overset{R_3}{\underset{|}{\overset{|}{C}}}}\!-\!R_2 \ + \ R_1\!-\!\underset{ONO}{\overset{R_3}{\underset{|}{\overset{|}{C}}}}\!-\!R_2$$

Haloalkane   Silver nitrite   Nitroalkane   Alkylnitrite

R_1 = H, Alkyl, Aryl
R_2 = H, Alkyl, Aryl
R_3 = H, Alkyl, Aryl
X = Cl, Br, I, etc.

Mechanism

Formation of nitroalkane: Tertiary haloalkanes reacts with silver nitrite to give nitroalkane with the elimination of silver halide.

Formation of alkylnitrite: Tertiary haloalkanes combines with silver nitrite to give alkylnitrite

Applications

a) *The reaction is widely useful for the synthesis of different substituted nitroalkanes. For example: Preparation of 1-nitropentane.*

1-bromopentane + AgNO$_2$ $\xrightarrow{\text{Ether, }\Delta}$ 1-nitropentane

b) *Synthesis of 1-nitro-4-(nitromethyl)benzene.*

1-(bromomethyl)-4-nitrobenzene + AgNO$_2$ $\xrightarrow{\text{Ether, }\Delta}$ 1-nitro-4-(nitromethyl)benzene

183. Vilsmeier-Haack Reaction

Principle

Vilsmeier and Haack in 1927 reported the formylation of an electron rich aromatics or heterocyclics using DMF as an acylating agent in presence of an activating reagent (phosphorus oxychloride) is referred as Vilsmeier formylation or Vilsmeier-Haack reaction. The combination of DMF and an activating reagent (phosphorus oxychloride) or the complex of iminium salt formed is known as Vilsmeier-Haack reagent.

The reaction is quite useful for aromatics which are more reactive than benzene, such as the benzenes activated by an *N,N*-dimethylamino group and heterocycles (pyrroles, carbazoles, indoles, ferrocenes, and acridines). Vilsmeier formylation has known to show regioselectivity as observed with *Friedel-Crafts acylation*, which depends on the electron density of the attacked site on the substrate. For example: among the annulated furans, thiophenes, and pyrroles, indoles are the most active heterocyles to undergo the formylation at the 3-position, unless it is blocked by an existing group. This is useful formylation method with advantage of monoformylation, though it has unavoidable drawbacks such as evolution of gaseous HCl and difficult work-up process.

General Reaction

a) Preparation of mixture of 4-methoxybenzaldehyde (34%) and 2-methoxybenzaldehyde (4%)

b) Thiophene on reaction with N,N-dimethylformamide and phosphorus oxychloride gives Thiophene-2-carbaldehyde and Dimethylamine.

c) Preparation of mixture of 4-(dimethylamino)benzaldehyde and Dimethylamine

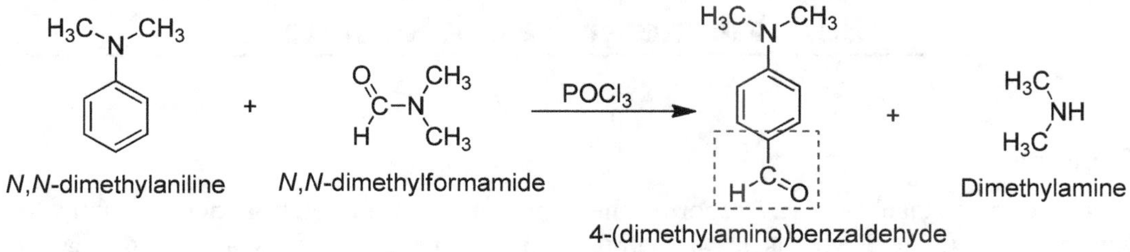

N,N-dimethylaniline   *N,N*-dimethylformamide

4-(dimethylamino)benzaldehyde

Dimethylamine

Mechanism

Step 1: Nucleophilic substitution reaction between N,N-dimethylformamide and phosphorus oxychloride.

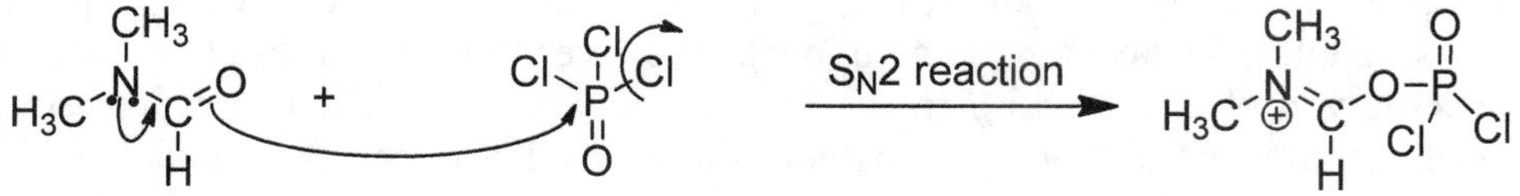

S_N2 reaction

Step 2: Preparation of Vilsmeir-Haack reagent.

Vilsmeir-Haack reagent

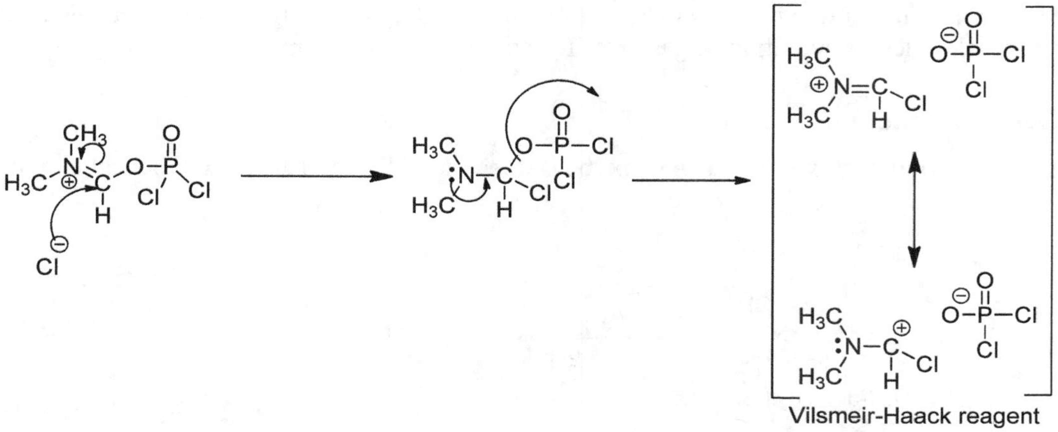

Step 3: Reaction between Vilsmeir-Haack reagent and anisole gives reactive intermediate.

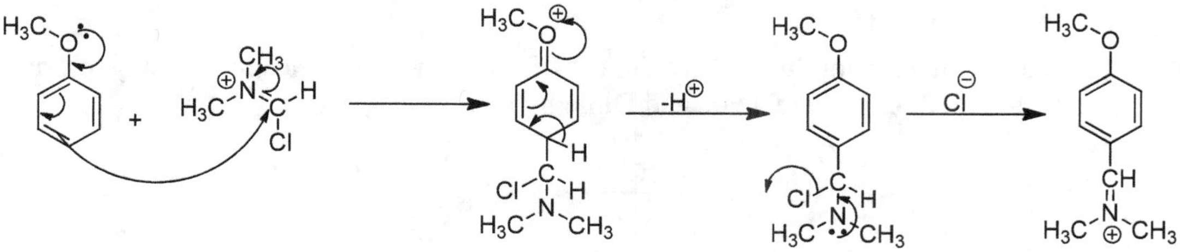

Step 4: Work up to form product.

Applications

a) *The reaction is very useful in monoformylation of different organic compounds. For example: Preparation of 2,4-dihydroxybenzaldehyde from resorcinol.*

Resorcinol + N,N-dimethylformamide $\xrightarrow[\text{NaOH}]{\text{POCl}_3}$ 2,4-dihydroxybenzaldehyde + Dimethylamine

b) *Synthesis of anthracene-9-carbaldehyde from anthracene.*

Anthracene + N,N-dimethylformamide $\xrightarrow[\text{NaOH}]{\text{POCl}_3}$ Anthracene-9-carbaldehyde + Dimethylamine

184. Wacker Oxidation

Principle

Phillips in 1894 invented an industrial production of acetaldehyde through oxidation of ethylene with oxygen in presence of aqueous acidic solution of palladium chloride and cupric chloride. But the reaction was named after work of Hoechst and Wacker Chemie GmbH, who implemented reaction for the synthesis of acetaldehyde known as Hoechst-Wacker process. The catalytic version of this reaction is used to transform the mono-substituted terminal olefins into methyl ketones with air as oxidant is known as Wacker oxidation. The mixture of palladium chloride and cupric chloride is Wacker catalyst. The reaction can be catalyzed by palladium chloride alone but palladium precipitates out during the reaction; hence requires presence of cupric chloride and hydrochloric acid. The reaction is highly chemoselective. Wacker oxidation can also be used to activate an allylic C-H bond.

General Reaction

Terminal olefins

R = H, Alkyl, Aryl

Methyl ketones

Mechanism

Applications

a) Wacker's process is widely useful in various organic synthetic reactions.

Methyl 7-allyl-6-hydroxy-5-oxo-1,2,3,5-
tetrahydroindolizine-8-carboxylate

Methyl 2-methyl-9-oxo-5,6,7,9-tetrahydrofuro
[2,3-*f*]indolizine-4-carboxylate

185. Walden Inversion

Principle

Walden in 1895 reported chemical phenomenon of inversion of configuration at a chiral center during the bimolecular nucleophilic substitution reaction (S_N2) that occurs at the chiral center. This inversion of configuration in nucleophilic substitution is known as Walden inversion. Occasionally, it is also referred to as the Walden inversion reaction. For example: Walden inversion is the conversion of L-malic acid into D-chlorosuccinic acid by action of phosphorus pentachloride, and the resulting L-chlorosuccinic acid can be turned into either D-malic acid or L-malic acid by treatment with potassium hydroxide and silver oxide, respectively.

Walden inversion may be due to backside attack of the nucleophile at the chiral center; hence, it is strongly influenced by steric effect so that blocking the backside of chiral center with large substituents halts the inversion. Walden inversion is supported by the following experimental evidence:

(a) Backside approach of a nucleophile to 2-chloropropionyl chloride is less hindered than that from 1,2-dichloro-1,2-difluoroethane or 2,3-dichlorobutane;

(b) Walden inversion never occurs at the bridgehead of bicyclic structures, For example: 1-bromotriptycene and bicyclo-[2.2.1]-heptane or bicyclo-[2.2.2]-octane with a substituent at the bridgehead.

On the other hand, the conversion of L-chlorosuccinic acid into L-malic acid by silver oxide is due to the double Walden inversion that completes the Walden cycle with retention of configuration, in addition, silver oxide also accelerates the formation of lactone. In addition, the reaction condition also affects the outcome of displacement. For example, the reaction between l-phenylchloroacetic acid with ammonia in water or alcohol gives D-phenylaminoacetic acid; the same reaction in acetonitrile or liquid ammonia gives l-phenylaminoacetic acid. D-α-chloro-phenylethane is tranformed into corresponding l-phenylethyl acetate on the second order when treated with tetraethylammonium acetate in acetone at 50°C, but is on the first order in anhydrous acetic acid at 50°C, accompanied by a higher extent of racemization, indicating a S_N1 reaction in anhydrous acetic acid.

General Reaction

L-(-)malic acid → (PCl₅) → D-(+)-chlorosuccinic acid → (KOH) → L-(-)malic acid / (Ag₂O) → D-(+)-malic acid

Mechanism

Applications

a) Walden inversion of configuration is widely useful in organic synthesis.

186. Wagner-Meerwein Rearrangement

Principle

Wagner accounted this reaction in 1899 and extended by Meerwein in 1914. 1,2-rearrangement of carbocation intermediates, from which the migratory group rearranges from an adjacent carbon atom to the carbocation center to form a more stable carbocation or to alleviate the ring strain; this rearrangement is known as Wagner-Meerwein rearrangement. Wagner-Meerwein rearrangement is occurred mainly in fused cyclic compounds; for example: bicyclic terpene systems. Rearrangement is also observed in monocyclic compounds (transformation of α-campholic acid to isolauric acid); noncyclic molecules [rearrangement of 2,2-dimethyl-3-hydroxy-3-phenyl propionic acid or ester to α-carboxyl (or ester)-β,β-dimethylstyrene in superacid solution]. The migratory groups consist of phenyl, vinyl, methylene, hydrogen, the electron-donating groups like alkyl, methyl, and isopropyl, and the electron-withdrawing groups (carboxyl and ester).

General Reaction

Tertiary alcohol → Alkene

a) Conversion of 3,3-dimethylbutan-2-ol into mixture of alkenes, namely 2,3-dimethylbut-2-ene (Major product) and 3,3-dimethylbut-1-ene (Minor product).

3,3-dimethylbutan-2-ol → 2,3-dimethylbut-2-ene (Major) + 3,3-dimethylbut-1-ene (Minor)

Mechanism

The rearrangement involves formation of cationic center, subsequent migration of a neighboring group to carbocation center. The migrating group differs, depending on the migratory aptitude and stereoelectronic effects.

Step 1: Formation of secondary carbocation with the elimination of water molecule.

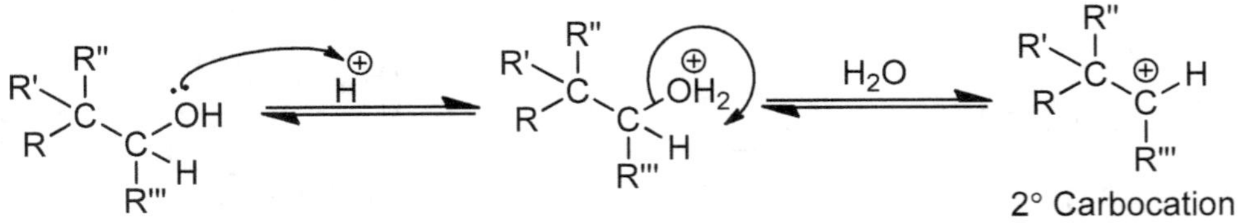

Step 2: 1,2-methyl shift rearranges secondary carbocation to tertiary carbocation.

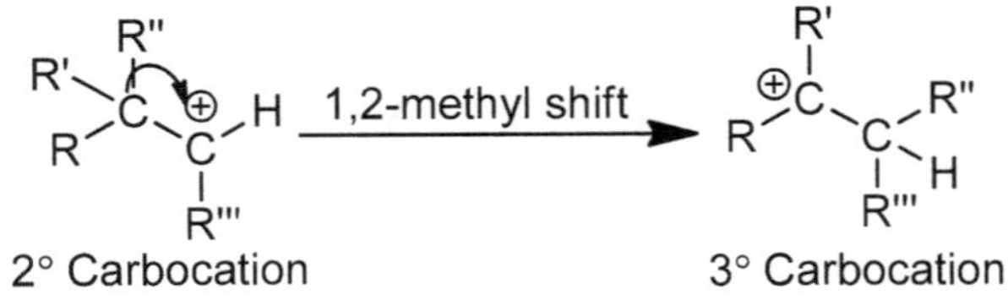

2° Carbocation   3° Carbocation

Step 3: Tertiary carbocation is converted to more stable alkene.

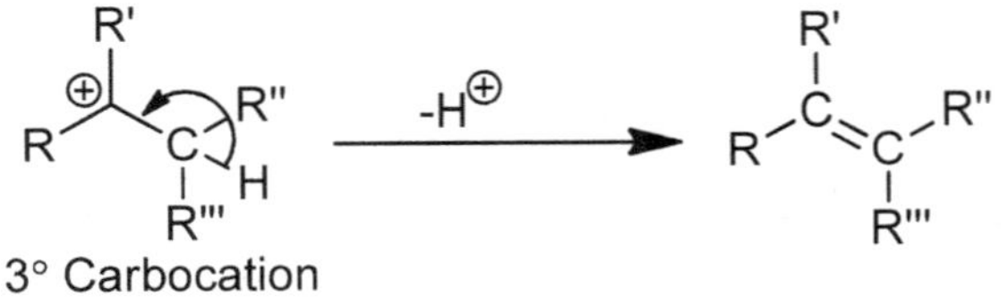

3° Carbocation

Applications

a) The rearrangement reaction is widely useful in organic chemistry. Rearrangement of 2,2-dimethylpropan-1-ol into 2-methylbut-2-ene (Major product) and 2-methylbut-1-ene (Minor product).

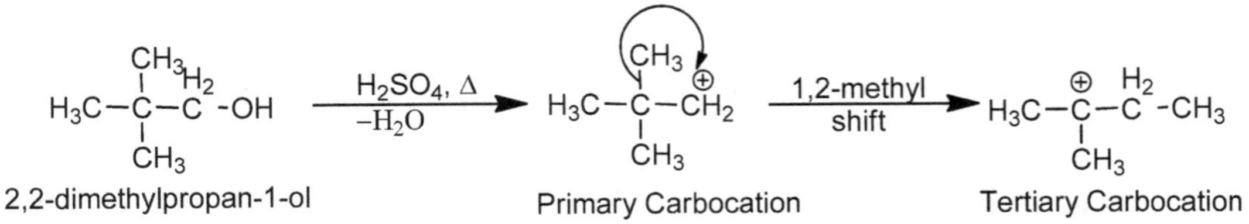

2,2-dimethylpropan-1-ol   Primary Carbocation   Tertiary Carbocation

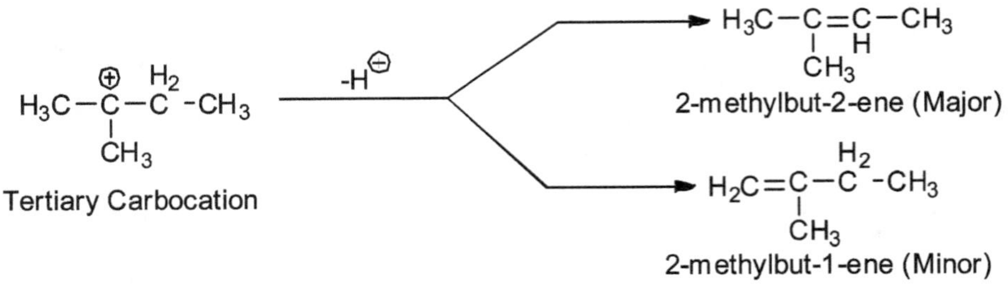

Tertiary Carbocation

2-methylbut-2-ene (Major)

2-methylbut-1-ene (Minor)

b) Rearrangement of (1R,2S,4S)-2-chloro-2,3,3-trimethylbicyclo[2.2.1]heptanes.

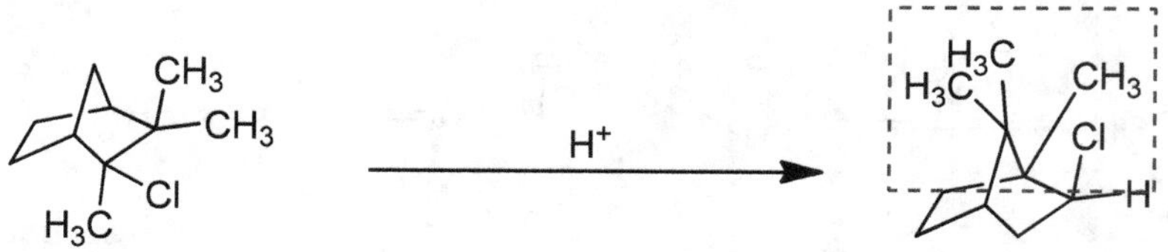

H$^+$

(1R,2S,4S)-2-chloro-2,3,3-
trimethylbicyclo[2.2.1]heptane

(1S,2S,4S)-2-chloro-1,7,7-
trimethylbicyclo[2.2.1]heptane

187. Wilkinson's Catalyst

Principle

Wilkinson and colleagues in 1965 described utility of catalyst chemically known as tris(triphenylphosphine) chlororhodium [Rh(PPh$_3$)$_3$Cl]). Hence this catalyst is commonly known as Wilkinson's catalyst. The Wilkinson's catalyst is a square planar, air and water-stable, 16-electron complex in the form of a red-violet crystalline and can be simply prepared from RhCl$_3$.H$_2$O in EtOH with an excess amount of PPh$_3$ or from [Rh(cod)Cl]$_2$ and PPh$_3$ in CHCl$_3$ or CH$_2$Cl$_2$. Organic reactions utilizing Wilkinson's catalyst are (a) Catalytic hydroboration of alkenes using catechol borane and pinacol borane, (b) 1,4-reduction of α,β-unsaturated carbonyls in combination with Et$_3$SiH, (c) Decarbonylation of aldehydes, (d) Hydroformylation, (e) Isomerization of olefin (allyl ether to vinyl ether), (f) Hydrosilylation, (g) Dehydrogenative silylation of olefins to vinylsilanes, (h) Annulation of aromatic imines in which the alkene is joined *meta* to imine, and (k) Catalytic oxidation of styrene into benzaldehyde and formaldehyde.

General Reaction

Wilkinson's Catalyst
(Chlorotris(triphenylphosphine)rhodium (I) complex or Tris(triphenylphosphine)rhodium(I) chloride)

Mechanism

Wilkinson's catalyst is useful for hydrogenation of olefin through dissociation of one PPh$_3$ molecule from the catalyst to give a 14-electron complex, oxidative addition of hydrogens to rhodium metal, π-complexation of olefin, intramolecular olefin insertion, isomerization, and reductive elimination of alkane.

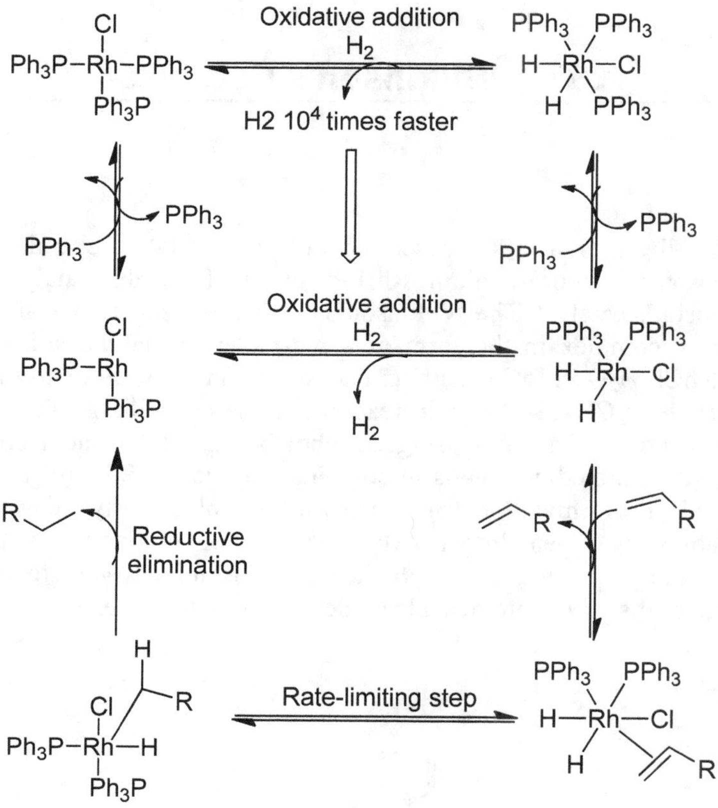

Applications

The catalyst is widely applicable in various organic synthetic reactions.

a) Synthesis of 7, 8, 9, 10-tetraphenyl fluoranthene from 1,8-bis (phenylethynyl)naphthalene in presence of Wilkinson's catalyst.

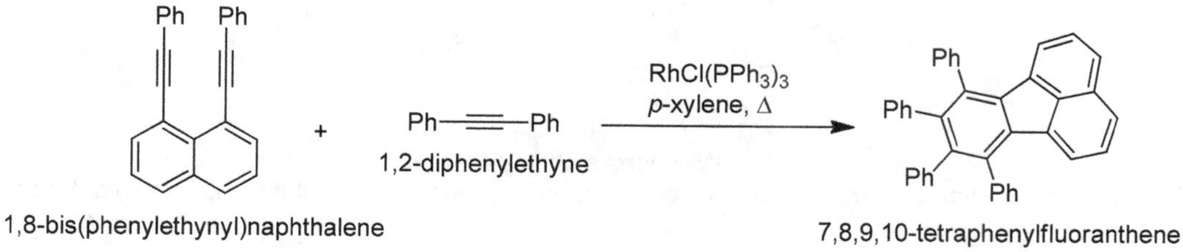

1,8-bis(phenylethynyl)naphthalene

1,2-diphenylethyne

$RhCl(PPh_3)_3$
p-xylene, Δ

7,8,9,10-tetraphenylfluoranthene

b) Preparation of cyclohexanone.

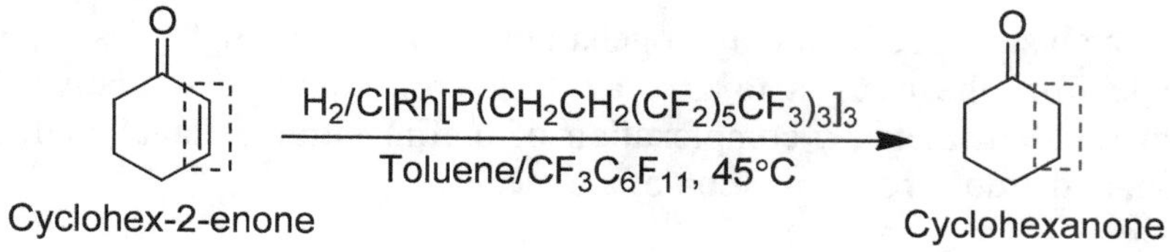

$H_2/ClRh[P(CH_2CH_2(CF_2)_5CF_3)_3]_3$
Toluene/$CF_3C_6F_{11}$, 45°C

Cyclohex-2-enone

Cyclohexanone

188. Willgerodt-Kindler Reaction

Principle

Willgerodt in 1887 explained and reported synthesis of phenylacetamide by reaction of styrene with ammonium polysulfide at high temperatures is known as Willgerodt reaction. Kindler extended reaction for the synthesis of thionamide from acetophenone and anhydrous primary or secondary amine is referred as Kindler modification. As reaction mechanism of Willgerodt and Kindler variation is analogous; hence synthesis of thionamide is generally known as Willgerodt-Kindler reaction. In this reaction α,β-unsaturated acid affords thionamides with one less carbon atom, whereas the reaction of methyl phenylethynyl ketone occurs only at the terminal carbon atom, and a reaction with equal amounts of benzaldehyde, morpholine and sulfur leads to the isolation of *N*-(thionbenzoyl)morpholine.

General Reaction

Propiophenone

TsOH, S_8, Δ

1-morpholino-3-phenylpropane-1-thione

Mechanism

Step 1:

morpholine   octathiocane

4-heptasulfanylmorpholine

1,2-dimorpholinodisulfane   hexasulfane

Step 2:

thiirene

Step 3:

Applications

a) *Synthesis of 5-phenylpentanamide from 1-phenylpentan-1-one.*

1-phenylpentan-1-one

1) S_8, Pyridine, $(NH_4)_2Sx$, NH_4OH, Δ

2) H_2O, Δ

5-phenylpentanamide

189. Williamson Synthesis of Ether

Principle

Williamson in 1851 explained nucleophilic substitution reaction of an alkali alkoxide with alkylating reagent to form ether and an inorganic salt is known as Williamson ether synthesis. Both symmetrical and asymmetrical ethers can be synthesized by using different alcohols and alkylating reagents. Commonly used alkylating reagents are haloalkanes, some other alkylating reagents such as dialkyl sulfates, alkyl mesylates, and tosylates are also used in this reaction. Compounds such as esters of weak acids (For example: acetic acid and benzoic acid), dialkyl ethers, alkyl aryl ethers, and epoxides can be used as alkylating reagents. Because alkali alkoxide is also strong base, the Williamson ether synthesis is often accompanied by elimination as side reaction, if a secondary or tertiary haloalkanes are used; where steric hindrance blocks the backside attack from an alkoxide. Rate of Williamson synthesis depends on solvent, temperature, leaving group on alkylating reagent, nucleophilicity of alkoxide.

General Reaction

$$R-O^{\ominus}Na^{\oplus} \; + \; X-R' \xrightarrow{\Delta} R-O-R' \; + \; NaX$$

Sodium alkaoxide   Haloalkane   Ether

a) Preparation of ethoxyethane

$$H_3C-\overset{H_2}{C}-O^{\ominus}Na^{\oplus} \; + \; I-\overset{}{\underset{H_2}{C}}-CH_3 \xrightarrow{\Delta} H_3C-\overset{H_2}{C}-O-\underset{H_2}{C}-CH_3 \; + \; NaI$$

Sodium ethoxide   Iodoethane   Ethoxyethane

b) Preparation of methoxyethane

$$H_3C-\overset{H_2}{C}-O^{\ominus}Na^{\oplus} \; + \; I-CH_3 \xrightarrow{\Delta} H_3C-\overset{H_2}{C}-O-CH_3 \; + \; NaI$$

Sodium ethoxide   Iodomethane   Methoxyethane

c) Preparation of (methoxymethyl)benzene

$$H_3C-O^{\ominus}Na^{\oplus} \; + \; \overset{Br}{\underset{}{H_2C}}{-}\bigcirc \xrightarrow{\Delta} H_3C-O-\overset{H_2}{C}-\bigcirc \; + \; NaBr$$

Sodium methoxide   (bromomethyl)benzene   (methoxymethyl)benzene

d) Preparation of ethoxybenzene

Sodium phenoxide   Bromoethane   Ethoxybenzene

Mechanism

Williamson synthesis includes two steps:

Step 1: The formation of alkali alkoxide by the treatment of alcohol or phenol with a strong base to generate nucleophile.

$$R-O^{\ominus} Na^{\oplus} \longrightarrow R-O^{\ominus} + Na^{\oplus}$$

Sodium alkaoxide   Nucleophile

Step 2: Nucleophilic substitution of the resulting alkoxide on an alkylating reagent to form ether.

$$R-O^{\ominus} + R'-X \xrightarrow[\text{substitution}]{\text{Nucleophilic}} R-O-R' + X^{\ominus}$$

Nucleophile   Haloalkane   Ether   Leaving group
                   (substrate)

Limitations:

2-bromo-2-   Ethoxide ion   2-ethoxy-2-methylpropane   Tertiary butoxide   Bromoethane
methylpropane

Applications

a) *Williamson ether synthesis is one of the oldest reactions widely useful for the synthesis of etheral analogues.*

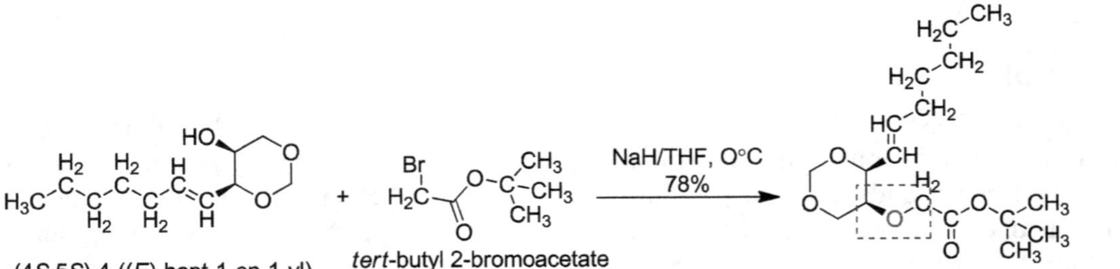

(4S,5S)-4-((E)-hept-1-en-1-yl)-
1,3-dioxan-5-ol

tert-butyl 2-bromoacetate

NaH/THF, 0°C
78%

tert-butyl 2-(((4S,5S)-4-((E)-hept-1-en-1-yl)-
1,3-dioxan-5-yl)oxy)acetate

190. Wittig Reaction

Principle

Wittig and Geissler in 1953 invented a new way to form a carbon-carbon double bond from a carbonyl group and a carbanion. The synthesis of an olefin by reaction between carbonyl compounds and phosphonium ylide, *via* either betaine and/or oxaphosphetane intermediate is generally known as Wittig reaction. Wittig's reaction uses a phosphorus stabilized carbanion to attack a ketone or an aldehyde. These phosphonium ylides are commonly known as Wittig reagents. Extensions of this reaction include aza-Wittig reaction, phospha-Wittig reaction, and thia-Wittig reaction.

Wittig reaction consists of following steps:

a) Generation of phosphonium salt by reaction of triphenylphosphine with haloalkanes;

b) Deprotonation of the corresponding phosphonium salt to form phosphorus ylide (phosphorane);

c) Condensation of phosphorane and carbonyl compound;

d) Rapid decomposition of the condensation intermediate (either betaine or oxaphosphetane) to give olefin and phosphine oxide.

Besides the carbonyl compounds (aldehydes and ketones); other carbonyl derivatives can also be used for the coupling with Wittig reagents, such as include esters, isocyanates, lactols, lactones, aminolactol.

General Reaction

a) Preparation of methylenecyclohexane from cyclohexanone in presence of Wittig reagent

b) Preparation of (Z)-1,2-diphenylethene from benzaldehyde in presence of Wittig reagent

Preparation of Wittig's reagent:

The phosphorus stabilized carbanion is an ylide. A molecule that bears no overall charge but has a negatively charged carbon atom bonded to positively charged heteroatom. Phosphorus ylides are prepared from triphenylphosphine and alkyl halide in two step process.

Step 1: The first step involves a nucleophilic attack by triphenylphosphine on an unhindered haloalkane (more often primary haloalkane). The product obtained is an alkyltriphenyl phophonium salt.

Step 2: In second step, the phosphonium salt is treated with a strong base (usually butyl lithium) to abstract a proton from the carbon atom bonded to phosphorus.

Butan-1-id-1-yllithium(II)

Alkyl triphenylphosphonium salt

Phopsphorus ylide

Phosphorus ylide has two resonating structures:

I) One with a double bond between carbon and phosphorus.

II) Another with charges on carbon and phosphorus.

The double bonded resonating structure requires ten electrons in the valence shell of phosphorus, using one of the phosphorus d-orbital. The π bond between carbon and phosphorus is relatively weak; charged structure is major contributor. The carbon atom actually bears a partially negative charge balanced by corresponding positive charge on phosphorus atom.

Mechanism

Step 1: Because of carbanion character of Wittig's reagent, the ylide carbon is strongly nucleophilic. It attacks on a carbonyl group to give a charge separated intermediate called a betaine. A betaine is unusual compound because it contains negatively charged oxygen and positively charged phosphorus on adjacent carbon atoms. Phosphorus and oxygen form a very strong bond; the attraction of the opposite charges promotes the fast formation of a four membered oxaphosphetane ring.

Phosphorus ylide   Aldehyde/Ketone
(Wittig reagent)

Betaine

Oxaphosphetane

Step 2: The four memebered ring quickly collapses to give desired alkenes and triphenylphosphine oxide. Triphenylphosphine oxide is an exceptionally stable compound and the conversion of triphenylphosphine to triphenylphosphine oxide provides the driving force for the formation of alkene in Wittig's reaction.

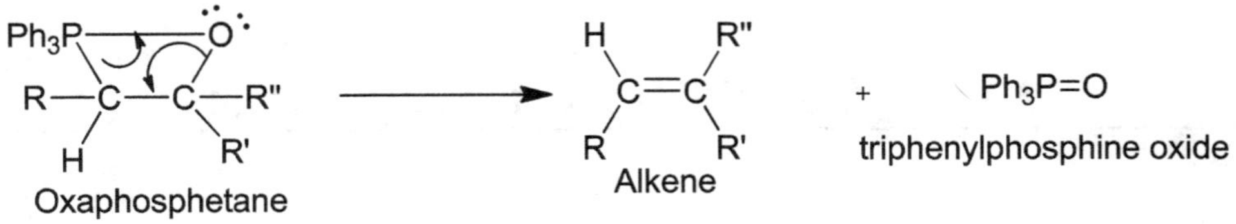

Ph3P
O
R
C
C
R"
H
R'
Oxaphosphetane
H
R"
C=C
R
R'
Alkene
+
Ph3P=O
triphenylphosphine oxide

Applications

a) Preparation of olefins; for example: formation of prop-1-en-2-ylbenzene.

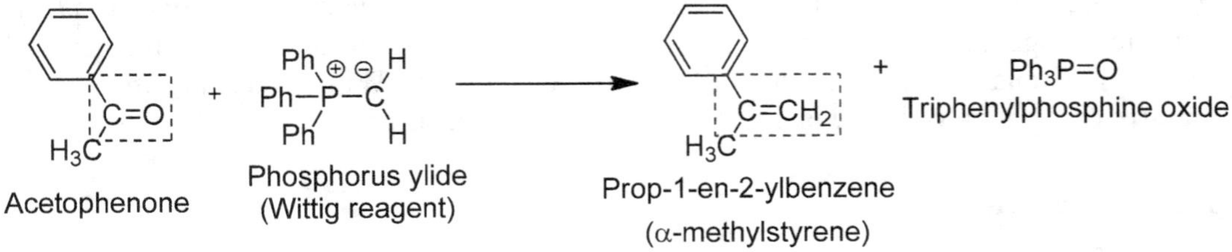

C=O
H3C
Acetophenone
+
Ph
Ph—P—C
Ph
H
H
Phosphorus ylide
(Wittig reagent)
C=CH2
H3C
Prop-1-en-2-ylbenzene
(α-methylstyrene)
+
Ph3P=O
Triphenylphosphine oxide

b) Formation of exocyclic double bonds; transformation of cyclohexanone into methylenecyclohexane.

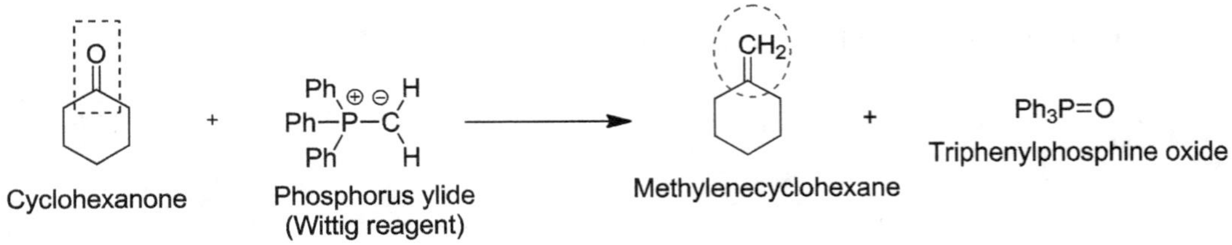

O
Cyclohexanone
+
Ph
Ph—P—C
Ph
H
H
Phosphorus ylide
(Wittig reagent)
CH2
Methylenecyclohexane
+
Ph3P=O
Triphenylphosphine oxide

191. [1,2]-Wittig Rearrangement

Principle

Schlenk and Bergmann in 1928 reported transformation of an ether into a secondary alkoxide *via* [1,2]-carbon shift by treatment of ether with a strong base such as alkyl or aryl lithium, subsequently leads to formation of secondary alcohol is known as [1,2]-Wittig rearrangement. The driving force for rearrangement is instability of α-oxygenated carbanion. Transformation of thioether into thiol is thio-Wittig rearrangement, and similarly conversion of imine is imino 1,2-Wittig rearrangement.

General Reaction

$$R-\underset{\underset{O-R'}{|}}{\overset{H_2}{\overset{|}{C}}} \quad + \quad R-\underset{Li}{\overset{H_2}{\overset{|}{C}}} \quad \longrightarrow \quad R-\underset{\underset{H}{|}}{\overset{R'}{\overset{|}{C}}}-OH$$

Ether   Alkyl Lithium   Alcohol

a) Preparation of Phenoxy(phenyl)methanol

(benzyloxy)benzene   +   Phenyllithium   Ether   Phenoxy(phenyl)methanol

Mechanism

Step 1: Generation of carbanion by deprotonation with the help of alkyl lithium.

Deprotonation

$+ \quad R-CH_3$

Step 2: [*1,2*]-sigmatropic rearrangement of carbanion followed by acidic workup to give alcohol.

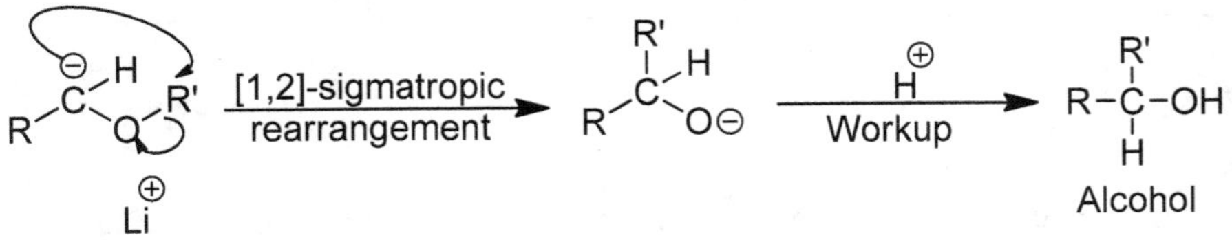

Applications

a) The reaction is useful for the synthesis of secondary alcohols.

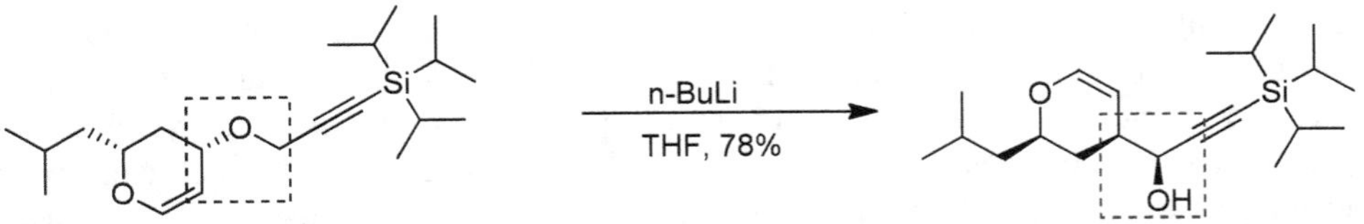

(3-(((2R,4S)-2-isobutyl-3,4-dihydro-2H-pyran-
4-yl)oxy)prop-1-yn-1-yl)triisopropylsilane

(S)-1-((2R,4S)-2-isobutyl-3,4-dihydro-
2H-pyran-4-yl)-3-(triisopropylsilyl)prop-2-yn-1-ol

192. [2,3]-Wittig Rearrangement

Principle

A highly regioselective [2,3]-sigmatropic rearrangement of the conjugate bases of allylic ethers (or benzylic ethers) to alcohols with carbon-carbon bond formation is known as [2,3]-Wittig sigmatropic rearrangement. In 1978, rearrangement was extended by Still and Mitra for α-alkoxystannes in which the carbanion was generated regioselectively *via* a tin-lithium exchange; hence referred as Still-Wittig [2,3]-sigmatropic rearrangement. Similar rearrangement of an allylic amine in presence of strong base is called *aza*-2,3-Wittig sigmatropic rearrangement. Different from [1,2]-Wittig Rearrangement, [2,3]-Wittig rearrangement obeys the Woodward-Hoffmann's orbital symmetry conservation theory, and proceeds *via* a concerted route in both the gas and the liquid phase, involving a five-membered envelope-like transition state. It has been proposed that the breaking carbon-oxygen bond and the formation carbon-carbon bond in transition state are eclipsed, and the steric effect and electronic interaction at this point results in a high stereoselectivity.

General Reaction

R' = Alkynyl, Alkenyl, Ph, COR, CN, etc.

Alkyl Lithium

Homoallylic alcohol

a) Synthesis of (1R,2R)-2-methyl-1-phenylbut-3-en-1-ol

(E)-((but-2-en-1-yloxy)methyl)benzene   Ethyllithium

(1R,2R)-2-methyl-1-phenylbut-3-en-1-ol

Mechanism

Step 1: The driving force for this rearrangement is the conversion of stabilized carbanion to a more stable alkoxide; as a result, an early transition state is expected for this exothermic reaction.

Step 2: [2,3]-sigmatropic rearrangement of carbanion followed by acidic workup to give alcohol.

Applications

a) Rearrangement has been applied in synthesis of different alcohols.

(Z)-isopropyl 4-butoxy-2-(((Z)-4-butoxybut-2-en-1-yl)oxy)but-2-enoate

(2S,3S)-isopropyl 3-(butoxymethyl)-2-((E)-2-butoxyvinyl)-2-hydroxypent-4-enoate

193. Wohl-Ziegler Reaction

Principle

A bromination of unsaturated organic compounds at allylic or benzylic position (or at α-position of the side chain of other aromatic compounds) using *N*-bromosuccinimide (NBS) as brominating agent reported by Wohl in 1919 and improved by Ziegler in 1942. The bromination by *N*-bromoacetamide is known as Wohl reaction, whereas the bromination by NBS is known as Wohl-Ziegler bromination. NBS is feasible for the bromination under basic conditions, including the bromination of methoxy benzoic acid in aqueous sodium hydroxide solution and *ortho*-bromination of phenol in the presence of di-isopropylamine.

General Reaction

a) Formation of 3-bromocyclohex-1-ene with release of succinamide by reaction of Cyclohexene with Ziegler reagent.

Cyclohexene + N-bromosuccinamide (Ziegler reagent) $\xrightarrow[\text{CCl}_4]{\text{Peroxide}}$ 3-bromocyclohex-1-ene + Succinamide

b) Preparation 3-bromoprop-1-ene from prop-1-ene

Prop-1-ene + N-bromosuccinamide (Ziegler reagent) $\xrightarrow[\text{CCl}_4]{\text{Peroxide}}$ 3-bromoprop-1-ene + Succinamide

Allylic hydrogen

Allylic hydrogen atom

Mechanism

Step 1: Generation of free radical by homolytic cleavage.

Free radical

Step 2: Abstraction of proton by free radical to generate allyl free radical.

Abstraction of proton

Allyl free radical

Step 3: NBS reacts with allyl free radical to give 3-bromocyclohex-1-ene.

Allyl free radical

3-bromocyclohex-1-ene

Applications

a) *The reaction is used in synthesis of allylic and benzylic bromides. For example: Synthesis of 1-(4-(bromomethyl)phenyl)ethanone.*

1-(*p*-tolyl)ethanone

N-bromosuccinamide
(Ziegler reagent)

CH_3CN, Δ

1-(4-(bromomethyl)phenyl)ethanone

b) *Synthesis of 9-(bromomethyl)anthracene by reaction with N-bromosuccinamide.*

9-Methylanthracene

N-bromosuccinamide
(Ziegler reagent)

CCl_4, Δ, 98%

9-(bromomethyl)anthracene

Principle

Wolff and Kishner reported base catalyzed conversion of carbonyl compounds into corresponding hydrocarbons through the decomposition of hydrazone intermediates is referred as Wolff-Kishner reduction. Both Wolff's protocol and Kishner's method are inconvenient and not much practical for the laboratory transformation, because the preformed hydrazones were added to a hot mixture of solid KOH and a platinized porous plate in Kishner's protocol, whereas in Wolff's method, hydrazones were heated with NaOEt in ethanol at 160-200 °C. Huang Minlon extended this reaction with important modification in which carbonyl compounds and hydrazine hydrate in presence of base are heated with diethylene glycol; upon the completion of hydrazone formation, the temperature is raised to 190-200 °C to drive off water and excess hydrazine hydrate. Under the conditions of Huang-Minlon modification, a reasonable reaction rate can be reached only when the temperature is raised above 190 °C.

General Reaction

$$
\underset{\substack{\text{Carbonyl} \\ \text{compounds}}}{\overset{R}{\underset{R'}{>}}C=O} + \underset{\substack{\text{Hydrazine} \\ \text{hydrate}}}{NH_2NH_2} \longrightarrow \underset{\text{Hydrazone}}{\overset{R}{\underset{R'}{>}}C=NNH_2} \xrightarrow[\Delta]{\text{Base}} \underset{\text{Alkane}}{\overset{R}{\underset{R'}{>}}CH_2} + N\equiv N
$$

a) Preparation of ethane.

$$
\underset{\text{Ethanal}}{\overset{H_3C}{\underset{H}{>}}C=O} + \underset{\substack{\text{Hydrazine} \\ \text{hydrate}}}{NH_2NH_2} \longrightarrow \underset{\text{Hydrazone}}{\overset{H_3C}{\underset{H}{>}}C=NNH_2} \xrightarrow[\Delta]{\text{Base}} \underset{\text{Ethane}}{\overset{H_3C}{\underset{H}{>}}CH_2} + N\equiv N
$$

b) Preparation of butane.

$$
\underset{\text{Butanone}}{\overset{H_3C}{\underset{H_3C-CH_2}{>}}C=O} + \underset{\substack{\text{Hydrazine} \\ \text{hydrate}}}{NH_2NH_2} \longrightarrow \underset{\text{Hydrazone}}{\overset{H_3C}{\underset{H_3C}{>}}C=NNH_2} \xrightarrow[\Delta]{\text{Base}} \underset{\text{Butane}}{\overset{H_3C}{\underset{H_3C-CH_2}{>}}CH_2} + N\equiv N
$$

c) Formation of aromatic hydrocarbon.

$$\underset{\text{Aromatic ketone}}{Ph-\overset{|}{\underset{R}{C}}=O} \ + \ \underset{\substack{\text{Hydrazine} \\ \text{hydrate}}}{NH_2NH_2} \longrightarrow \underset{\text{Hydrazone}}{Ph-\overset{|}{\underset{R}{C}}=NNH_2} \xrightarrow[\Delta]{\text{Base}} \underset{\text{Aromatic hydrocarbon}}{Ph-\overset{|}{\underset{R}{C}}H_2} \ + \ N\equiv N$$

d) Formation of toluene.

$$\underset{\text{Benzaldehyde}}{Ph-\overset{|}{\underset{H}{C}}=O} \ + \ \underset{\substack{\text{Hydrazine} \\ \text{hydrate}}}{NH_2NH_2} \longrightarrow \underset{\text{Hydrazone}}{Ph-\overset{|}{\underset{H}{C}}=NNH_2} \xrightarrow[\Delta]{\text{Base}} \underset{\text{Toluene}}{Ph-\overset{|}{\underset{H}{C}}H_2} \ + \ N\equiv N$$

Mechanism

Step 1: Wolff-Kishner reduction involves the formation of hydrazone in a manner analogous to the formation of imine between a carbonyl compound and a primary amine.

$$\underset{R}{}\overset{O}{\underset{}{\underset{}{C}}}\underset{R'}{} \ + \ :NH_2NH_2 \longrightarrow \left(\underset{R}{}\overset{HO \quad HN-NH_2}{\underset{}{\underset{}{C}}}\underset{R'}{}\right) \xrightarrow{-H_2O} \underset{\text{Hydrazone}}{R-\overset{N-NH_2}{\underset{R'}{\underset{}{C}}}}$$

Step 2: Successive deprotonation on hydrazone eventually results in the evolution of nitrogen and the formation of hydrocarbon.

Loss of N_2

Carbanion

Step 3: Work-up

Applications

a) Conversion of camphor to camphene.

Camphor   Hydrazone   Camphene

b) Synthesis of 1,5-dimethyl-8-oxabicyclo-[3.2.1]-octane.

1,5-dimethyl-8-oxabicyclo-[3.2.1]-octan-3-one   1,5-dimethyl-8-oxabicyclo-[3.2.1]-octane

195. Wolff Rearrangement

Principle

A thermal, photochemical or catalytic transformation of α-diazoketones into ketenes is reported and explained by Wolff in 1902 is commonly referred as Wolff rearrangement. Wolff rearrangement carried out by photo illumination is known as photochemical Wolff rearrangement. The Wolff rearrangement for transformation of β,γ-unsaturated α-diazoketones into γ,δ-unsaturated esters is called vinylogous Wolff rearrangement. The resulting ketenes from the Wolff rearrangement can be simply converted into carboxylic acids or esters in presence of water or alcohols, respectively.

General Reaction

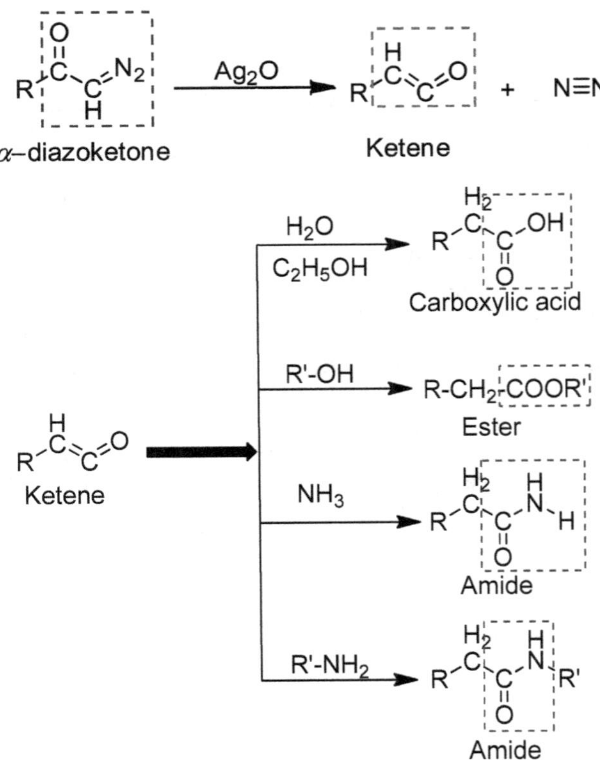

Mechanism

Step 1: Formation of α-ketocarbene from α-diazoketone with the elimination of nitrogen molecule.

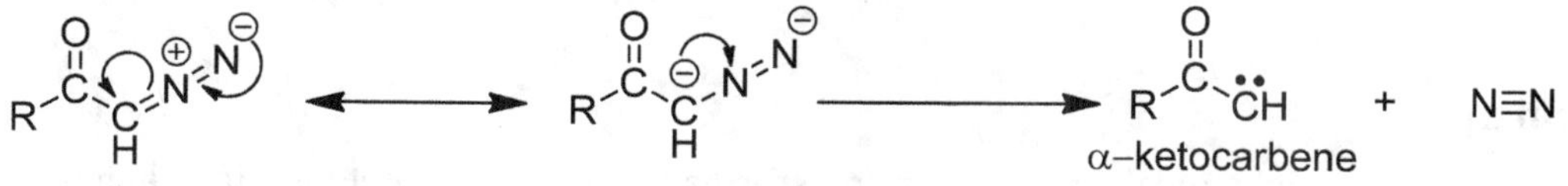

α-ketocarbene + N≡N

Step 2: 1,2-shift of alkyl group to produce ketene.

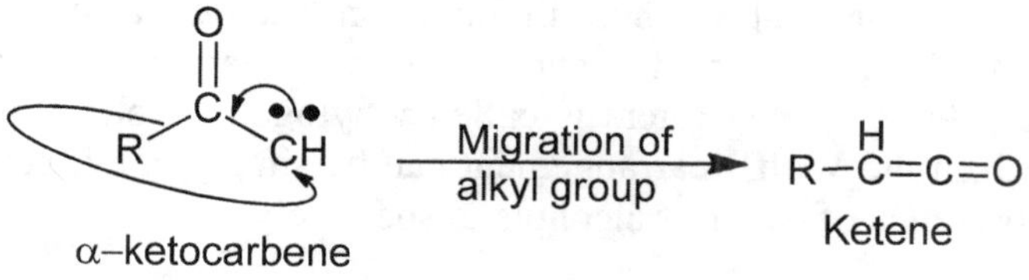

α-ketocarbene

Migration of alkyl group

R–C=C=O

Ketene

Step 3: Ketene on reaction with primary amines gives amide.

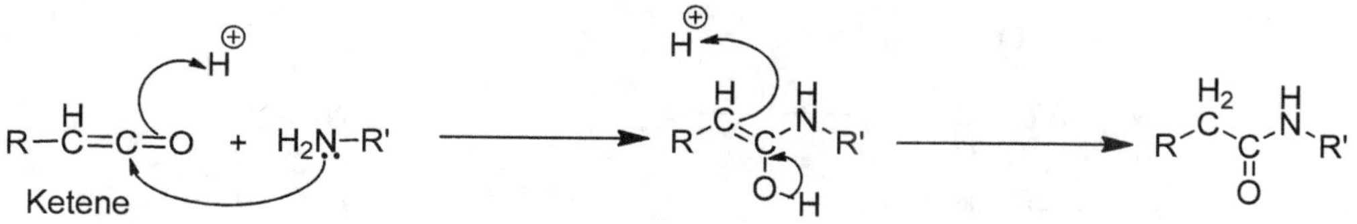

Ketene + H₂N–R'

Applications

The reaction is applicable for the synthesis of ketene, which on treatment with different replaceable hydrogen containing compounds gives acid derivatives.

a) Formation of amides; For example: Synthesis of N-methyl-2-phenylacetamide.

α–diazoketone

Ketene

N-methyl-2-phenylacetamide

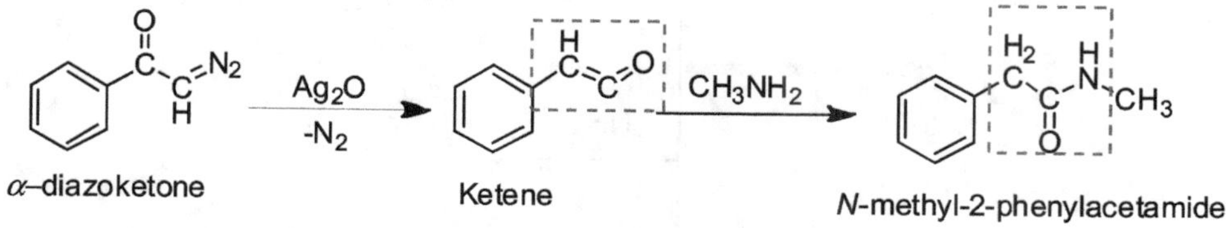

b) *Synthesis of methyl 8-(benzyloxy)-4-(2-(methyl(phenethyl)amino)-2-oxoethyl)-2-naphthoate.*

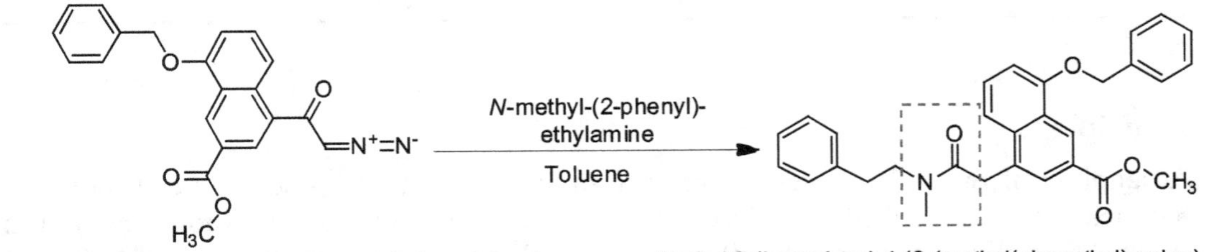

Methyl 8-(benzyloxy)-4-(2-diazoacetyl)-2-naphthoate

Methyl 8-(benzyloxy)-4-(2-(methyl(phenethyl)amino)-2-oxoethyl)-2-naphthoate

196. Wurtz Reaction

Principle

A coupling of two haloalkanes in presence of sodium metal to produce higher order of hydrocarbon is described by Wurtz in 1855; hence the reaction is known as Wurtz synthesis or Wurtz reaction. Similarly, the coupling between haloalkane and Grignard reagent is known as Grignard-Wurtz coupling. In this reaction, it was assumed that coupling of two alkyl radicals takes place; because sodium has an odd electron and can donate one electron to haloalkane to form an alkyl radical and sodium halide; in addition, the formation of a saturated alkane with the same number of carbon atoms is attributed to the proportion of the alkyl radical.

General Reaction

$$R\text{--}X \ + \ 2Na \ + \ X\text{--}R \ \xrightarrow{\text{Dry ether}} \ R\text{--}R \ + \ 2NaX$$

Haloalkane   Haloalkane   Alkane

a) Preparation of butane.

$$CH_3\text{-}CH_2\text{-}Br \ + \ 2Na \ + \ Br\text{-}CH_2\text{-}CH_3 \ \xrightarrow{\text{Dry ether}} \ CH_3\text{-}CH_2\text{-}CH_2\text{-}CH_3$$

Bromoethane   Bromoethane   Butane

b) Preparation of 2,3-dimethylbutane.

$$\underset{\overset{|}{CH_3}}{CH_3\text{-}CH}\text{-}I \ + \ 2Na \ + \ I\text{-}\underset{\overset{|}{CH_3}}{CH}\text{-}CH_3 \ \xrightarrow{\text{Dry ether}} \ \underset{\overset{|}{CH_3}\ \overset{|}{CH_3}}{CH_3\text{-}CH\text{-}CH\text{-}CH_3}$$

2-iodopropane   2-iodopropane   2,3-dimethylbutane

c) Formation of mixture of alkanes by reaction between two unsymmetrical haloalkanes.

$$\left.\begin{array}{c} CH_3\text{-}Br \\ + \\ CH_3\text{-}CH_2\text{-}Br \end{array}\right\} \xrightarrow[\text{Dry ether}]{Na}$$

$CH_3\text{-}CH_3$ (Unfavoured product)
Ethane
+
$CH_3\text{-}CH_2\text{-}CH_3$ (Favoured product)
Propane
+
$CH_3\text{-}CH_2\text{-}CH_2\text{-}CH_3$ (Unfavoured product)
Butane

Mechanism

Step 1: Formation of organosodium compound in presence of dry ether

$$\bullet Na \; + \; R\text{-}X \; + \; \bullet Na \xrightarrow{\text{Dry ether}} \; :\overset{\ominus}{R}\text{-}\overset{\oplus}{Na} \; + \; \overset{\oplus}{Na}\text{-}\overset{\ominus}{X}$$

Haloalkane   Organosodium compound

Step 2: Organosodium compound combine with second molecule of haloalkane to generate higher alkane.

$$:\overset{\ominus}{R}\text{-}\overset{\oplus}{Na} \; + \; \overset{+\delta}{R}\text{-}\overset{-\delta}{X} \longrightarrow R\text{-}R \; + \; NaX$$

Alkane

Applications

The reaction is widely applicable in synthesis of carbon-based polymers, such as poly(1,4-phenylene-1,1-diphenylethylene), silicons, tin-based polymers, including polysilanes, poly(dibutylstanne), organostannane-organosilane copolymer. In addition, tetraarylsilanes, diphenylalkanes, and tetrakis(p-dimethylaminophenyl)ethylene are also synthesized by Wurtz reaction.

a) Conversion of (chloromethyl)trimethylsilane into mixture of products by Wurtz reaction.

(chloromethyl)trimethylsilane

Tetramethylsilane (18.3%)

Methane (1%)

1,2-bis(trimethylsilyl)ethane (40.5%)

Ethyldimethyl((trimethyl silyl)methyl)silane (11.8%)

b) Transformation of (4-chloro-2-methylbutan-2-yl)benzene into mixture of hydrocarbons.

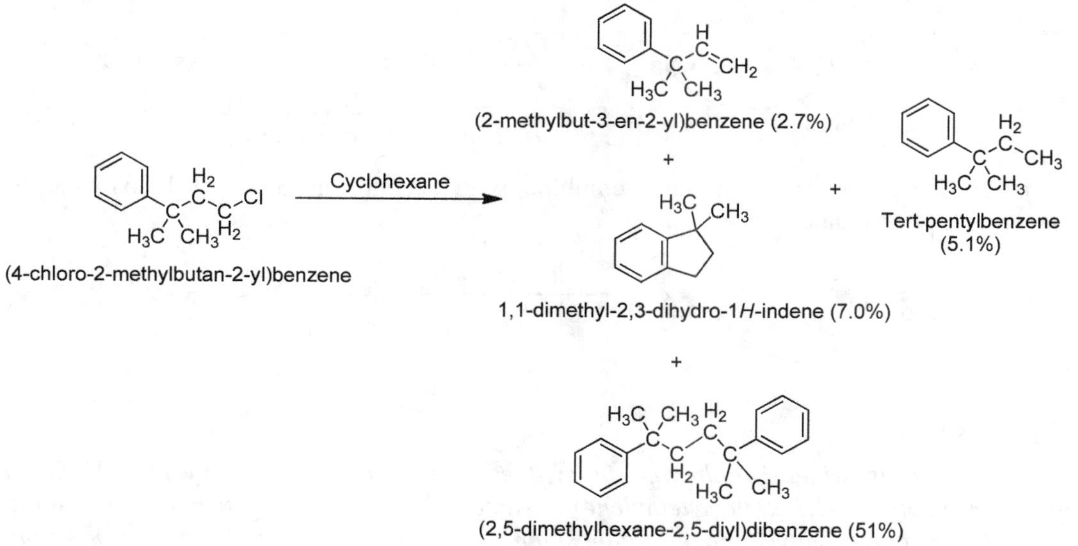

(4-chloro-2-methylbutan-2-yl)benzene

Cyclohexane

(2-methylbut-3-en-2-yl)benzene (2.7%)

+

1,1-dimethyl-2,3-dihydro-1*H*-indene (7.0%)

+

Tert-pentylbenzene (5.1%)

+

(2,5-dimethylhexane-2,5-diyl)dibenzene (51%)

197. Wurtz-Fittig Reaction

Principle

The extension of Wurtz synthesis for the preparation of alkylated aromatics by sodium-mediated coupling of an alkyl and an aryl halide was reported by Fittig in 1864 commonly known as Wurtz-Fittig reaction or Wurtz-Fittig synthesis. Similarly, sodium-mediated coupling between alkenyl halide and silyl halide to obtain alkenylsilane is known as Wurtz-Fittig silylation. It has been reported that ω-diaryl alkanes is obtained very easily by Wurtz-Fittig reaction from ω-dihaloalkanes, even though such coupling is affected factors such as size of sodium particle; distance between two halogen atoms in alkanes, and solvents.

General Reaction

Aryl halide   Alkyl halide   Alkylated aromatic compounds

R = Alkyl, X = Cl, Br, I
X' = Cl, Br, I

a) Preparation of ethylbenzene.

Bromobenzene   Bromoethane   Ethylbenzene

b) Preparation of 1-ethyl-4-methylbenzene.

1-bromo-4-methylbenzene   Bromoethane   1-ethyl-4-methylbenzene

Mechanism

Step 1: Formation of organosodium compound in presence of dry ether

Haloalkane   Organosodium compound

Step 2: Organosodium compound combine with molecule of haloalkane to generate alkylated aromatics.

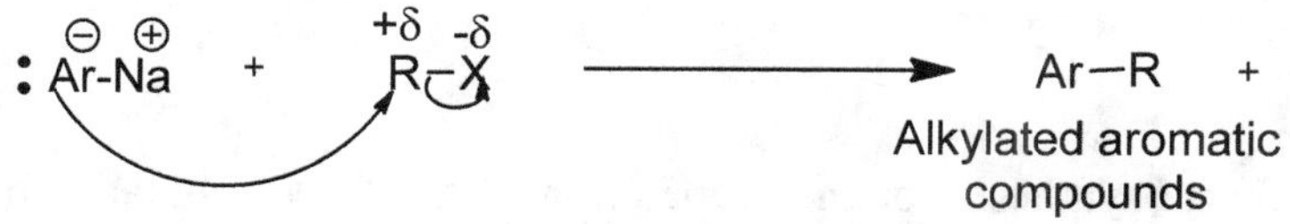

$$: \overset{\ominus}{Ar} - \overset{\oplus}{Na} \quad + \quad \overset{+\delta}{R} - \overset{-\delta}{X} \quad \longrightarrow \quad Ar-R \quad + \quad NaX$$

Alkylated aromatic
compounds

Applications

a) Direct methylation of aromatic compounds; For example: Preparation of 1,3-diphenylpropane from bromobenzene and 1,3-dibromopropane.

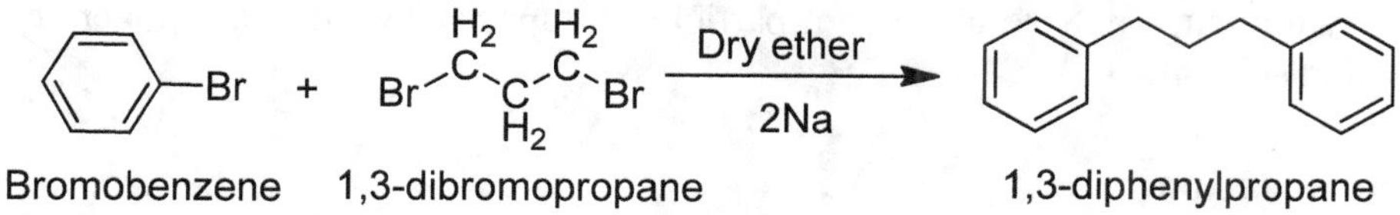

Bromobenzene   1,3-dibromopropane   1,3-diphenylpropane

Principle

The regioselective synthesis of highly functionalized esters from carboxylic acids and alcohols using 2,4,6-trichlorobenzoyl chloride as condensation reagent in presence of a stoichiometric amount of 4-N,N-dimethylamino pyridine (DMAP) was initially reported by Yamaguchi in 1979 is known as Yamaguchi esterification. 2,4,6-trichlorobenzoyl chloride is Yamaguchi reagent. The use of 2,4,6-trichlorobenzoyl chloride ensures that aromatic moiety of mixed anhydride is not accessible to other nucleophiles due to steric hindrance at aromatic counterparts, and thus the regioselectivity of esters can be anticipated. As a result, the mixed anhydride reacts with nucleophiles once it forms. Owing to the high reactivity of the mixed anhydride, which is thermally unstable, different results might be obtained at various temperatures as well as with a different order of mixing reactants.

General Reaction

The reaction involves two steps:

a) Formation of a mixed anhydride between a carboxylic acid and the Yamaguchi reagent in the presence of Et_3N.

b) Esterification of alcohol from the mixed anhydride with DMAP.

Mechanism

Yamaguchi esterification involves the formation of a mixed anhydride between an aliphatic acid and 2,4,6-trichlorobenzoyl chloride, and the regioselective reaction between alcohol and mixed anhydride in the presence of DMAP. Formation of acylium ion in presence of triethylamine. Removal of good leaving group (Cl⁻) from 2,4,6-trichlorobenzoyl chloride to generate intermediate which later combines with acylium ion to give TCBz⁻.

Formation of ester

Applications

a) The reaction is applicable in synthesis of esters and large sized lactones.

3-((4R,5R)-5-((R)-6-hydroxyheptyl)-
2,2-dimethyl-1,3-dioxolan-4-yl)propanoic acid

(3aR,8S,13aR)-2,2,8-trimethyloctahydro-3aH-
[1,3]dioxolo[4,5-e][1]oxacyclododecin-6(8H)-one

199. Ziegler-Hafner Azulene Synthesis

Principle

The synthesis of azulenes *via* condensation of cyclopentadienyl anion with the intermediate arising from the nucleophilic addition of dimethylamine onto an activated pyrinium salt was initially given by Ziegler and Hafner in 1955. The reaction is referred as Ziegler-Hafner azulene synthesis. It is the most versatile method for synthesis azulene analogues with substituents at the seven-membered ring. In this reaction, 2,4-dinitrochlorobenzene is usually used to activate the pyridine moiety; however, when the pyridine ring connects to other structures that reduce the basicity of the pyridine moiety; hence 2,4-dinitrochlorobenzene is not suitable for the activation purpose probably because the nitrogen atom of the pyridine ring does not smoothly undergo nucleophilic substitution. Trifluoromethanesulfonic anhydride is a good alternative for 2,4-dinitrochlorobenzene for activation of pyridine moiety.

General Reaction

Pyridine

1) DNCB; 2) $(CH_3)_2NH$, $NaOCH_3$; 3) $C_2H_5^-Na^+$

(4Z,6Z,8Z)-3a*H*-cyclopenta[8]annulene

DNCB = 2,4-dinitrochlorobenzene
C_2H_5-Na^+ = Sodium cyclopentadiene

Mechanism

Step 1: Nucleophilic substitution of the chloro species from 2,4-dinitrochlorobenzene by pyridine, affording an activated pyridinium salt.

Step 2: Nucleophilic addition of the dimethylamine anion to give a dihydropyridine intermediate, which undergoes *retro*-electrocyclization to yield a conjugated arylimino enamine.

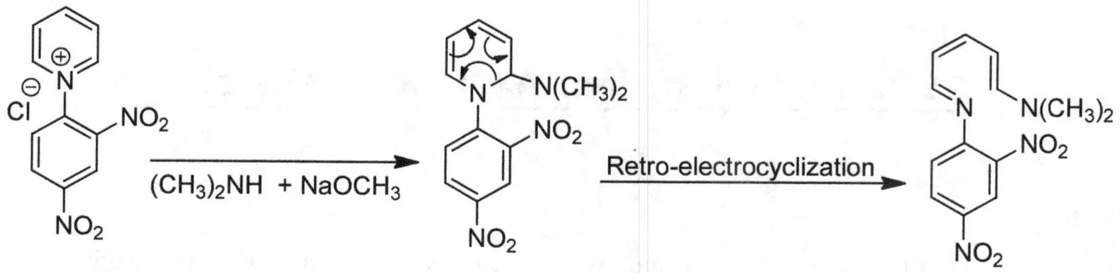

Step 3: Further addition of dimethylamine to the imino group and elimination of arylamine to afford the conjugated iminium salt.

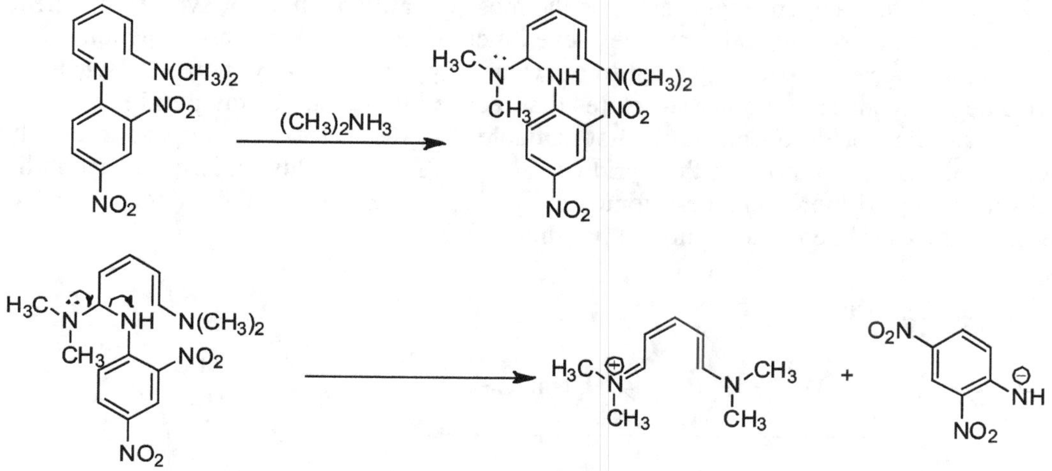

Step 4: Addition of cyclopentadienyl anion to the iminium moiety and elimination of dimethylamine to form the butadienyl conjugated fulvene scaffold.

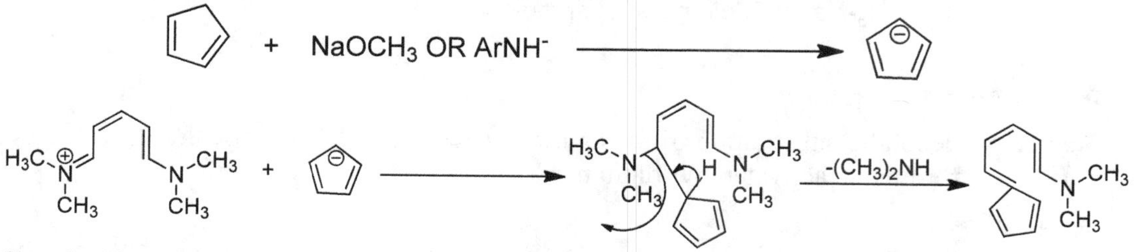

Step 5: 10-π electrocyclization followed by elimination of dimethylamine to produce the azulene nucleus.

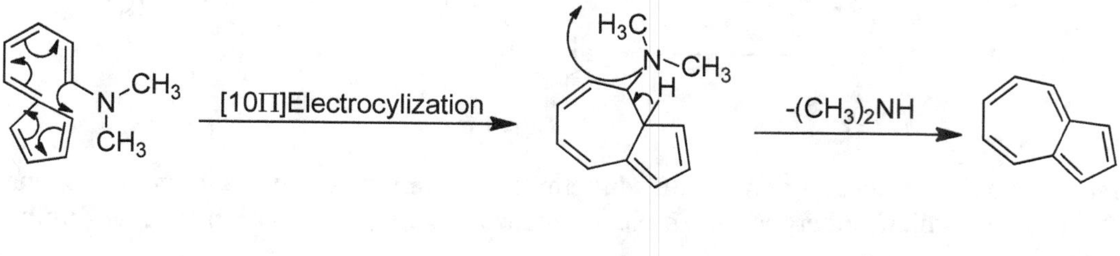

Applications

a) *The reaction is useful in synthesis of azulene analogues, containing substituents at the seven-membered moiety. For example: Formation of 5,6-dimethylazulene.*

$$CH_3I, C_2H_2OH \quad \text{then} \quad NaOC_2H_5/\text{Cyclopentadiene} \quad C_2H_5OH$$

3,4-dimethylpyridine   1,3,4-trimethylpyridin-1-ium iodide   5,6-dimethylazulene

200. Ziegler-Natta Polymerization

Principle

Ziegler and Natta in 1954 reported polymerization of vinyl monomers in presence of trialkylaluminum to give stereoregulated or tactic polymers known as Ziegler-Natta polymerization. The polymerization requires lewis acid of an early transition metal (Such as Ti, Zr, V) to catalyze polymerization. The combination of $TiCl_4$ and Et_3Al or α-$TiCl_3/Et_2AlCl$ used mostly for heterogeneous polymerization is referred as Ziegler-Natta catalyst. Ziegler-Natta polymerization has been modified extensively, and several generations of Ziegler-Natta catalysts have been developed. The first generation of Ziegler-Natta catalyst is chromium and zirconium used at initial development stage of Ziegler-Natta polymerization; second generation of Ziegler-Natta catalysts is $TiCl_4/Et_3Al$ or $TiCl3/Et_2AlCl$. The first two generations of Ziegler-Natta catalysts have little use in industries due to their low activities; however, the magnesium chloride supported $TiCl_4$ developed in the 1960 is third generation of Ziegler-Natta catalyst was implemented for production of polymers. Major limitation of reaction is the intolerance of Ziegler-Natta catalysts to Lewis bases. For example, the catalysts will be poisoned by the heteroatom of an ether, ester, amino and carboxyl functional group in the monomers. In addition, polyvinyl chloride as well as polyacrylates cannot be made by the Ziegler-Natta polymerization, due to the fact that vinyl chloride undergoes radical polymerization and acrylates precede the anionic polymerization; both can be initiated by the Ziegler-Natta catalysts.

General Reaction

$$\text{Vinyl monomers} \xrightarrow[\text{or } Cp_2TiCl_2/MAO/\text{electron donor}]{TiCl_4/MgCl_2/Et_3Al} \text{stereoregulated or tactic polymers}$$

R = H, CH_3, C_2H_5, C_6H_5

Mechanism

Step 1:

Step 2:

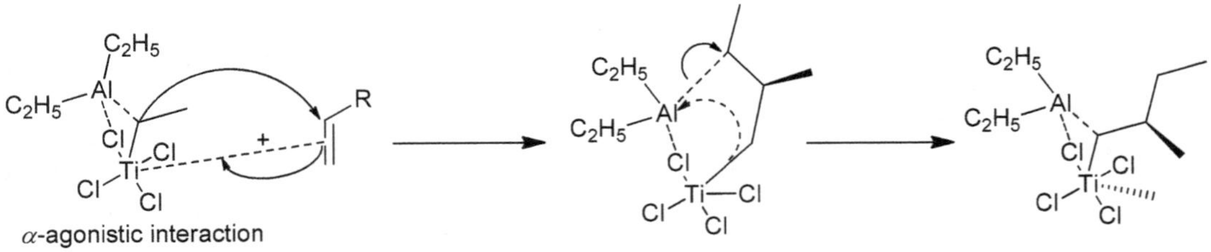

α-agonistic interaction

Step 3:

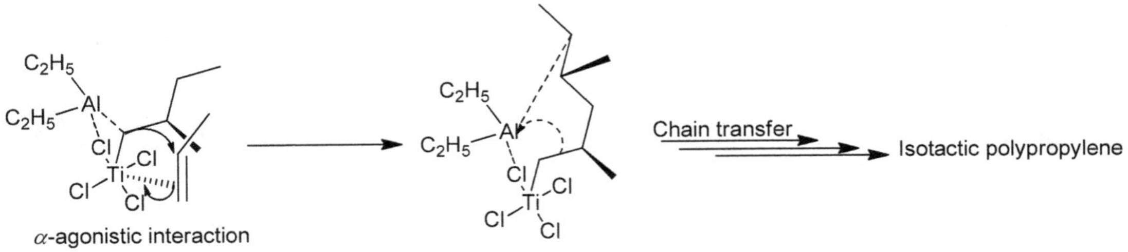

α-agonistic interaction

Chain transfer → Isotactic polypropylene

Applications

a) *The reaction has been widely used in industry for the synthesis of different polymers in bulk quantity; For example: Synthesis of isotactic polymer.*

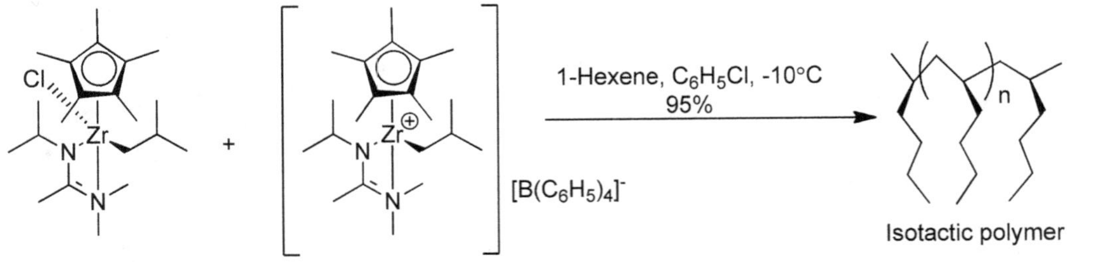

1-Hexene, C_6H_5Cl, -10°C
95%

$[B(C_6H_5)_4]^-$

Isotactic polymer

Acronym/Abbreviations

Sr. No.	Acronym/ Abbreviations	Name	Chemical Formula or Functionality
1	Ac	Acetyl	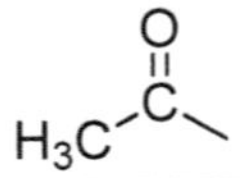
2	AcOH (HOAc)	Acetic acid	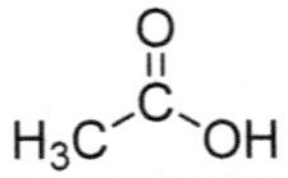
3	AIBN	2,2'-Azobisisobutyronitrile	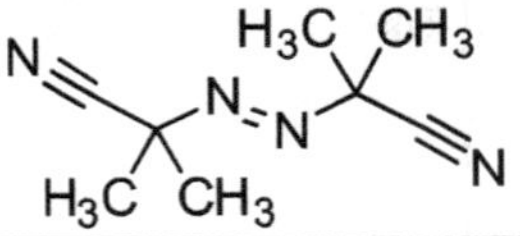
4	ACN	1,1'-Azobis-1-cyclohexanenitrile	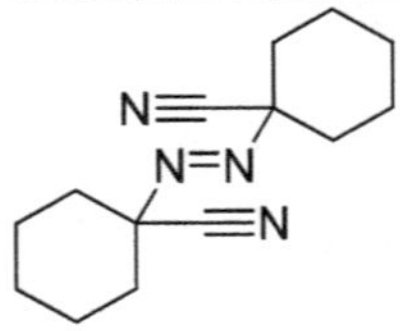
5	9-BBN	9-Borobicyclo[*3.3.1*]nonane	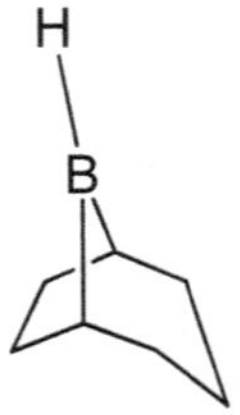
6	BINAP	2,2'-Bis(diphenylphosphino)-1,1'-binaphthyl	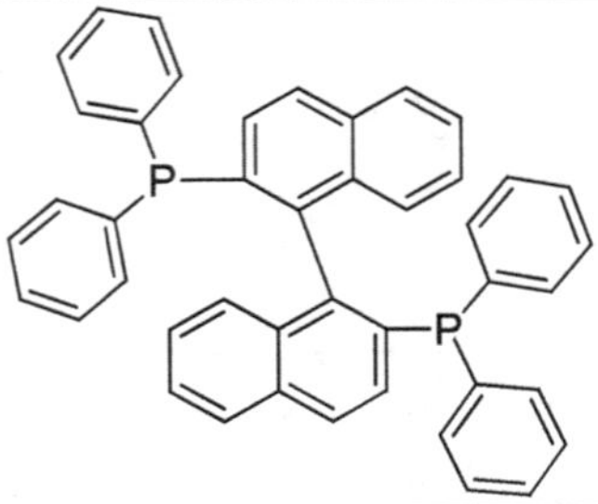
7	BINOL	[1,1'-binaphthalene]-2,2'-diol	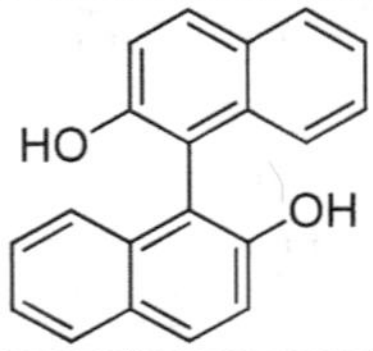

8	BMDA	Bromomagnesium Diisopropylamide	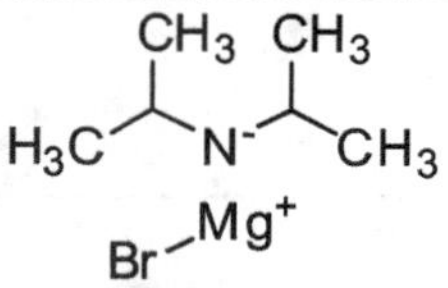
9	BMS	Borane Dimethylsulfide	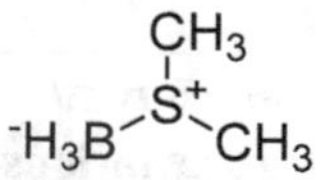
10	Bn-	Benzyl	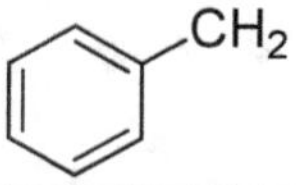
11	BOC- (t-Boc)	t-Butoxycarbonylchloride	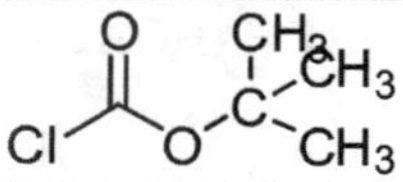
12	Bs	Brosylate	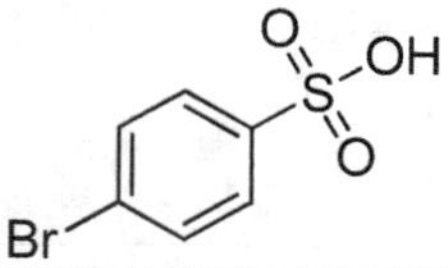
13	Bz	Benzoyl	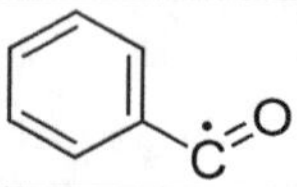
14	CAN	Ceric ammonium nitrate	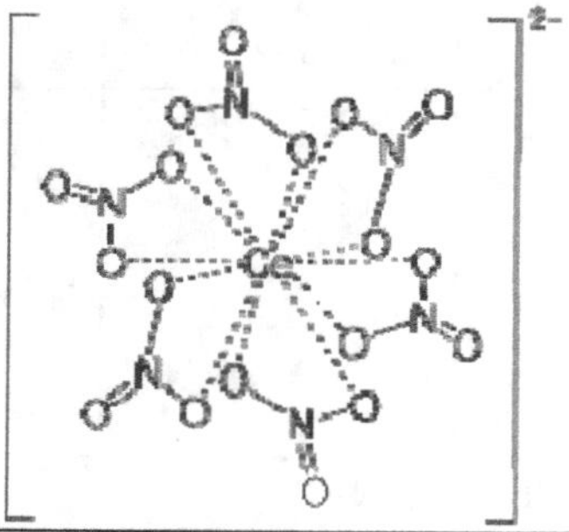
15	CAS	Ceric ammonium sulfate	$(NH_4)_4Ce(SO_4)_4 \cdot 2\ H_2O$
16	Cbz-	Carbobenzyloxy	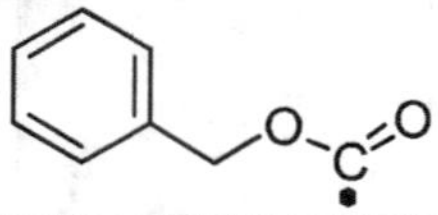
17	Cetyl	Hexadeca- $C_{16}H_{33}$	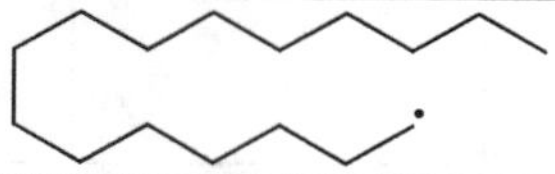
18	DABCO, TED	1,4-Diazabicylo[*2.2.2*]octane; TED, triethylenediamine	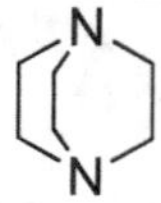 or $(N_2(C_2H_4)_3)$
19	DBN	1,5-Diazabicyclo[*4.3.0*]non-5-ene	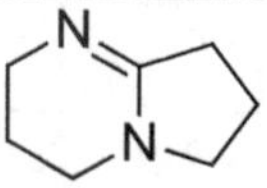

20	DBU	1,8-Diazabicyclo[*5.4.0*]undec-7-ene	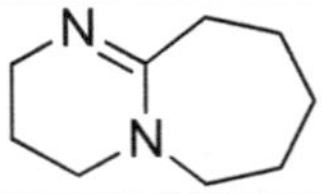
21	DCC	*N,N'*-dicyclohexylcarbodiimide	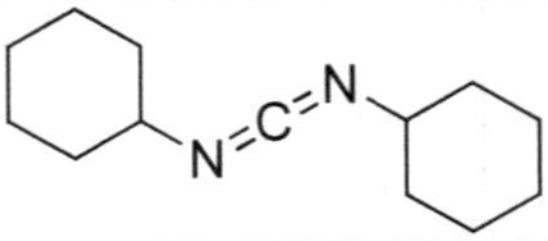
22	DDQ	2,3-Dichloro-5,6-dicyano-1,4-benzoquinone	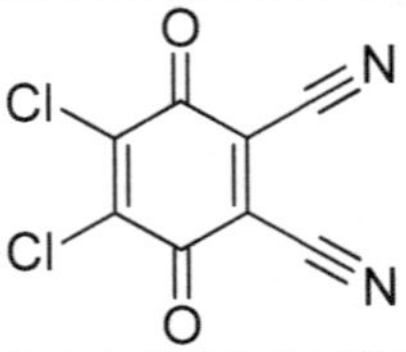
23	DEAD	Diethyl Azodicarboxylate	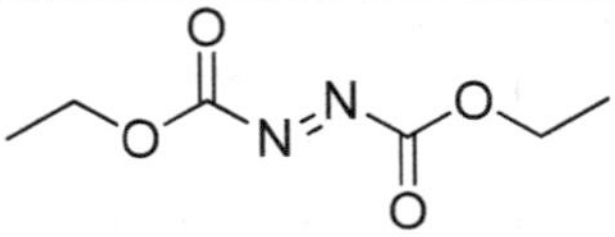
24	DET	Dietkyl tartrate	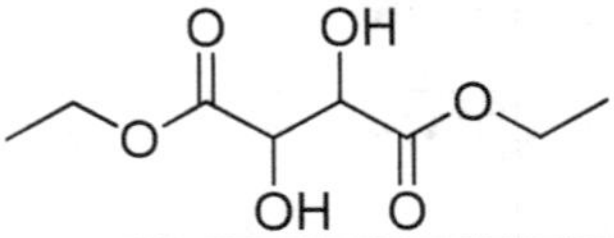
25	DZBAL DZBAL-H	Disobutylaluminum hydride	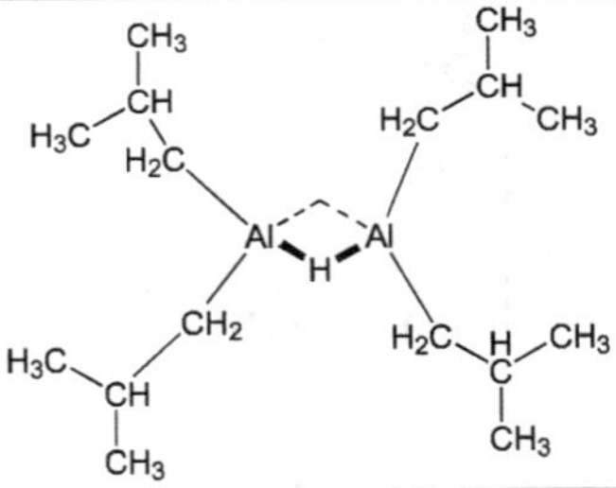
26	DIPEA	Diisopropylethylamine (Hunip's base)	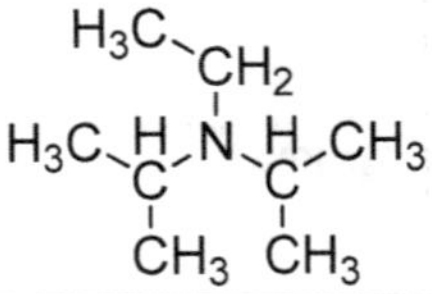
27	DIPT	Diisopropyl tartrate	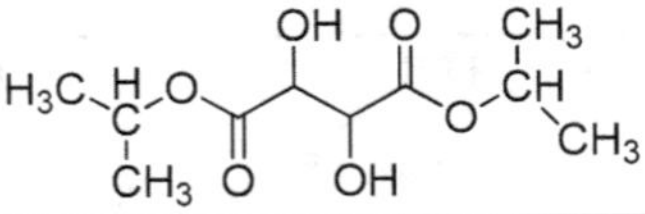
28	Diglvme	Diethylene glycol dimethyl ether	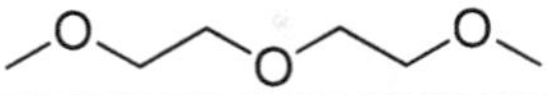
29	DMAP	4-(Dimethylamino)pyridine	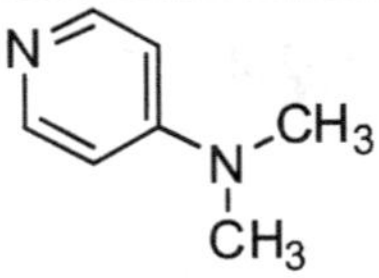
30	DME	1,2-Dimethoxyethane or Glyme	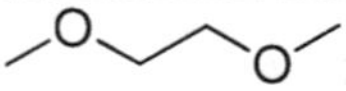

31	DMIPS	Dimethylisopropylsilyl	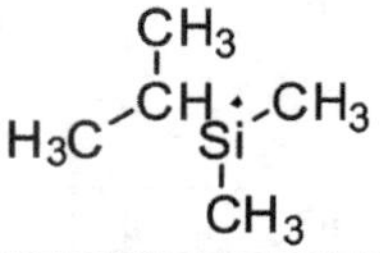
32	DMF	Dimethylformamide	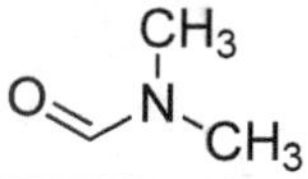
33	DMP	Dimethylpyrazole	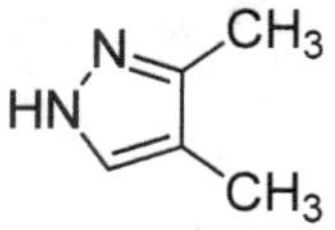
34	DMPU	N,N'-Dimethylpropyleneurea	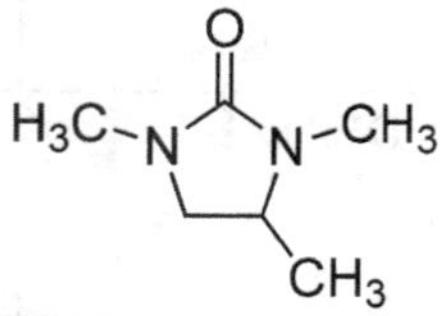
35	DMS	Dimethylsulfide	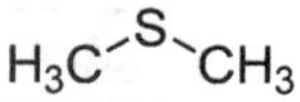
36	DMSO	Dimethylsulfoxide	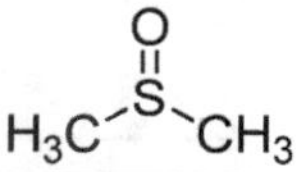
37	DNP	2,4-dinitrophenyl	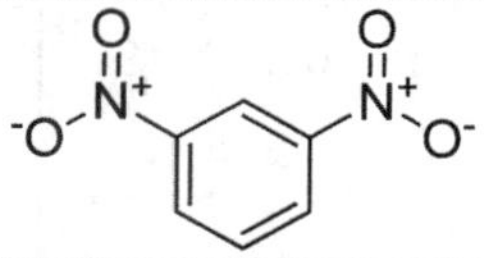
38	HMPT/HMPA	Hexamethylphosphoric triamide	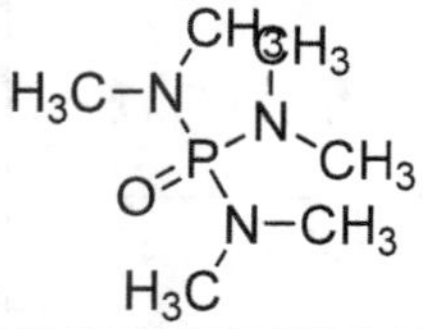
39	HMTA	Hexamethylenetetramine	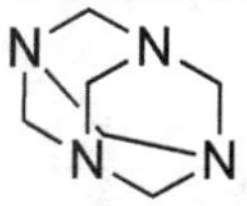
40	HTIB	Hydroxy(tosyloxy)-iodobenzene (Koser's Reagent)	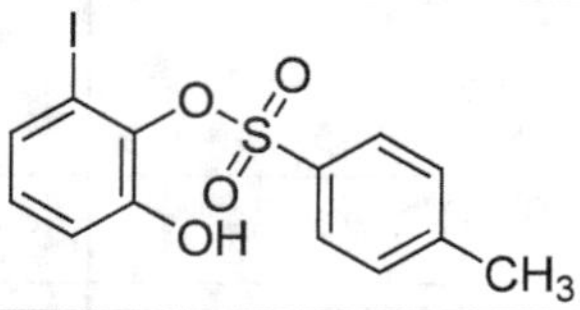
41	Icp2BH	Diisopinocampheylborane	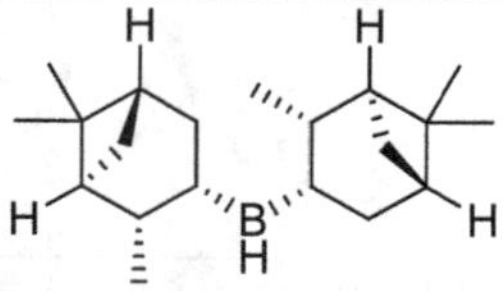
42	LTA	Lead tetraacetate	$Pb(C_2H_3O_2)_4$

43	LTMP	1-Lithio-2,2,6,6-tetramethylpiperidine	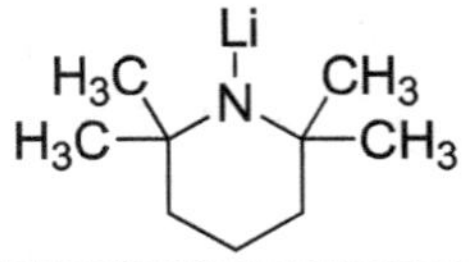
44	MCPBA	*m*-Chlorperoxybenzoic acid	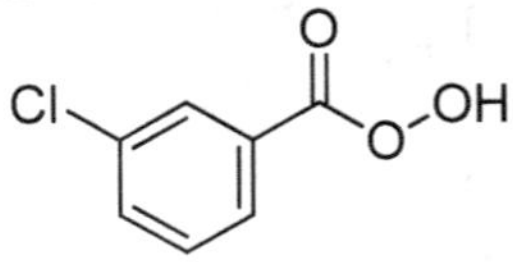
45	MeCN	Acetonitrile	
46	MEM-	2-Methoxyethoxymethyl	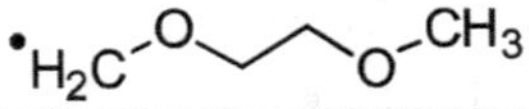
47	Ms	Mesyl, Methanesulfonyl	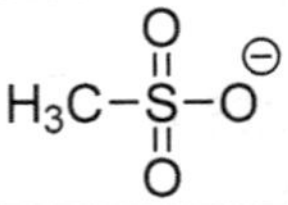
48	MTM	Methylthiomethyl	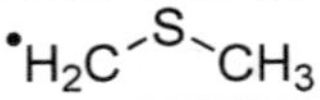
49	MVK	Methyl Vinyl Ketone	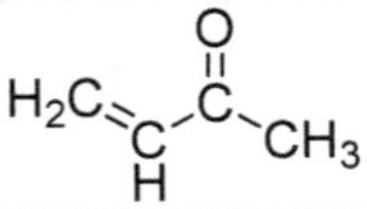
50	NBS	*N*-Bromosuccinimide	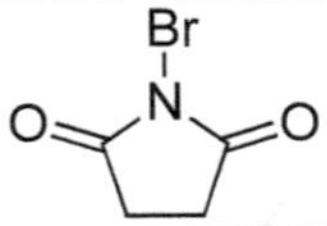
51	NCS	*N*-Chlorosuccinimide	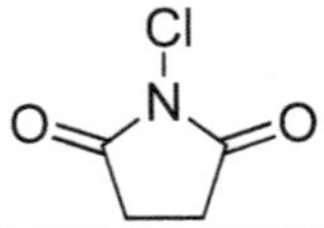
52	NMM	4-Methylmorpholine	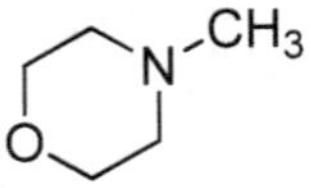
53	PCC	Pyridinium chlorochromate (Corey's Reagent)	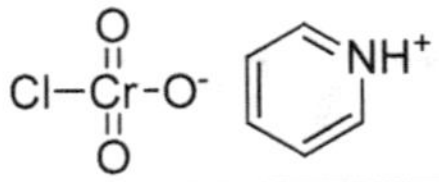
54	PDC	Pyridinium dichromate (Cornforth reagent)	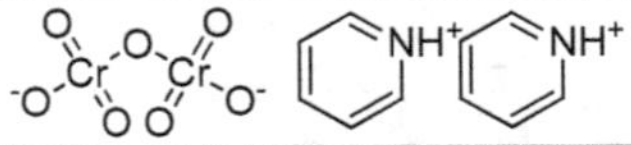
55	[Pd$_2$(dba)$_3$]	Tris(dibenzylideneacetone)dipalladium(0)	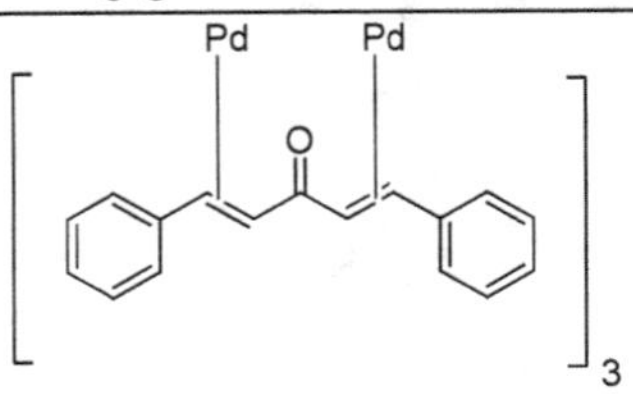
56	PMB	*p*-Methoxybenzyl	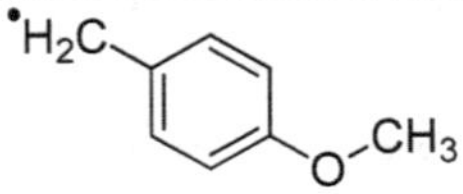

57	PNB	*p*-Nitrobenzoyl	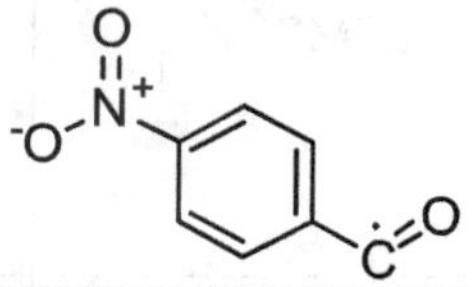
58	PPA	Polyphosphoric Acid	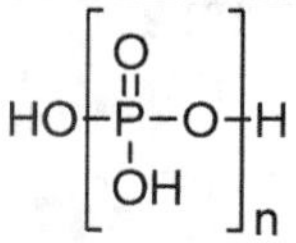
59	pTT {PTAB)	Phenyltrimethylammonium Tribromide	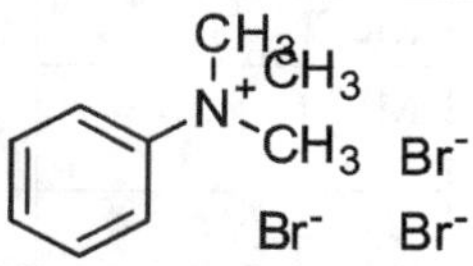
60	PPTS	Pyridinium *p*-toluenesulfonate	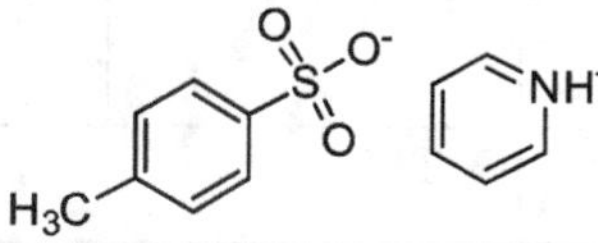
61	PTSA	*p*-Toluenesulfonic acid or Tosic acid	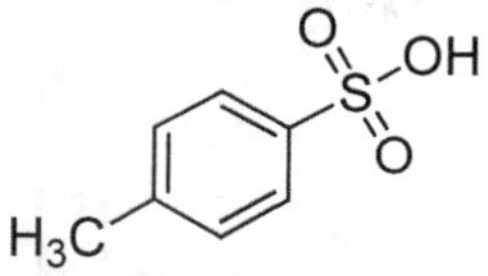
62	Py	Pyridine	
63	RAMP	(*R*)-1-Amino-2-methoxymethylpyrrolidine	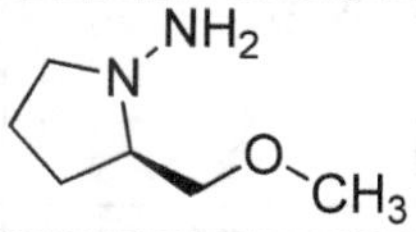
64	SAMP	(S)-1-Amino-2-methoxymethylpyrrolidine; (Ender's Reagent)	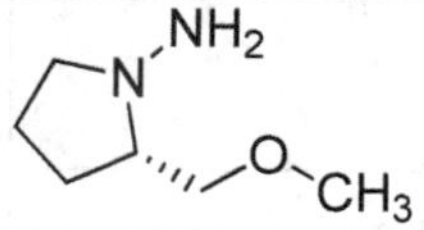
65	SEM	2-Trimethylsilylethoxy-methoxy	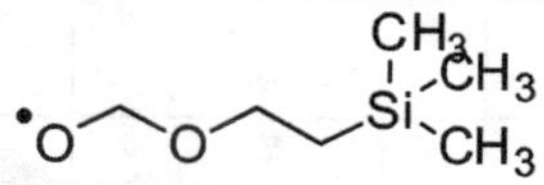
66	TBAF	Tetrabutylammonium fluoride	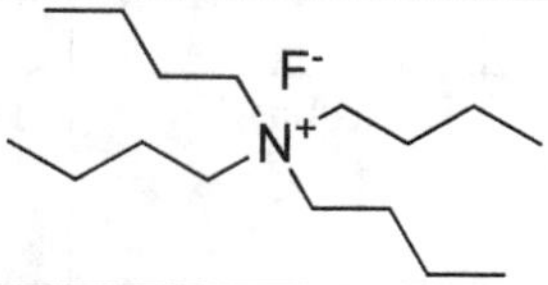
67	TBDPS	*tert*-Butyldiphenylsilyl	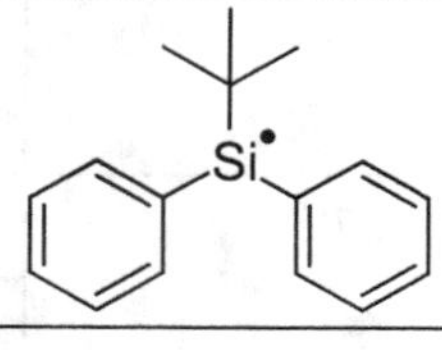

68	TBHP	*tert*-Butyl hydroperoxide	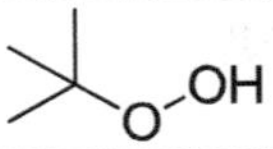
69	TBS/TBDMS	*tert*-Butyldimethylsilyl	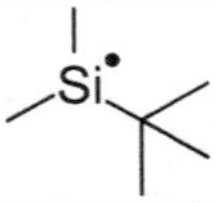
70	TEA	Triethylamine	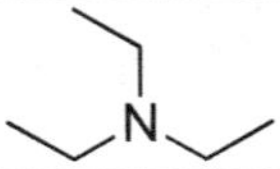
71	TEBA/TEBAC	Benzyltriethylammonium chloride	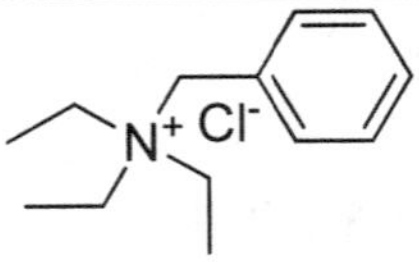
72	TEMPO	2,2,6,6-Tetramethylpiperidin-1-oxyl	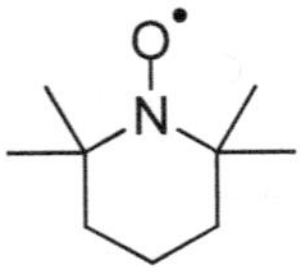
73	TES	Triethylsilyl	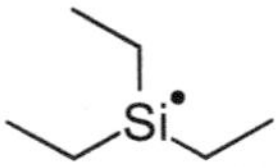
74	Tf	Triflate	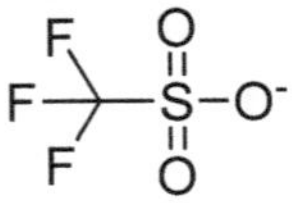
75	THF	Tetrahydrofuran	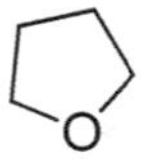
76	THP	Tetrahydropyranyl	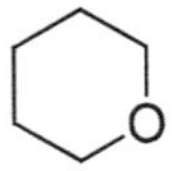
77	TIPS	Triisopropylsilyl	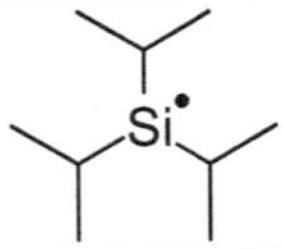
78	TPAP	Tetrapropylammonium perruthenate	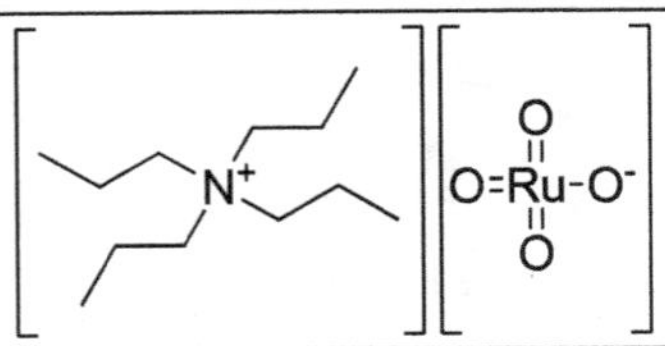
79	TPP	Triphenyl phosphine	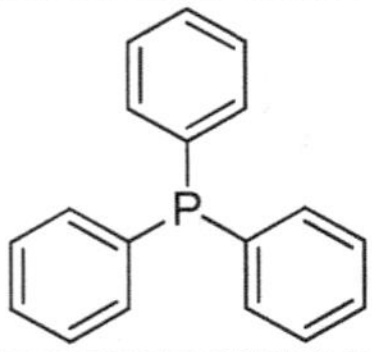

80	TMS	Trimethylsilyl	
81	TPS	Triphenylsilyl	
82	Ts-; TOS-	Tosyl; *p*-toluenesulfonyl	